# Sustainable Energy: Green Technologies

# Sustainable Energy: Green Technologies

Edited by Marrianne Fox

SYRAWOOD
PUBLISHING HOUSE

New York

Published by Syrawood Publishing House,
750 Third Avenue, 9th Floor,
New York, NY 10017, USA
www.syrawoodpublishinghouse.com

**Sustainable Energy: Green Technologies**
Edited by Marrianne Fox

International Standard Book Number: 978-1-68286-471-5 (Hardback)

**Cataloging-in-Publication Data**

Sustainable energy : green technologies / edited by Marrianne Fox.
    p. cm.
Includes bibliographical references and index.
ISBN 978-1-68286-471-5
1. Renewable energy sources. 2. Clean energy. 3. Energy development--Technological innovations.
4. Biomass energy.  I. Fox, Marrianne.
TJ808 .S87 2017
621.042--dc23

Printed in the United States of America.

# TABLE OF CONTENTS

**Permissions**

**List of Contributors**

**Index**

# PREFACE

This book on sustainable energy discusses the various technologies and methods that have been made popular for energy-efficient consumption. Sustainable energy that promotes environmental sustainability is a very effective source of energy generation. Energy generation must concentrate on reducing pollution and other emissions to combat detrimental effects like greenhouse gas emissions and ozone layer depletion. Contents included in this text deal with coherent data that explains the effectiveness of green energy technologies and reduced emissions. This book is a compilation of chapters that discuss the most vital concepts and emerging trends in the field of sustainable energy. A number of latest researches have been included to keep the readers up-to-date with the global concepts in this area of study. This book is an essential guide for both academicians and those who wish to pursue this discipline further.

This book unites the global concepts and researches in an organized manner for a comprehensive understanding of the subject. It is a ripe text for all researchers, students, scientists or anyone else who is interested in acquiring a better knowledge of this dynamic field.

I extend my sincere thanks to the contributors for such eloquent research chapters. Finally, I thank my family for being a source of support and help.

**Editor**

# Performance enhancement of PV array based on water spraying technique

**Salih Mohammed Salih, Osama Ibrahim Abd, Kaleid Waleed Abid**

Renewable Energy Research Center, University of Anbar, Ramadi, Iraq

**Email address:**
dr_salih_moh@ieee.com (S. M. Salih), osama_eng21@yahoo.com (O. I. Abd), kaleidwaleed@yahoo.com (K. W. Abid)

**Abstract:** This paper experimentally presents water spraying technique to improve photovoltaic (PV) array efficiency and enhance the net power saving. A forced-water spraying and cooling technique with constant flow rate of water on PV array surface is designed and implemented. The decreasing rate in the panel surface temperature has a direct proportional relation with PV efficiency. Simultaneously, the output hot water is very beneficial for houses, buildings etc., as water heating system, specifically in the remote areas. The electrical performance of PV array was also studied. The cooling rate of panel surface for 5 min.= 4 in midday. The electrical performance of PV array also was studied. As a final point, the economical results were achieved as result of the power saving increases 7w/degree at midday.

**Keywords:** PV Cooling, Temperature Effect, Water Spraying, RERC, Cooling Rate

## 1. Introduction

Performance of a solar-photovoltaic (PV) system not only depends on its basic electrical characteristics; maximum power, tolerance rated value %, maximum power voltage, maximum power current, open-circuit voltage (Voc), short-circuit current (Isc), maximum system voltage, but also is negatively influenced by several obstacles such as ambient temperature, relative humidity, dust storms and suspension in air, shading, global solar radiation intensity, spectrum and angle of irradiance [1, 2].

The operating temperature plays an essential role in the PV energy conversion process. The electrical performance of a PV module which involves both the electrical efficiency and the power output depends linearly on the operating temperature [3-5]. Temperature affects how electricity flows through an electrical circuit by changing the speed at which the electrons travel. This is due to an increase in resistance of the circuit that results from an increase in temperature. Likewise, resistance is decreased with decreasing temperatures, i.e., the efficiency of photovoltaic cells decreases as temperature increases. All previous investigations agreed that the performance of PV panels reduces with increasing temperatures [6-10]. Therefore, most panels do not operating under ideal conditions due to different weather conditions or real ones. Since PV panels are more efficient at lower temperatures, PV systems have to

design with active and passive cooling. Cooling the PV panels allows them to function at a higher efficiency and produce more power. Panels can be cooled actively or passively. An active system requires some external power source to run. Many researchers have investigated and proposed different techniques of active cooling, consisting of forced air or water-cooling, to increase total energy output of the PV modules [10].

In this study, to improve the electrical efficiency of solar panels that operate in non-optimal conditions, an active water cooling system has been built on top of PV array to spray (pump) a cool water on front of the panel surface, to pull away heat, keep the panel cool and keep the panel within certain temperatures, i.e., within ambient temperature range.

## 2. Experimental Methodology

In this experimental study a PV-array cooled by a thin continues film of water flowing on the front of the panels has been considered. Cooling technique was utilized by flowing and spraying a film of water on the PV-array front. Due to the rapid flow of the water there should be only a slight increase in water temperature.

The main water source is divided to two sources for homogenous work over the array that consists of twenty PV panels which are connected in parallel to increase the electric current flow. Moreover, the evaporating water should further decrease the panel surface temperature, so result in increased power output, therefore, gaining better electrical efficiency due to decreasing the reflection loss (refractive index of water is 1.3, which is intermediate between glass, with 1.5, and air, with 1.0) [10,11].

### 2.1. System Configuration

Figure (1) shows the cooling of PV system which built in this study, at the Renewable Energy Research Center (RERC) /University of Anbar Campus. PV system consists of five units, connected in parallel with a capacity of each unit is 216 watt), therefore, the total capacity of the PV array is 1.080kw. It is fixed at 33.33° south facing. The DC power produced from the PV array is sent to the control system which consists of five convertors of NAPS NS77 type, ten batteries of 105Ah for each one, and one inverter of AJ sinewave type.

### 2.2. Measurements

1-Plastic tube (Reservoir)
2-Measurment devices
3-Water Sprayer Part
4-Pumping system and Flow Meter
5- Measurement devices
6-Control valve
7-Spraying view

*Figure 1. Cooling of PV system which built in this study*

*Table 1. Technical Specifications of PV modules*

| Parameter | Values |
| --- | --- |
| Vmax (Volt) | 17.4 |
| Imax (Ampere) | 3.11 |
| Voc (Volt) | 21.7 |
| Isc (Ampere) | 3.31 |
| Pout (Watt)/ panel | 54 |
| Rated capacity of each PV array (Watt)(parallel) | 54*4 = 216 |
| Total installed capacity (Watt) | 216*5=1080 |
| Dimensions of PV array | 1.5 m * 7.4 m *1.6m |
| Area required for the installation of five PV unit | 11.1 m2 |
| PV model | Kyocera 54 W |

The measurements were recorded during a clear day at the RERC in Ramadi city on the 15th of April 2014 and 1st of July

2014. The power of the pump for spraying and circulation of water is 0.25 hp. For practical design, the power of water pump should be less than the obtained gain of power from PV due to using the spraying technique. The flow rate of water is measured by an ULTRASONIC FLOMETER-TDS-100H flow meter. The Solar irradiance was measured by a Pyranometer at the same incident plane of the modules. Wind speed was recorded by weather station at RERC which was less than 1m/sec. Ambient temperature was measured in the shade space. Two temperature sensors (Thermocouple Type K) were installed on the front of the module surface (upper and lower surface) and average surface temperature was calculated. The temperatures of water inlet and water outlet, were also measured to calculate the temperature difference ($\Delta$Tw=heating rate). Current and voltage were measured by NAPS type millimeter with accuracy of 1 mA and 1 mV respectively. The water collected at the bottom of the panel passes through plastic tube used as a heat exchanger with environment and produces a constant low water temperature. Therefore when the water is return back to the feeding tube it would be at a desired temperature level to flow on the panel surface. Pumping system and the heat exchanger which are used in the combined system are shown in Fig. 1. Humidity was measured during the testing time and it was between 20% to 30%. The power output from each of the PV array was recorded every 25 min, and after that the spraying system is done for 5 minutes with recording the output power. More than one reading is taken at the same time in order to get smoothing curves, and this is due to difficulties in getting such smooth curves due to fluctuation in solar irradiance at the experimental site. The testing was done at clear day in order to get a maximum sun radiation with highest temperature on the PV surfaces.

## 3. Results and Discussion

### 3.1. Measurements on 15th of April 2014

Figure (2) shows the total system power, solar irradiance, and the average surface temperature for PV system. From this figure, the solar irradiance increases from 9:30am to 12:25pm, where it has a maximum value at 12:25pm is 1055w/m$^2$, this high value of solar irradiance causes the output power of PV system to be 1135W at 12:30pm after spraying the surface of PV system, this value is more than the standard total value of system power at 1000w/m$^2$ (i.e. 1080W). The surface temperature increases from 24.7°C at 7:30am to 60.5°C at 12:55pm. The reduction of temperature due to applying the spraying technique is 28.5°C at 1:00pm, where the panels' surface temperature is 32°C. The maximum power gain is 215W (1135W at 12:30pm - 920W at 12:55pm=215W). The total obtained power during the testing period is 18858W (with spraying, i.e. summation of 20 reading during the day) and is 16283w without spraying. So, the obtained gain for the whole period of test in one day is 2575W. The variation of power per time can give an indication of temperature effect on the output power which is:

$\Delta p/\Delta t=(18858\text{-}16283)/(947.1\text{-}584.3)= 7.097W/^{o}C$ (for 20 PV panel)

The values 947.1 and 584.3 present the sum of 20 reading of temperature without and with spraying respectively. The effect of increasing the temperature on each model is:

7.097/20 panel=0.354 W/ $^{o}$C.

The reduction of electrical power output within the range of 0.4W/$^{o}$C–0.5W/$^{o}$C for mono and multicrystalline silicon solar cells, respectively (which are used in most power applications) [11]. Note that the above value (i.e. 0.354W/$^{o}$C) is less than the range loss due to increasing the surface temperature of PV system, also it is calculated at non STC for 1000w/m$^{2}$.

The power with and without spraying is summarized in fig. (3). The average power with water spraying technique is 943W from 7:30am to 5:00pm in corresponding to 818W without spraying. Additional power can be obtained if the temperature of PV surface is kept below 32 $^{o}$C (the PV surface temperature at the midday in fig. (2)) even it is difficult due to the increasing in ambient temperature up to 36 $^{o}$C at the midday. This difference comes from the fact that in fig. (2) there are many temperature points on the curve have values above the standard test condition (25 $^{o}$C) after spraying the PV system by water, so the obtained power will also be affected by such increment of PV surface temperature. The efficiency and electrical power decrease with increases the operating temperature.

The output power increases as the solar irradiance (G) increases. From the same figure the output power with and without spraying will approach each other at the start and end of day which is considered as a normal case due to less increasing in surface temperature of PV system and the decreasing of ambient temperature at that time.

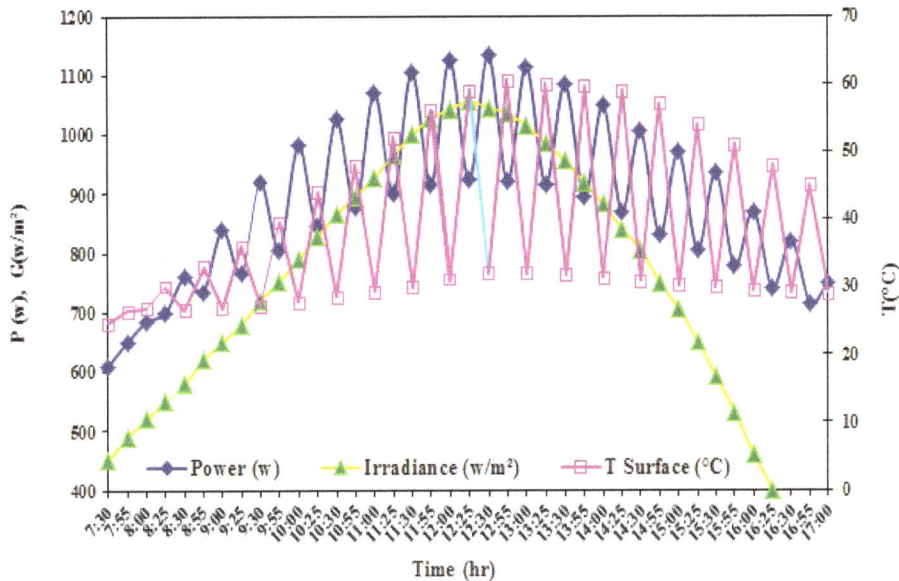

Figure 2. System output power, Irradiance, and models surface temperature versus time (15$^{th}$ of April 2014).

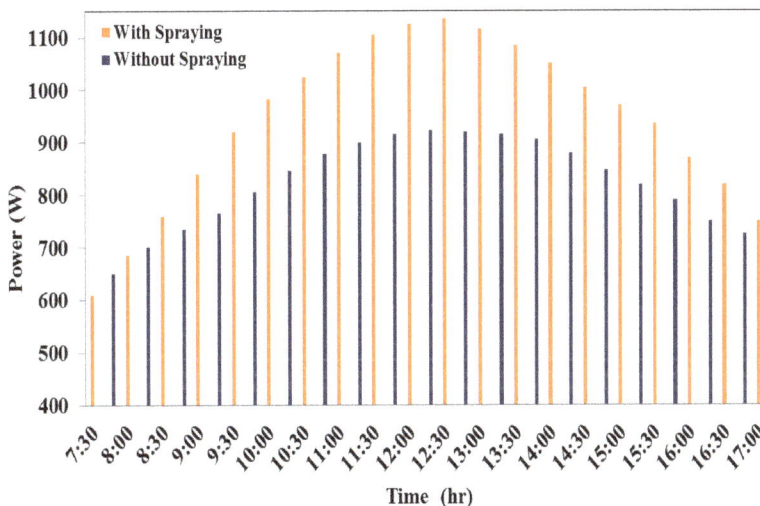

Figure 3. System output power with and without spraying (15$^{th}$ of April 2014).

### 3.2. Measurements on 1st of July 2014

Figure (4) illustrates the PV system performance with and without spraying in 01/July/2014. Since the day time is around 14 hour in July, the graph axis is started from 6:00am to 7:00pm. The ambient temperature at this month is high as shown in fig. (5). The output power increases gradually up to 9:00am. The effect of temperature at this period is not clear due to increasing in solar insulation values which substitute the negative effect of temperature on the generated power. The fluctuation of output power increases by time till the midday. The solar insolation has maximum values in the period (12:30-14:00)pm, which are around 1080w/m$^2$. Since the ambient temperature is around 47 $^{\circ}$C (from 1:10pm to 4:30pm) this will cause the output power to be reduced even the solar irradiance is high at this period. The PV surface temperature is about 35 $^{\circ}$C in the period (12:30 to 4:30)pm with water spraying model. The maximum reduction of temperature is about 38.7$^{\circ}$C (74.2 $^{\circ}$C at 1:55pm -35.5 $^{\circ}$C at 2:00pm=38.7 $^{\circ}$C). The average power for the whole day is 800W with spraying technique and 735W without spraying. The obtained average power in April is higher than the corresponding value in July due to high ambient temperature values in July which have direct effect on the out power. The PV surface temperature is staying over 50 $^{\circ}$C at 7:00pm, where the ambient temperature is 40 $^{\circ}$C at the same time. The variation of power per time is:

$$\Delta p/\Delta t = (21585-19848)/(1560.1-840.2)= 2.1414\text{W}/^{\circ}\text{C (for 20 PV panel)}$$

The values 21585 and 19848 present the sum of 28 reading of power with and without spraying technique. The other values in brackets are for the temperature on same period. The effect of increasing the temperature on each model is:

$$2.1414/20 \text{ panel}=0.107 \text{ W}/ ^{\circ}\text{C}$$

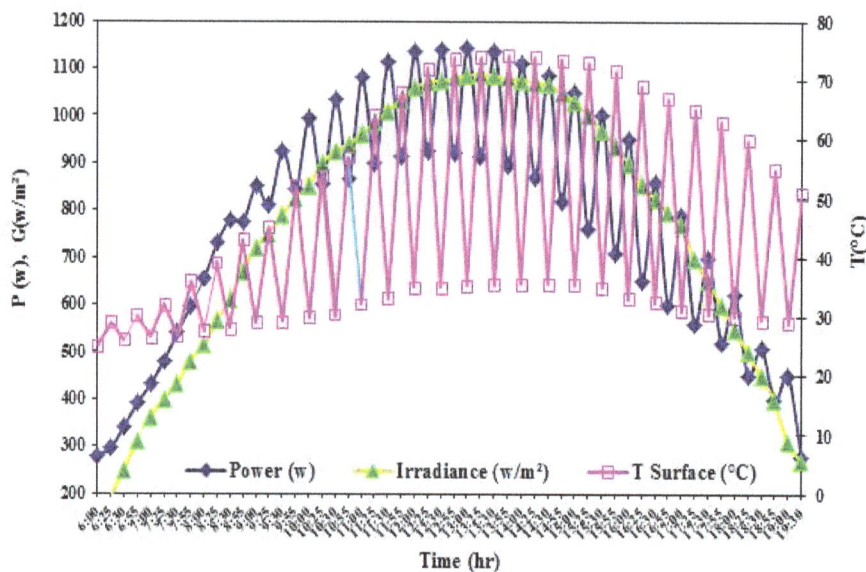

***Figure 4.*** *System output power, Irradiance, and models surface temperature versus time (1$^{st}$ of July, 2014).*

Figure (5) presents the ambient temperature at the middle of April-2014 and the first day of July-2014.The measurement is based on thermal sensor for Soly2 sun tracker system which already works at the same site of this practical project. Also the measurement is automatically registered by data logger device for each two minutes. From this figure, the ambient temperature increases with time from 6:00am to 1:00pm. The maximum temperatures are 36 $^{\circ}$C and 47 $^{\circ}$C at 2:00pm in these two days. The temperature reduces slowly after the midday which verifies the question of why the surface temperature reduces slowly in fig. (2) and fig. (4) after the midday.

Using spraying of water on the front of panel as a coolant cools it cause the PV to generate more power than the system without water spraying. Approximately, the cooling rate for 5 min.= 4 $^{\circ}$C/min in midday, as could be shown in Table 2. and Fig. 6. A water coolant system for solar panels may help the solar panels to cool and increase energy output during clear, sunny days.

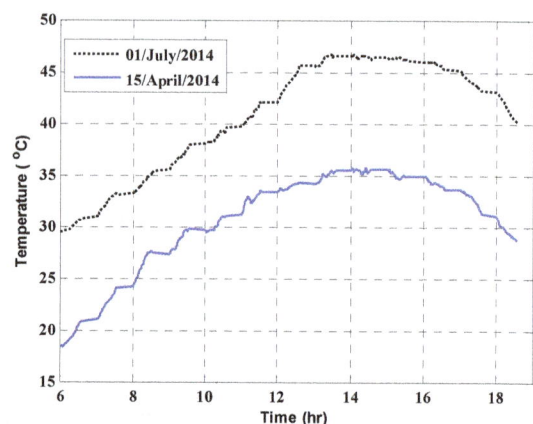

***Figure 5.*** *The ambient temperature versus time at the RERC-University of Anbar (15th of April and 1st of July, 2014)*

**Table 2.** *Cooling rate of Panel surface along one day in October 2014.*

| Time (hr) | dT/time (5 min.) =Cooling rate (°C/min.) |
|---|---|
| 9:55-10:00 | 1.87 |
| 10:25-10:30 | 2.71 |
| 10:55-11:00 | 3.59 |
| 11:25-11:30 | 3.61 |
| 11:55-12:00 | 2.71 |
| 12:25-12:30 | 4.93 |
| 12:55-13:00 | 4.76 |

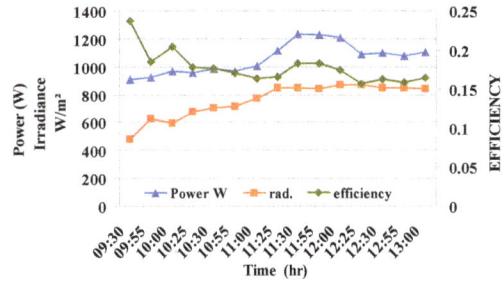

**Figure 8.** *The relationship between power, radiation and efficiency versus time along one day in April 2014.*

## 4. Conclusions

Increasing the efficiency and the power output depends largely on reducing the PV surface temperature, which allows the greatest benefit from the whole system. Experimental results show that the PV cells power has increased due to spraying of water over the PV front surface. The cooling rate for 5 min.= 4 °C/min in midday. This can significantly increase the system efficiency. The average value of efficiency for spraying system along one day was 17.8 %. Besides, the reduction in installation area for solar panels was acquired as result of power saving.

## Acknowledgements

This work is supported by the University of Anbar-Iraq /Renewable Energy Research Center with Grant No. RERC-PP23.

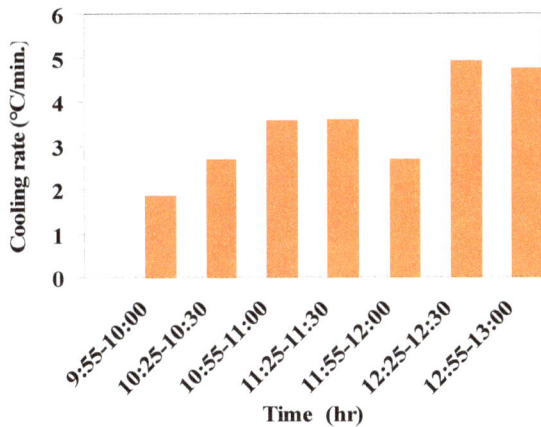

**Figure 6.** *Cooling rate of Panel surface along one day in April 2014.*

There is another application of spraying system which is water heating, so spraying system considered as hybrid system. Temperature difference of water along one day or heating rate $\Delta T_W$ is shown in Fig. 7. Higher heating rate acquired at mid day was 9.8°C, and the average of heating rate of water along one day was 6.83°C.

The efficiency is determined by dividing the product of voltage and battery current by average global solar radiation as shown in Figs. 7 and 8. The top values of efficiency were observed in midday. The average value of efficiency of spraying system along one day was 17.8 %.

## References

[1] Shafiqur Rehman and Ibrahim El-Amin, "Performance evaluation of an off-grid photovoltaic system in Saudi Arabia", *Energy* 46, pp. 451-458, 2012.

[2] K.E. Park , G.H. Kang, H.I. Kim, G.J. Yu and J.T. Kim, "Analysis of thermal and electrical performance of semi-transparent photovoltaic (PV) module", *Energy*, pp. 2681–2687, 2012.

[3] Omubo-Pepple V B, Israel-Cookey C and Alaminokuma G I, "Effects of Temperature, Solar flux and Relative Humidity on the Efficient Conversion of Solar Energy to Electricity", *European Journal of Scientific Research*, vol.35 (2), pp. 173-180, 2009.

[4] Kawamura T, Harada K, Ishihara Y,  Todaka T, Oshiro T, Nakamura H, and Imataki M, stics in Photovoltaic power system", *Solar Energy Materials and Solar Cells*, vol.47, pp. 155-165, 1997.

[5] E. Skoplaki and J.A. Palyvos, "On the temperature dependence of photovoltaic module electrical performance: A review of efficiency/power correlations", *Solar Energy* 83, pp. 614–624, 2009.

[6] Ben Richard Hughes, Ng Ping SzeCherisa, and Osman Beg, "Computational Study of Improving the Efficiency of Photovoltaic Panels in the UAE", World Academy of Science, *Engineering and Technology* 49, pp. 278-287, 2011.

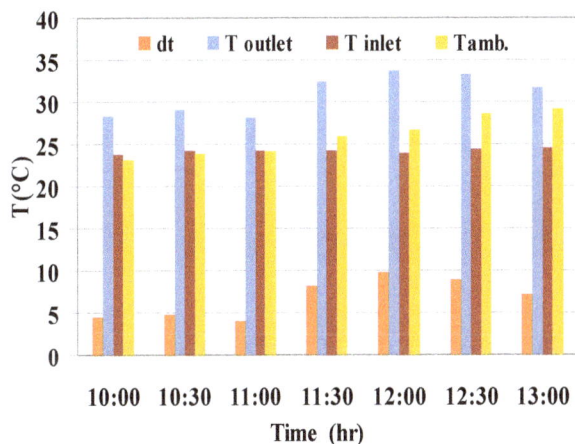

**Figure 7.** *Temperature difference of water along one day in April 2014 (Heating rate ΔTW).*

[7]   Sandstorm JD., "A method for predicting solar cell current-voltage curve characteristics as a function of incident solar intensity and cell temperature", National Aeronautics and Space Administration, series technical report; pp. 32-1142, 1967.

[8]   Osterwald CR, Glatfelter T and Burdick J., "Comparison of the temperature coefficients of the basic I-V parameters for various types of solar cells" In: Proceedings of the 19th IEEE photovoltaic specialists conference; pp. 188-193, 2008.

[9]   Makrides G, Zinsser B, Georghiou GE, Schubert M. and Werner JH. "Outdoor effi-ciency of different photovoltaic systems installed in Cyprus and Germany", the 33th IEEE photovoltaic specialists conference;11-16 May 2008, pp. 1-6, 2008.

[10]  L. Dorobanţu, M. O. Popescu, C. L. Popescu, and A. Crăciunescu, "Experimental Assessment of PV Panels Front Water Cooling Strategy", International Conference on Renewable Energies and Power Quality (ICREPQ'13) Bilbao (Spain), 20th to 22th March, 2013.

[11]  M. Abdolzadeh, M. Ameri, "Improving the effectiveness of a photovoltaic water pumping system by spraying water over the front of photovoltaic cells", *Renewable Energy*, vol. 34, no. 1, pp. 91–96, January 2009.

# Effect of silver nano-particle blended biodiesel and swirl on the performance of diesel engine combustion

**Nagaraj Banapurmath[1], T. Narasimhalu[1], Anand Hunshyal[2], Radhakrishnan Sankaran[3], Mohammad Hussain Rabinal[4], Narasimhan Ayachit[5], Rohan Kittur[1]**

[1]Department of Mechanical Engineering, B. V. B. College of Engineering and Technology, Hubli-580031, India
[2]Department of Civil Engineering, B. V. B. College of Engineering and Technology, Hubli-580031, India
[3]Nehru College of Engineering and Research Centre Pampady, Trissure (Dist), Kerala, India
[4]Department of Physics, Karnatak University, Dharwad, Karnataka, India
[5]Department of Physics, Rani Channamma University, Belgaum, Karnataka, India

**Email address:**

nr_banapurmath@rediffmail.com (N. Banapurmath), narasimhalu.t.n@gmail.com (T. Narasimhalu), amhunashyal@bvb.edu (A. Hunshyal), ngpcradhakrish@yahoo.co.in (R. Sankaran), mkrabinal@rediffmail.com (M. H. Rabinal), (nhayachit@gmail.com (N. Ayachit), rohan_kittur@yahoo.co.in (R. Kittur)

**Abstract:** Increased energy requirement in sectors like transportation, power generation and others coupled with depletion of high energy non-renewable energy resources like petroleum products and their harmful tail pipe emissions has led to search for new alternative and renewable energy resources. Different methods have been adopted to reduce tail pipe emissions and these include engine modification, fuel alteration, and exhaust gas treatment. Low emission characteristics and equivalent energy density of biodiesel are useful for replacement for petroleum fuels in internal combustion engines. Recently addition of catalytic reactivity materials like metal and oxide materials to biodiesel and their effect on engine performance has been reported in the literature. Due to their special properties like higher thermal conductivity, chemical and electrical properties enhanced properties of the base fuel diesel/biodiesel when these additives were used has been reported. In the present work both engine modification as well as fuel alteration techniques have been adopted to study their effect on diesel engine performance and emission characteristics. Engine modification involved provision of tangential slots on the piston crown surface. Fuel modification included addition of metal and metal oxide nano-particles to Honge biodiesel called Honge Oil Methyl Ester (HOME) as an alternative fuel for diesel engine applications. Experimental investigations were carried out to determine performance, emission, and combustion characteristics of diesel engine operated on diesel, HOME and HOME-silver nano-particles blended fuels. The biodiesel was prepared from honge oil called Honge Oil Methyl Ester [HOME]. The silver nano-particles were blended with HOME in the mass fractions of 25ppm and 50ppm using a mechanical homogenizer and an ultrasonicator. Subsequently, the stability characteristics of silver nano-particles blended–biodiesel fuels were analyzed under static conditions for their homogeneity. A considerable enhancement in the brake thermal efficiency with substantial reduction in the harmful pollutants from the engine for the nano-additive biodiesel blends was observed. Maximum brake thermal efficiency was obtained for HOME+ 50SILVER with reduced harmful pollutants compared to HOME+25SILVER blends. With swirl intended slots provided on the piston crown surface the performance was further improved using HOME+50SILVER in general and for 6.5mm slot on the combustion chamber in particular.

**Keywords:** Diesel Engine, Silver Nano-Particle, HOME, Biodiesel, Ultrasonicator, Combustion, Emission

## 1. Introduction

Depletion of fissile fuel resources, their harmful emissions when used in engines and increased energy requirement for sectors like transportation, energy generation has resulted in a need to search for new alternative energy sources. The increased environmental pollution and stringent emission norms are considered to be the main reason for need of new energy source. Many researchers have made attempts to increase the engine efficiency and decrease pollutants from engine exhaust using different methods of engine modification as well as fuel alteration.

### 1.1. Nano-Particles as Additives

Addition of some metal and metal oxide in the form of nano-powder to the base fuel may enhance the properties of the fuels. This is due to the interesting properties of nano-particles like higher specific surface area, thermal conductivity, catalytic activity and chemical properties as compared to their bulk form. Many researchers have used nano-particles as additives in diesel as well as biodiesel as new hybrid fuel blends (Williams & Van den Wildenberg 2005)1. They reported properties of nano-particles such as size, thermal conductivity and chemical properties affect the performance and emission characteristics of engine. Reduced size of the nano-particles increased specific surface, surface to volume ratio, and surface area improving catalytic reactivity and magnetic properties as compared to their bulk form. Hence metal and metal oxide nano-particles addition to biofuel will improve the performance as well as reduce the harmful gases from engine exhaust (Jones et al. 2011)2. Adding aluminium nano-particles to ethanol and diesel as well as to biodiesel may enhance the ignition properties, faster burning, reduced incomplete combustion and ignition delay with heat built up in the fuel due to reactive nature of aluminium nano-particles added (Kao et al. 2008; Allen et al. 2011; Tagi et al. 2008; Basha and Anand 2010)3,4,5,6. Addition of cerium oxide nano-particles to biodiesel may also be effective as they enhanced surface area to volume ratio. Flash point of biodiesel, which is an indication of the volatility, was found to increase with the inclusion of such additives making them safer fuels. The viscosity of biodiesel was found to increase with the addition of cerium oxide nano-particles to biodiesel. The viscosity and the volatility were found to hold direct relations with the dosing level of the nano-particles (Sajith et al. 2010; Selvan et al. 2009)7,8. Performance and emission characteristics of a diesel engine operated on CNT- diesel blends have been experimentally investigated. Substantial enhancement in the brake thermal efficiency and reduced harmful pollutants were reported for such nano-additive-biodiesel blends compared to jatropha biodiesel and diesel. This could be due to better combustion associated with such novel hybrid nano-liquid fuel blends when used in diesel engines (Sabourin et al. 2009)9. Therefore, nano-particles can function as both catalyst and an energy carrier when used in base fuels in diesel engines. In addition, small scale of nano-particles, also facilitate stability of fuel suspensions when used with base fuels (Basha and Anand 2011)10.

### 1.2. Swirl Effect in Combustion

Efficiency and harmful pollutants from internal combustion engines mainly depends upon the combustion of fuel occurring inside engine cylinder. To obtain a better combustion with lesser emissions in diesel engines, it is necessary to achieve a good spatial distribution of the injected fuel throughout the entire space (Risi et al. 2003)11. The swirl effect of air and resulting fluid motion can have a significant effect on air-fuel mixing, combustion, heat transfer, and emissions (Schapertons and Thiele 1986)12. At the time of air intake during suction stroke swirl motion is needed for proper mixing of fuel and air, while at the time of compression swirl effect will be decreased due to decreased angular momentum of air compared to intake time. When the piston moves towards Top Dead Centre (TDC), the slots (grooves) provided on the piston crown surface has a significant effect on air flow resulting in better mixing with injected fuel spray and better combustion. This cannot happen during the time of air intake. For a base line engine provided with tangential grooves on the piston and retarded injection timing decreased delay period and better mixing of air-fuel mixture was observed resulting in better performance with reduced emissions (Subba Reddy et al. 2013)13. Introduction of swirl by providing number of grooves viz., 3, 6, and 9 on the piston crown to study its effect on the performance of a direct injection diesel engine fuelled with karanja biodiesel has been reported (Bharathi et al. 2012)14. Increased brake thermal efficiency by 6.9% and reduction in smoke emission by 5.9%, NOx by 1.8%, and HC by 2.83% were obtained. The effect of swirl on combustion and emissions of a heavy duty-diesel engines has been investigated (Timoey 1985)15 and optimum level of air swirl minimizing soot has been obtained and reported to depend on engine running conditions. Over-swirling caused centrifugal action directing fresh air away from the fuel, resulting in incomplete combustion and there by increased soot formation has been reported (Tippelmen 1977)16.

## 2. Nano-Biodiesel Fuel Blends Preparation

For the present study Honge oil was selected as an alternative fuel for diesel as it was locally and abundantly available. It was then subsequently converted into its biodiesel using transesterification method. It is a well established method used for biodiesel production. The optimum parameters for better conversion were determined maximizing biodiesel yield (Banapurmath and Tewari 2008, 2010)17, 18.

### 2.1. Blending Methodology

The nano-particles blended honge biodiesel fuel is prepared by mixing the HOME and silver nano-particles with the aid of an ultrasonicator. The ultrasonicator technique is the best suited method to disperse the silver nano-particles in the base fuel, as it facilitates possible agglomerate nano-particles back to nanometre range. The nano-particles are weighed to a predefined mass fraction say 25ppm and dispersed in the HOME with the aid of ultrasonicator set at a frequency of 40 kHz for 30 minutes. The resulting nano-particles blended honge biodiesel is named as HOME+25SILVER. The same procedure is carried out for the mass fraction of 50ppm to prepare the silver nano-particles blended honge biodiesel fuel (HOME+50SILVER). For analyzing the stability characteristics of silver nano-particles blended HOME, the blends were kept in bottles under static conditions.

*Figure 1. Blended Fuels.*

*Table 1. Specifications of Silver nano-particles.*

| Sl. No | Parameters | Silver nano-particles |
|---|---|---|
| 1 | Manufacturer | Sigma Aldrich, Bangalore |
| 3 | Average particle size (APS) – nm | <150 nm |
| 4 | Surface area (SSA) m2/g | 385 |
| 5 | Purity - % | 99 |
| 5 | Thermal conductivity –W/mK | 429 |
| 6 | Density   g/cm3 (lit.) | 10.49 |

Table 2 below shows the Properties of HOME-silver nano-particles blends that are determined by appropriate apparatus.

*Table 2. Properties of HOME-silver nano-particles blended fuels.*

| Properties | Diesel | Home | Home+25silver | home+50silver |
|---|---|---|---|---|
| Density, kg/m3 | 840 | 875 | 895 | 900 |
| Calorific value, kJ/kg k | 42390 | 36100 | 35000 | 35500 |
| Viscosity at 40°C (cSt) | 4.59 | 5.6 | 5.8 | 5.8 |
| Flash point in °C | 75 | 187 | 160 | 158 |
| Cetane number | 45-55 | 40 | - | - |

## 2.2. Swirl Effect

*Figure 2. Piston crown surfaces showing slots of widths.*

Swirl as mentioned earlier assists in proper mixing of air and fuel inside the combustion chamber. In the present study swirl was induced by providing suitable tangential slots (grooves) of different widths on the existing hemispherical combustion chamber. The piston crown of 80 mm bore was suitably modified to accommodate four tangential grooves. These grooves were of different widths viz., 5.5mm, 6.5mm,

and 7.5mm and each were having a constant depth of 2 mm. Figure 2 shows the images of piston with tangential slots.

## 2.3. Heat Release Rate Calculations

The heat release rate of the fuel causes a variation of gas pressure and temperature within the engine cylinder, and strongly affects the fuel economy, power output and emissions of the engine. It provides a good insight into the combustion process that takes place in the engine. So finding the optimum heat release rate is particularly important in engine research. During this work a computer program was developed to obtain the heat release rate. The heat release rate at each crank angle was calculated by using a first law analysis of the average pressure versus crank angle variation obtained from 100 cycles using the following expression given below:

$$Qapp = \frac{\gamma}{\gamma-1}[PdV] + \frac{1}{\gamma-1}[Vdp] + Qwall$$

Where,

| | |
|---|---|
| $Q_{app}$ | Apparent heat release rate (J) |
| $\gamma$ | Ratio of specific heats Cp/ (Cp – $\bar{R}$) |
| $\bar{R}$ | Gas constant in (J / kmol-K) |
| $C_p$ | Specific heat at constant pressure (J / kmol – K) |
| V | Instantaneous volume of the cylinder (m3) |
| P | Cylinder pressure (bar) |
| $Q_{wall}$ | Heat transfer to the wall (J) |

The heat transfer was calculated based on the Hohenberg equation (Hohenberg 1979) given below and the wall temperature was assumed to be 7230 K (Hayes1986).

$$Qwall = h \times A \times [Tg - Tw]$$

$$h = C_1 V^{-0.06} P^{0.8} T^{-0.4} (V_P + C_2)^{0.8}$$

| | |
|---|---|
| h | Heat transfer coefficient in W/m2 K |
| $C_1$ & $C_2$ | Constants, 130 & 1.4 |
| V | Cylinder volume in m$^3$ |
| P | Cylinder pressure in bar |
| T | Cylinder gas temperature in K |
| $V_P$ | Piston mean speed in m/s |
| A | Instantaneous Area (m$^2$) |

# 3. Experimental Set Up

*Figure 3. Experimental Test Ring.*

*Table 3. Specifications of test rig.*

| Sl. No. | Engine specification | |
|---|---|---|
| | Parameters | Engine |
| 1 | Type of engine | Kirlosker make Single cylinder four stroke direct injection diesel engine |
| 2 | Nozzle opening pressure | 200 to 205 bar |
| 3 | Rated power | 5.2 kW (7 HP) @1500 rpm |
| 4 | Cylinder diameter (Bore) | 87.5 mm |
| 5 | Stroke length | 110 mm |
| 6 | Compression ratio | 17.5 : 1 |

The experimental investigations were conducted in two phases. In the first phase, the various physicochemical properties of modified biodiesel blends were determined and compared with those of the base fuels. In the second step, extensive experiments were conducted to determine performance, combustion and emission characteristics of a single cylinder four stroke direct injection compression ignition engine fuelled with modified and base fuels. Figure 3 shows the schematic experimental set up used. The engine was always operated at a rated speed of 1500 rev/min. The engine had a conventional fuel injection system. Injector was provided with three holes each having an orifice diameter of 0.3 mm. The injector opening pressure and the static injection timing as specified by the manufacturer was 205 bar and $23^0$ BTDC respectively. Eddy current dynamometer was used for loading the engine. The fuel flow rate was measured on the volumetric basis using a burette and stopwatch method. A piezoelectric pressure transducer was mounted with the cylinder head surface to measure the cylinder pressure. The emission characteristics were measured by using HARTRIDGE smoke meter and 5 gas analyzer during the steady state operation. The tests were conducted with diesel, HOME and HOME–silver nano-particle blends. The specification of the compression ignition (CI) engine is given in Table 3. Injection timing and injection pressure for diesel were maintained at 23°bTDC and 205 bar, while for HOME and HOME-Nanoparticle blended fuels it was kept at 19°bTDC, 230 bar. Engine was in turn run with the selected fuel combinations for different slotted pistons. Table 3 shows specifications of the engine used for present work.

# 4. Results and Discussion

This section explains the performance, emission and combustion characteristics of the diesel engine fuelled with novel hybrid fuel blends. Effects of nano-particle additives to HOME and swirl on the diesel engine performance is presented.

*Effects of nano-particle additives to HOME:*

### 4.1. Variation of Brake Thermal Efficiency

Figure 4 shows variation of brake thermal efficiency for diesel, HOME and HOME-silver nanoparticle blended fuels. The HOME resulted in inferior performance due to its higher

viscosity (nearly twice diesel) and lower volatility and lower calorific value. However the brake thermal efficiency of the HOME-silver nano-particle blended fuels improved compared to neat HOME operation. This could probably be attributed to the better combustion characteristics of HOME-silver nanoparticle blends. In general, the nano-sized particles possess high surface area and reactive surfaces that contribute to higher chemical reactivity and act as potential catalyst. In this perspective, the catalytic activity of HOME-silver nano-particle could have improved due to the existence of higher surface area and active surfaces prevailing. The silver nano-particles provide higher surface area and higher thermal conductivity as compared to HOME. Moreover, in case of HOME+50SILVER the catalytic activity may be enhanced due to the high dosage of silver nano-particles compared to that of HOME+25SILVER. An increase in surface area in liquid fuel droplets due to the possibility of droplets somehow being formed on the nano-particles could also be responsible for this observed trend. These properties provide increased reactivity and faster burning rates of fuel. Due to this effect, the brake thermal efficiency was higher for HOME+50SILVER compared to that of HOME+25SILVER.

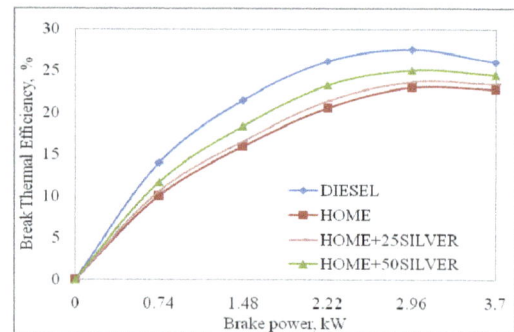

*Figure 4. Variation of Brake Thermal Efficiency.*

### 4.2. Variation of HC Emission

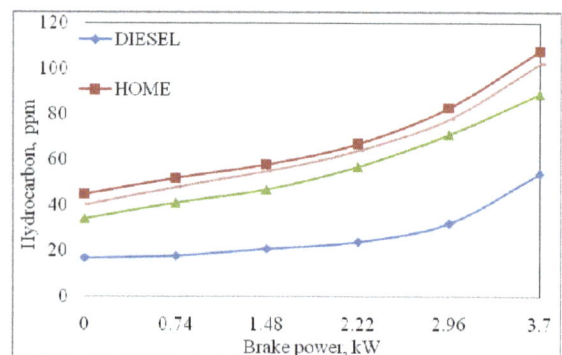

*Figure 5. Variation of HC Emission.*

The un-burnt HC emission variations for HOME and HOME-silver nanoparticle blended fuels are shown in Figure 5. The HC emission for HOME operation was higher compared to diesel due to its lower brake thermal efficiency resulting from incomplete combustion. However HC emissions were marginally lower for the HOME-silver blended fuels compared to HOME alone operation. This could be due to increased catalytic activity and improved combustion characteristics of silver NPs which lead to improved combustion. HOME+50SILVER showed better performance with comparatively lower HC as compared to HOME+25SILVER due to the increased dosing level of silver nano-particles that provided higher surface area resulting in improved combustion characteristics.

### 4.3. Variation CO Emission

The CO emissions for diesel, HOME, HOME-silver blended fuel are shown in Figure 6. The CO emission for HOME operation was higher compared to diesel due to its lower thermal efficiency resulting in incomplete combustion. However CO emissions were marginally lower for the HOME-silver blended fuels than HOME. The higher catalytic activity and improved combustion characteristics of silver NPs and leading to improved combustion could be the reason for this performance. HOME+50SILVER showed the better results as compared to HOME+25SILVER fuel due to the increased dosing level of silver nano-particles that facilitates complete combustion of fuel inside the engine.

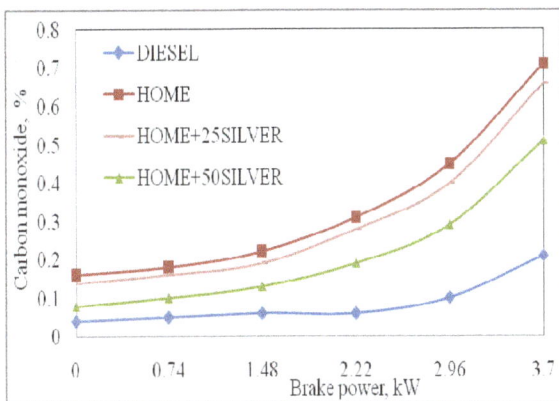

Figure 6. Variation of CO Emission.

### 4.4. Variation of Nox Emission

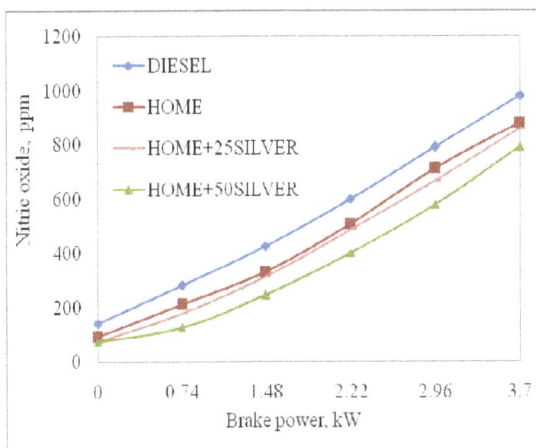

Figure 7. Variation of Nitric oxides

Figure 7 shows variation of NOx emission for diesel, HOME, HOME-silver blended fuels. HOME shows lower NOx emissions compared to diesel operation. Heat release

rates of HOME were lower during premixed combustion phase, with lower peak temperatures prevailing inside the combustion chamber. Nitrogen oxides formation strongly depends on the peak temperature, which explains the observed phenomenon. Furthermore, HOME-silver nano-particles blended fuels produced lower NOx emission compared to that of HOME. This could also be due to higher premixed combustion observed with HOME- silver nano-particles blends. The HOME+50SILVER showed lower NOx compared to the HOME+25SILVER.

### 4.5. Cylinder Pressure

Figure 8 shows the variation in pressure with crank angle for HOME and HOME-silver nanoparticle blended fuels. Combustion started later in comparison to diesel with biodiesel and biodiesel silver nano-particle blended fuels. However increased catalytic activity observed with silver nano-particle blended biodiesel results in reduced delay period with combustion starting earlier as well.

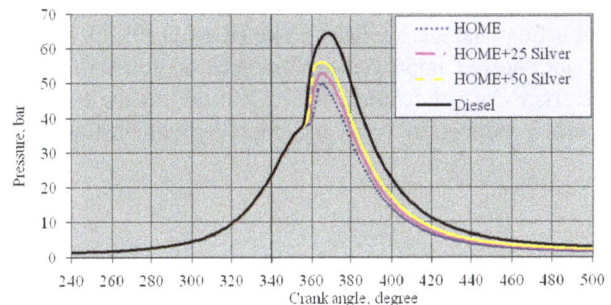

Figure 8. Variation of cylinder pressure with crank angle for 80% load

### 4.6. Heat Release Rate

Figure 9 shows heat release rate for nano-particle biodiesel fuel blends tested. It follows that for the biodiesel a shorter premixed heat-release portion occurs, in spite of their increased ignition delay. The heat release rate for HOME and HOME-silver nano-particles were lower compared to diesel operation. The reduced heat release rate during premixed combustion phase and increased heat release rate were observed during diffusion combustion phase for both HOME and HOME-silver nano-particles. This leads to increased exhaust gas temperature. With blend of silver nano-particles in HOME premixed combustion increased compared to HOME due to their increased catalytic activity, thermal conductivity and surface area.

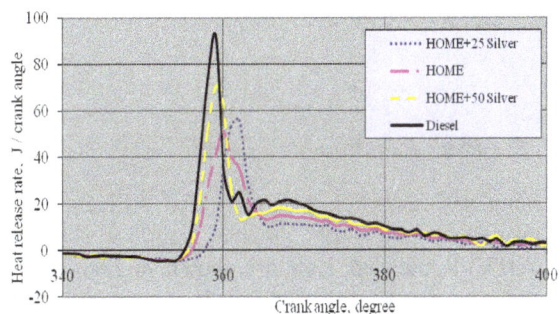

*Figure 9. Variation of heat release rate with crank angle for 80% load.*

# 5. Effect of Swirl on Diesel Engine Fuelled with Novel Hybrid Blends

In order to further improve the performance of diesel engine fuelled with HOME+50SILVER nano-particles, the swirl was created by inserting suitable slots on the hemispherical piston surface. Subsequently diesel engine performance and emission characteristics were obtained.

## 5.1. Variation of Brake Thermal Efficiency

Figure 10 shows variation of brake thermal efficiency for different slot widths provided on the piston surface using HOME+50SILVER blended fuels. The HOME+50SILVER with plain piston surface slot resulted in inferior performance. However brake thermal efficiency of the HOME+50SILVER blended fuels with swirl assisted by SLOTS improved compared to neat HOME+50SILVER operation. This could probably be attributed to the better mixing of fuel and air in the combustion chamber cylinder. The HOME+50SILVER+ 6.5mm slot showed improved performance as compared to the other two slots of 5.5mm and 7.5mm due to the better mixing effect induced by swirl and improved combustion of fuel.

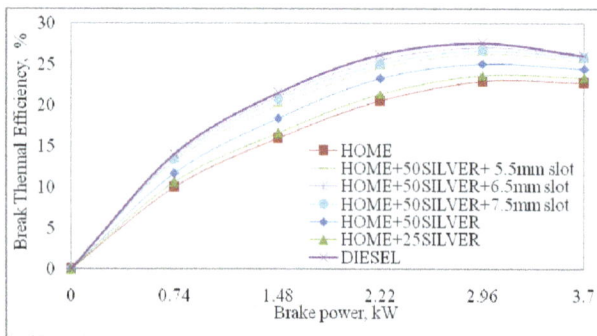

*Figure 10. Variation of Brake Thermal Efficiency*

## 5.2. Variation of HC Emission

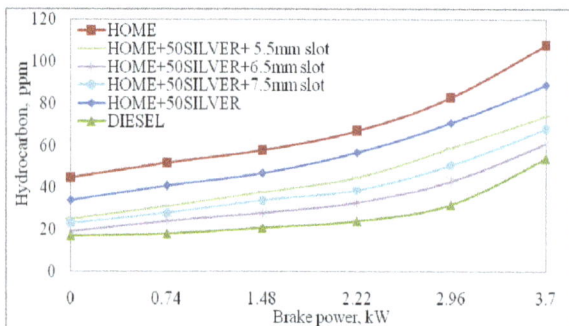

*Figure 11. Variation of HC Emission.*

The unburnt HC emission behaviour for different slot widths provided on the piston surface using HOME+50SILVER blended fuels are shown in Figure 11. The HC emission for HOME+50SILVER operation in all

modes is higher compared to diesel due to its lower thermal efficiency. However HC emissions are marginally lower for the HOME+50SILVER+SLOTS blended fuels than HOME. This could be due to increased catalytic activity and improved combustion characteristics of silver NPs and better mixing of air-fuel in combustion chamber, which lead to improved combustion. The HOME+50SILVER+ 6.5mm slot showed lower UBHC emission as compared to the other two slots of 5.5mm and 7.5mm.

## 5.3. Variation CO Emission

The CO emission variations for HOME+50SILVER fuel combinations with and without slots on the piston surface are shown in Figure 12. The CO emission for HOME+50SILVER operation is higher compared to HOME+50SILVER+SLOTS due to its lower thermal efficiency with incomplete combustion. However CO emissions are marginally lower for the HOME+50SILVER+SLOTS blended fuels than HOME+50SILVER. The higher catalytic activity and improved combustion characteristics of silver NPs combined with swirl induction probably lead to improved mixing of fuel and hence in the resulting behavior. The CO emission of HOME+50SILVER+ 6.5mm slot was much lower than that of as compared to the other two slots of 5.5mm and 7.5mm.

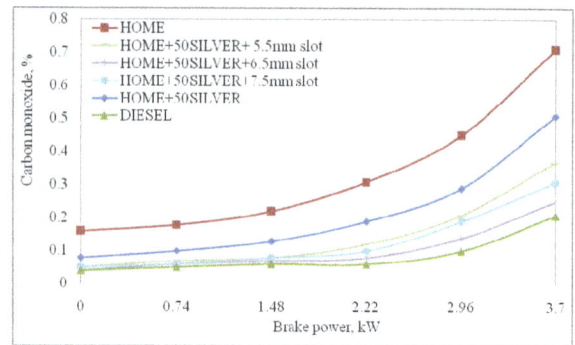

*Figure 12. Variation of CO Emission.*

## 5.4. Variation of NOx Emission

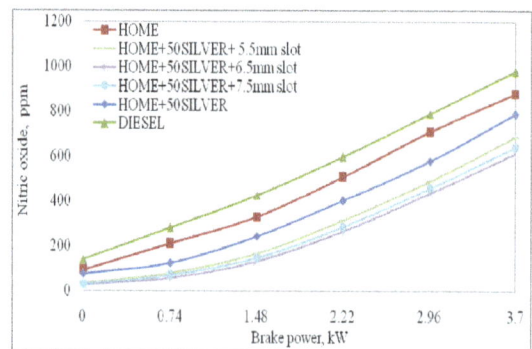

*Figure 13. Variation of NOx emission.*

Figure 13 compares variation of NOx emission for HOME+50SILVER fuel combinations with and without slots on the piston surface. The HOME+50SILVER showed lower NOx emissions compared to diesel operation. NOx emission of HOME+50SILVER+ 6.5mm slot was much higher than that observed for 5.5mm and 7.5mm slots. This could be due to more heat release rate obtained during premixed combustion for 6.5mm slot operation.

### 5.5. Cylinder Pressure

Figure 14 shows the variation in pressure with crank angle for HOME-silver nanoparticle blended fuels considering the effect of swirl. Combustion started later in comparison to diesel with biodiesel and biodiesel silver nano-particle blended fuels. However increased catalytic activity observed with silver nano-particle blended biodiesel coupled with swirl inducted resulted in reduced delay period with combustion starting earlier as well.

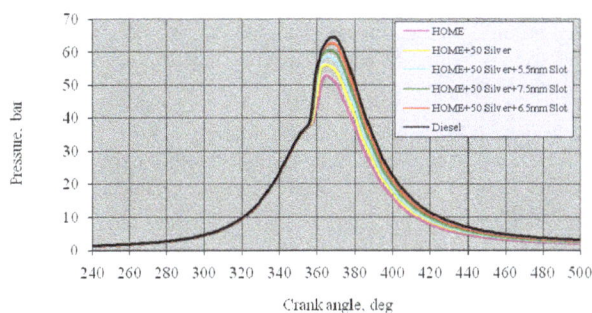

*Figure 14. Variation of cylinder pressure with crank angle for 80% load.*

### 5.6. Heat Release Rate

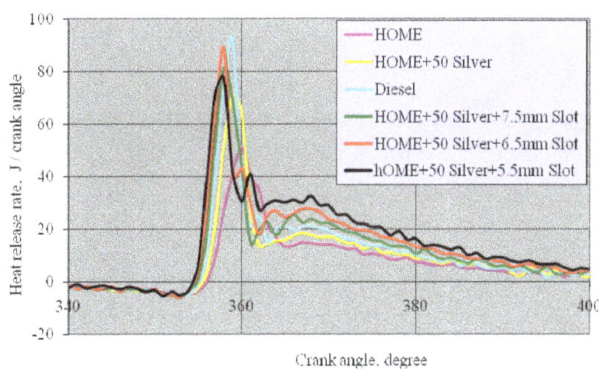

*Figure 15. Variation of heat release rate with crank angle for 80% load.*

Figure 15 shows heat release rate for nano-particle biodiesel fuel blends tested. It follows that for the biodiesel a shorter premixed heat-release portion occurs, in spite of their increased ignition delay. The heat release rate for HOME and HOME-silver nano-particles were lower compared to diesel operation. The reduced heat release rate during premixed combustion phase and increased heat release rate were observed during diffusion combustion phase for both HOME and HOME-silver nano-particles. This leads to increased exhaust gas temperature. With blend of silver nano-particles in HOME premixed combustion increased compared to HOME due to their increased catalytic activity, thermal conductivity and surface area. However increased catalytic activity observed with silver nano-particle blended biodiesel coupled with swirl inducted resulted in reduced delay period

with higher heat release rates.

## 6. Conclusions

The performance, and the emission characteristics of HOME, HOME-silver nano-particles blended fuels with and without the effect of swirl were investigated in a single-cylinder, constant speed, direct-injection diesel engine. Based on the experimentation data, the following conclusions were drawn.

1. HOME resulted in poor performance in terms of reduced brake thermal efficiency. However HOME performance was enhanced with silver nano-particle additives. Performance was further improved with higher dosing level of silver nano-particles in biodiesel.

2. Increased HC and CO emissions were observed for HOME alone operation. Emission reduced drastically with silver nano-addition. Further reduction in these emissions obtained with increased dosage of nano-particles to HOME. NOx emissions were lower for nano-particle blended HOME.

3. Effect of swirl with tangential slots provision on the piston surface showed better results and reduced emissions. 6.5 mm slot was found to be optimum.

4. HOME+50SILVER+SLOTS showed lowered NOx emission.

## References

[1]  Williams & van den Wildenberg, Roadmap Report on Nano-particles, November 2005.

[2]  J.Matthew, H.Calvin Li, Abdollah Afjeh, G.P. Peterson, Experimental study of combustion characteristics of nanoscale metal and metal oxide additives in bio fuel (ethanol), Nanoscale Research Letters, 2011; 6:246.

[3]  Kao Mu-Jung, Chen-Ching Ting, Bai-Fu Lin, and Tsing-Tshih Tsung, Aqueous Aluminum Nanofluid Combustion in Diesel Fuel, Journal of Testing and Evaluation, 36(2).

[4]  C.Allen, G.Mittal, C.Sung, E.Tulson and T.Lee, An Aerosal rapid compression machine for studying energetic-nanoparticle-enhanced combustion of liquid fuels, 2011; 33(2):3367-3374.

[5]  H.Tagi, P.E.Phelan, R. Prasher, R. Peck., T. Lee, J.R. Pacheco, and P. Arentzen, Increased Hot-Plate ignition Probability for Nanoparticle-Laden Diesel Fuel, Nano Lett, 2008; 8(5):1410-1416.

[6]  B. Sadhik and R.B. Anand, Effects of Alumina Nano-particles Blended Jatropha Biodiesel Fuel on Working Characteristics of a Diesel Engine, International Journal of Industrial Engineering and Technology, 2010; 2(1): 53-62: ISSN 0974-3146.

[7]  V. Sajith, C.B.Sobhan and G.P.Peterson. Experimental Investigations on the Effects of Cerium Oxide Nanoparticle Fuel Additives on Biodiesel, Advances in Mechanical Engineering Volume (2010), Article ID 581407, 6 pages doi:10.1155/2010/58140.

[8]    V. Arul Mozhi Selvan, R.B. Anand and M.Udayakumar, Effects of cerium oxide nanoparticle addition in diesel and diesel-biodiesel-ethanol blends on the performance and emission characteristics of a C.I engine, ARPN Journal of Engineering and Applied Sciences, September 2009; 4(7).

[9]    L.Sabourin Justin, M. Daniel Dabbs, A. Richard, Yetter, L. Frederick Dryer, and A. Ilhan Aksay, Functionalized Graphene Sheet Colloids for Enhanced Fuel/Propellant Combustion, 2009, 3(12) : 3945-3954.

[10]   J. Basha Sadhik and R.B. Anand, An experimental investigation in a diesel engine using carbon nanotubes blended water–diesel emulsion fuel, Journal of Power and Energy 2011,Vol 225:279-288.

[11]   Risi Arturo de, T. Donateo and D. Laforgia. Optimization of the Combustion Chamber of Direct Injection Diesel Engines, SAE 2003; 01(1064).

[12]   S. Herbert and T. Fred, Three Dimensional Computations for Flow Fields in D I Piston Bowls, SAE 1986; 60463.

[13]   C.V. Subba Reddy, C. Eswara Reddy, K. Hemachandra Reddy, Effect of Tangential Grooves on Piston Crown Of D.I. Diesel Engine with Retarded Injection Timing, International Journal of Engineering Research and Development, 2013, 5(10): 01-06.

[14]   V.V. Bharathi, Prathibha and Dr. Smt.G. Prasanthi, Experimental Investigation on the Effect of Air Swirl on Performance and Emissions Characteristics of a Diesel Engine Fueled with Karanja Biodiesel, International Journal of Engineering Research and Development, 2012, 2(8) : 08-13

[15]   D.J.Timoey, SAE paper 1985,851543,

[16]   G.A. Tippelmen. New method of investigation of swirl ports, S A E paper 1977, 770404.

[17]   N. R. Banapurmath, P. G. Tewari, 2008,"Combustion and emission characteristics of a direct injection CI engine when operated on Honge oil, Honge oil methyl ester (HOME) and blends of Honge oil methyl ester (HOME) and diesel", *International Journal of Sustainable Energy*, June 1 (2):80–93.

[18]   N. R. Banapurmath, and P. G. Tewari, 2010, "Performance, combustion, and emissions characteristics of a single-cylinder compression ignition engine operated on ethanol–biodiesel blended fuels", Proc. IMechE, Vol. 224, Part A: *J. Power and Energy, 533-543*.

[19]   T.K. Hayes, L.D.Savage, and S.C. Soreson (1986), Cylinder Pressure Data Acquisition and Heat Release Analysis on a Personal Computer. Society of Automotive Engineers, Paper No. 860029, USA.

# 3

# Construction and Performance Study of an Underground Air Heating and Cooling System

**Amit Shor, Md. Nazmus Sakib Khan, GM. Ahteshamul Haque**

Dept. of Mechanical Engineering, Rajshahi University of Engineering & Technology (RUET), Rajshahi, Bangladesh

**Email address:**

Shor.amit26@gmail.com (A. Shor), Sakib2161@yahoo.com (Md. N. Sakib Khan), Ahteshampulok@gmail.com (GM. A. Haque)

**Abstract:** This is the time of searching alternative energy sources because of increasing demand of energy day by day against our limited resources. Now-a-days finding and developing alternative energy sources is the important responsibility and duty for human being and it is also demand of time. However, it is not easy to find an efficient alternative energy source. Thus, we can concentrate in finding a solution which can ensure a reduced load on the conventional energy sources. This project is about an underground room heating and cooling system which takes a very little energy to run itself. The system is totally environment friendly, does not produce any harmful exhaust or gases which could pollute the atmosphere. Moreover, it can be highly economical so that everyone can install this at their houses and also in educational organizations and offices. Four types of pipes of different materials were installed underground to cause heat transfer between ambient air passed through the pipes and the soil in contact with the pipe. Heating and cooling action in winter and summer respectively can be achieved by this system. The length of the pipes used is 30 feet and dug 8 feet deep. An inlet air flow at 4.5 m/s was provided. The air flow rate was found to be reduced at the exit and a change in temperature was also found. The cooling rates varied depending upon the temperature of the inlet air.

**Keywords:** Alternative Energy, Energy Efficient, Environment Friendly Energy System, Underground Heating & Cooling System, Economical Energy System

## 1. Introduction

In tropical country like Bangladesh, cooling of indoor air is growing due to increasing comfort expectations. Air conditioning is the most widely used cooling system for indoor air in Bangladesh.

The main component used in an air conditioner is compressor mainly driven by electricity. Electricity generation processes are fossil based and responsible for nitrogen dioxide, carbon, Sulphur and other GHG emission.

Refrigerants that are used in air conditioning have also negative impact on the environment. Freon like CFC and HCFC refrigerants are harmful to Ozone layer and are out of used now-a-days. CFC, HCFC and HFC (which used as replacement of CFC and HCFC) are all greenhouse gas.

Global warming in the impact of GHG gases responsible for average temperature rise worldwide. "The increase in temperature in the 20th century is likely to have been the largest in any century during the past 1000 years. 1990s were the warmest decade and 1998 was the warmest year. Global average land and sea surface temperature in May 2003 were the second highest since records began in 1880-WMO [1] in a press release 2 July 2003. All these happened due to consumption of fossil energy.

Bangladesh currently facing a trouble of energy distribution due to the lack of availability. Still now only 62% populations [8] under electrification as per govt. data. The demand supply gap of electricity is almost more than 500 MW based on the connected load. Considering the energy shortage attention has gone to the energy intensive domestic appliance (air conditioning) currently using in Bangladesh to comfort of indoor air. The underground assisted system to heat and cool the indoor air whenever needed to replace the use of a conventional air conditioning system & contributes in the global GHG mitigation.

The overall objective is to construct the underground assisted air heating and cooling system and its performance study. The specific objectives are -To construct the system, performance evaluation for winter season and to estimate the cost.

The regular air conditioners or room heating and cooling devices cost a fortune. The effect of the elements used in those are not also environment friendly. Thus a solar powered room heating and cooling device can offer in a lower cost as well as can be excellent environment friendly as it does not contain any harmful chemicals. The people all over the world can be benefitted from this device. Specially, in the developing countries, it could be a better choice.

## 2. Methodology

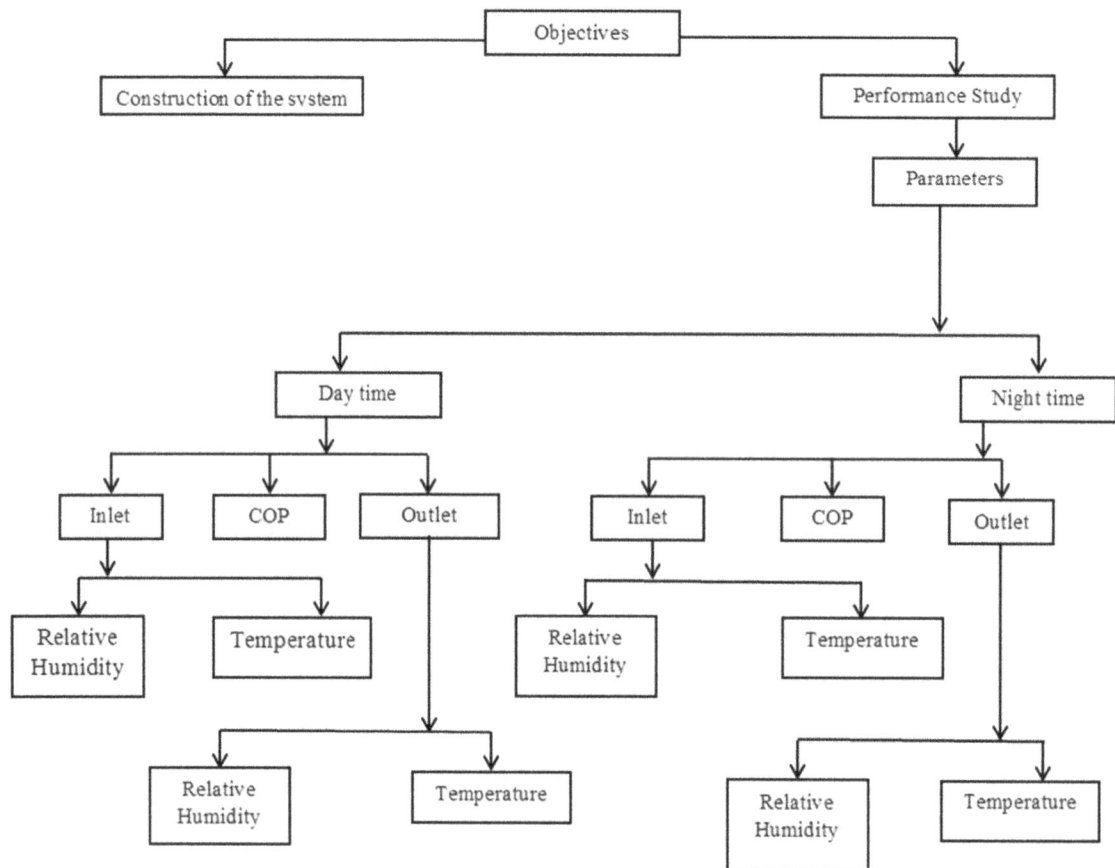

**Figure 1.** *Methodology.*

Figure 1 shows the methodology of the study. The objectives are in two dimensional named design, construction and performance study. The design of underground air passage includes depth, length, diameter, materials etc. The design of the sample house to be considered for heating and cooling includes volume of house, materials and set up arrangement etc. Then construction has been done assembling the entire materials. Finally the performance evaluation has been carried out using few performance indicators.

## 3. Construction

### 3.1. Technical Terms

Various technical terms are used to refer to earth tubes:
- Earth tubes
- Buried pipes
- Earth channels
- Air-to-soil heat exchanger
- Underground air pipe

- Earth-to-air heat exchanger (EAHX)
- Subsoil heat exchanger
- Earth-air tunnel system
- Ground tube heat exchanger
- Ground coupled heat exchanger

These terms all refer to the same kind of device: a pipe or series of pipes buried underground, and through which ventilation air is circulated. 'Earth-to-air heat exchanger' is probably the most technically accurate terminology, although 'earth tubes' also enjoys wide use. The systems can either be 'closed-loop' (i.e. recirculating the air from the building through the earth tubes), or 'open-loop' (i.e. drawing outside air through the pipes to ventilate the house). The system we have used in our set-up is an open loop system.

When a temperature gradient exists in a body, there is an energy transfer from the high temperature region to the low temperature region. The energy is transferred by conduction and that the heat transfer rate per unit area is proportional to the normal temperature gradient. [7]

$$q = kA\frac{dT}{dx} \tag{1}$$

Where,

q = the rate of heat transfer

k = Thermal conductivity of the material

A = Cross sectional area of the pipe

dT = The change in temperature

dx = Thickness of the pipe

$\frac{dT}{dx}$ = Temperature gradient

The conductivity of the different materials used is, [7]

Mild steel/ Carbon steel = 43 W/m °C

PVC = 0.09 W/m °C

Bamboo = 0.16-0.2 W/m °C

### 3.2. Pipe Length Selection

As the pipe length increases, the inlet air temperature decreases due to the fact that the longer pipe provides a longer path over which heat transfer between the pipe and the surrounding soil can take place given the same overall heat transfer coefficient of earth tube. Length can typically range from 10 to 100 m. Longer tubes correspond to more effective systems, but the required fan power and the cost also increase. [5]. For cost considerations, we have taken 30 feet for each of the four pipes.

### 3.3. Pipe Diameter Selection

As the pipe diameter increases, the earth tube outlet air temperature also increases due to the fact that higher pipe diameter results in a lower convective heat transfer coefficient on the pipe inner surface and a lower overall heat transfer coefficient of earth tube system. Smaller diameters are preferred from a thermal point of view, but they also correspond (at equal flow rate) to higher friction losses, so it becomes a balance between increasing heat transfer and lowering fan power. [5]. The pipes we have used has 1.5 inch diameter. For budget considerations, we only used similar diameter pipes for all the materials.

### 3.4. Pipe Depth Selection

As the pipe depth increases, the inlet air temperature decreases, indicating that the earth tube should be placed as deeply as possible. However, the trenching cost and other economic factors should be considered when installing earth tubes. Deeper positioning of the tubes ensures better performance. Typical depths are 1.5 to 3 m. The tubes can be positioned under the building or in the ground outside the building foundation. [5]

The depth of the pipes we used was 8 feet. The depth of the pipe required to heat or cool the air generally varies depending upon the geography of the set-up.

### 3.5. Material Selection

Four types of materials were used in our underground cooling and heating system.

1. Mild Steel pipe
2. PVC pipe
3. PVC filtered pipe
4. Bamboo

We used different types of materials to measure if there is any change in the cooling effect with respect to the change in materials. All materials mentioned above are available in local market and cost effective at the same time. The use of these materials with the above mentioned specifications would allow all class of people to install the system as household room heating and cooling device.

### 3.6. Flow Rate Selection

Lower flow rates are beneficial to achieve higher or lower temperatures, and also because they correspond to lower fan power. However, a compromise has to be made between pipe diameter, desired thermal performance, and flow rate. From literature review, we were suggested that the flow rate is preferable from 3-5 m/s. [5]. The flow rate we used was 4.5 m/s.

### 3.7. Fan

The circulating constant speed fan used at the inlet of the pipes to ensure the uniform air introduction into the system was a 1.92 watt fan. The power required to drive the fan or the input work is measured by the equation

$$P = VI \qquad (2)$$

Where,

P = Power (watt)

V = Voltage (volt)

I = Current (ampere)

The current was measured by connecting an ammeter to the multimeter circuit with the circulating fan.

$$P = 12 \times 0.16 = 1.92 \text{ watt}$$

*Figure 2. Circulating fan.*

### 3.8. Battery Capacity

The battery used to run the fan was a 12 Volt battery. Solar panel could also be used to run the fan.

### 3.9. Digital Hygro-Thermometer

A digital Hygro-thermometer is a versatile instrument which is used to measure the temperature and moisture content of air. This instrument was used to measure the inlet and exit temperature and humidity [9].

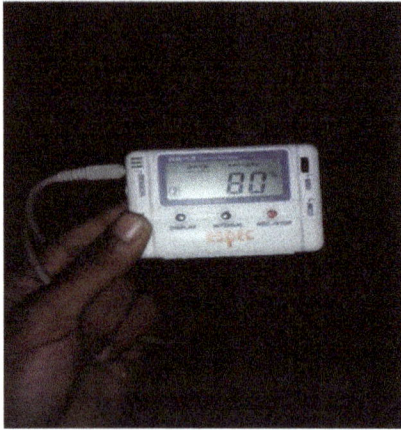

*Figure 3. Digital Hygro-Thermometer.*

### 3.10. Anemometer

An anemometer or wind meter is a device used for measuring wind speed, and is a common weather station instrument. Anemometers can be divided into two classes: those that measure the wind's speed, and those that measure the wind's pressure; but as there is a close connection between the pressure and the speed, an anemometer designed for one will give information about both [10].

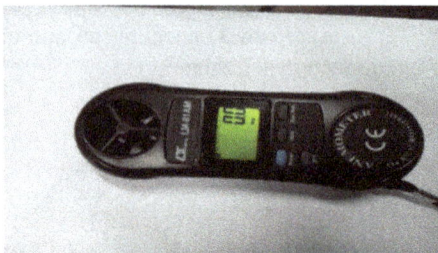

*Figure 4. Anemometer.*

### 3.11. Construction

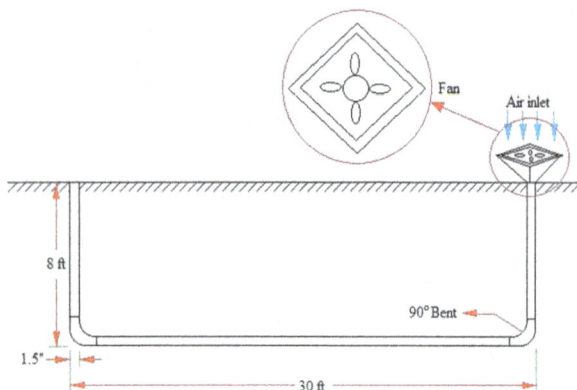

*Figure 5. Underground Air Heating and Cooling System.*

The above figure shows our basic design for the heating and cooling system for each pipes. The pipe depth is 8 feet, whereas the pipe length is 30 feet. The diameter of the pipes selected is 1.5 inch each.

*Figure 6. The setup of underground air heating and cooling system behind Boiler Lab.*

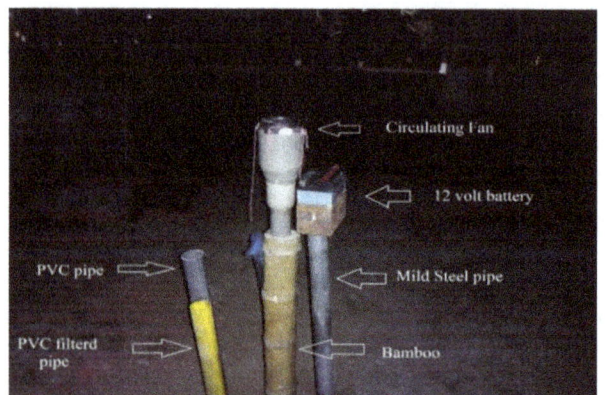

*Figure 7. The inlet port of the pipes with battery and fan.*

## 4. Result and Discussions

### 4.1. Description

The data required from the underground air conditioning system is categorized below as per their respective material. The data was recorded from 6.00 PM, 16th December, 2014 to 3.30 AM, 17th December, 2014 for the night time condition. And from 10.00 AM- 2.00 PM December 21st, 2014.

### 4.2. Variation of Inlet Temperature with Time (Night & Day)

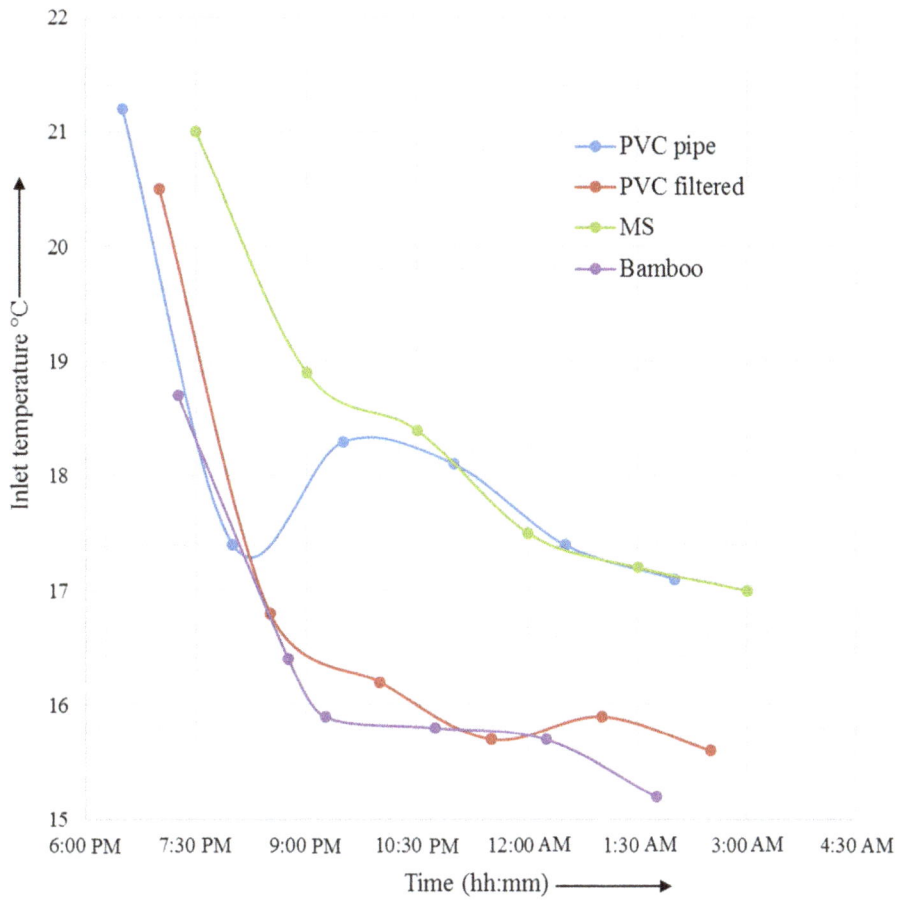

**Figure 8.** *Variation of inlet temperature with time (night).*

As the time passed, the temperature reduced. This is caused by the change in atmospheric condition. The lowest temperature recorded was 15.1 degree Celsius. The data was taken by digital hygro-thermometer.

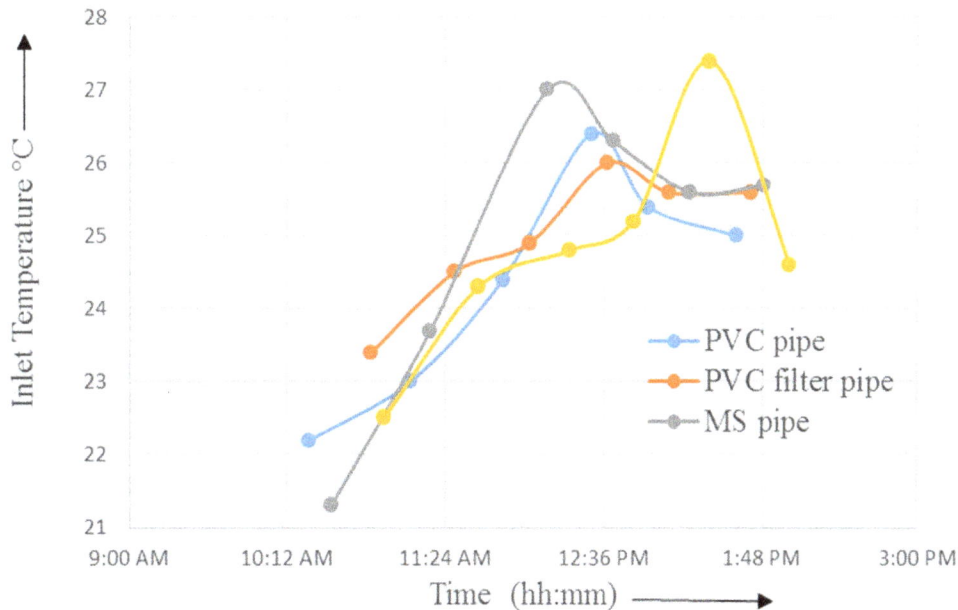

**Figure 9.** *Variation of inlet temperature with time (day).*

During the day time, the change in temperature was not gradual. Rather it was random. However, as there were cloud in the sky, the hygro thermometer had problem sensing the temperature. The highest temperature found was 27.4°C.

### 4.3. Variation of Inlet Humidity with Time (Night & Day)

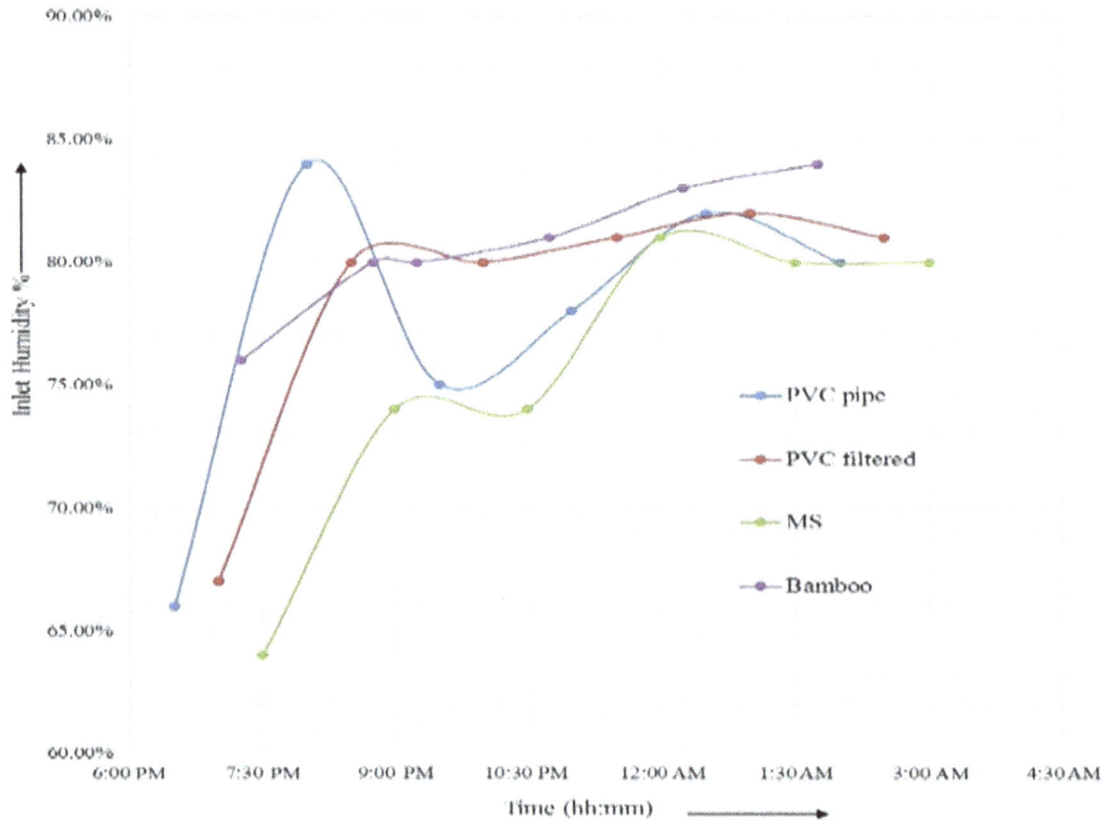

*Figure 10. Variation of inlet humidity with time (night).*

As the night grew older, there was a considerable increase in the moisture content of the air. The highest humidity was found 84%.

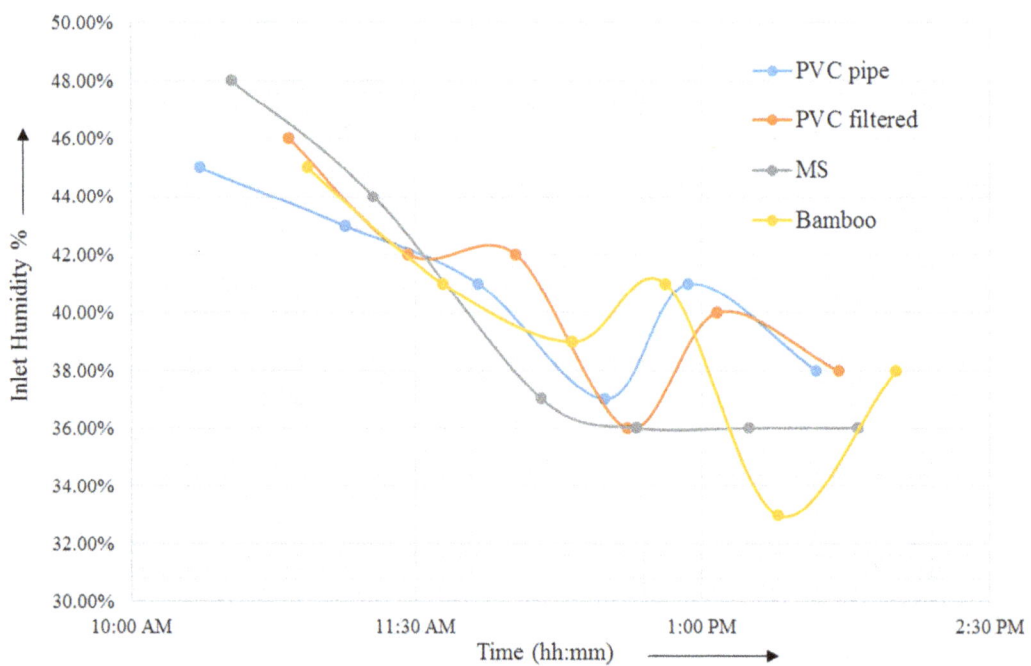

*Figure 11. Variation of inlet humidity with time (day).*

The humidity level at the day time was found to be considerably less than the humidity during the night. As time passed, it gradually decreased. The highest humidity was found 48%.

### 4.4. Variation of Temp. and RH for PVC with Time Under Inlet and Exit Condition (Night & Day)

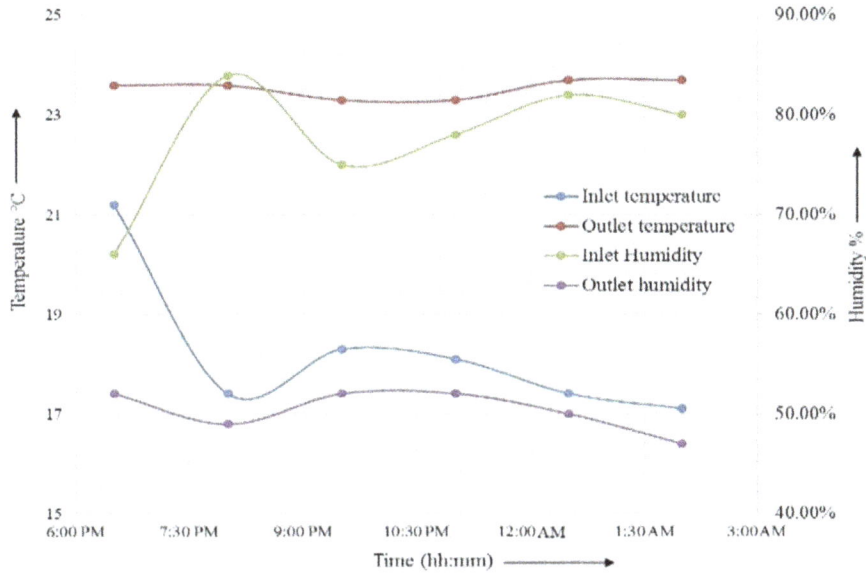

*Figure 12. Variation of temp. and RH for PVC with time under inlet and exit condition (night).*

This curve gives a qualitative comparison of the inlet and exit air and helps us judging the air properties. The outlet temperature for all the observations were around 23.5°C. We get a considerable amount of air heating by this set of observations.

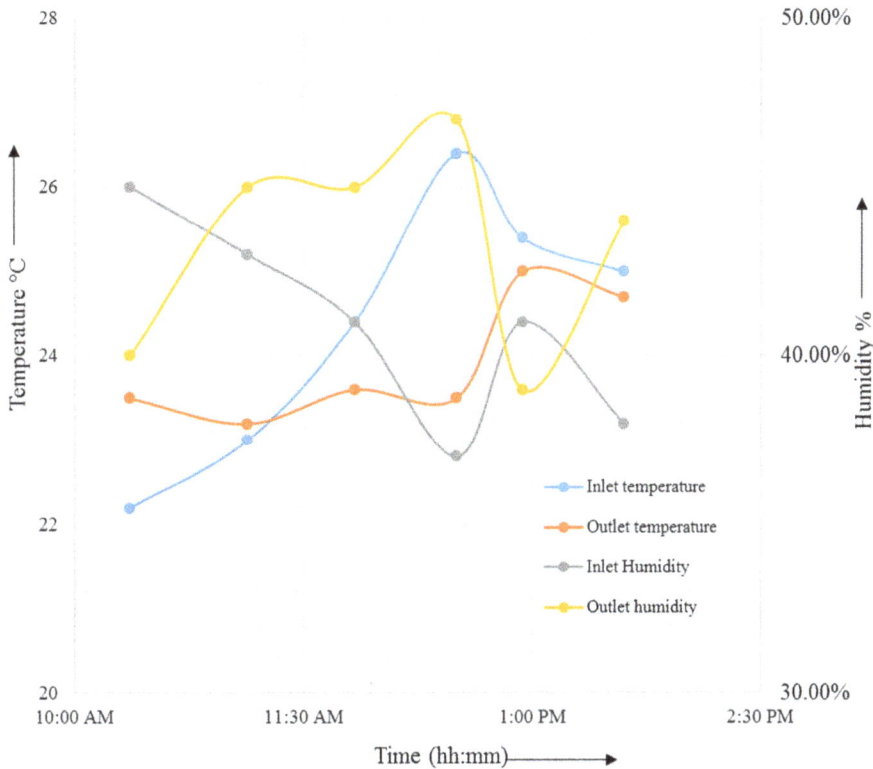

*Figure 13. Variation of temp. and RH for PVC with time under inlet and exit condition (day).*

The variation of temperature and RH was found to be different than night in the day time. Sometimes it has shown cooling character and sometimes it has shown heating character. The lowest outlet temperature found was 23.2°C.

### 4.5. Variation of Temp. and RH for MS with Time Under Inlet and Exit Condition (Night & Day)

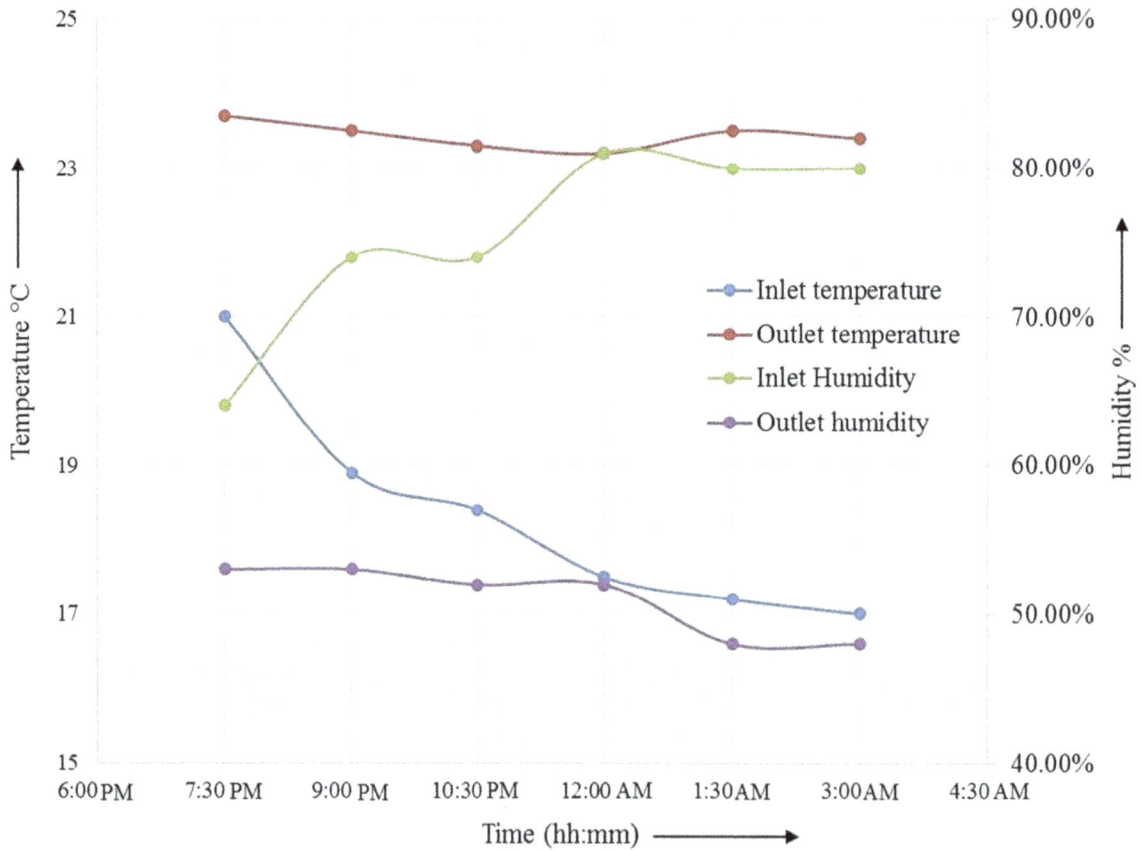

**Figure 14.** *Variation of temp. and RH for MS with time under inlet and exit condition (night).*

Mild Steel pipe has had a very stable heating property. The outlet temperatures for all the observations were around 23.5°C. We get a considerable amount of air heating by this set of observations.

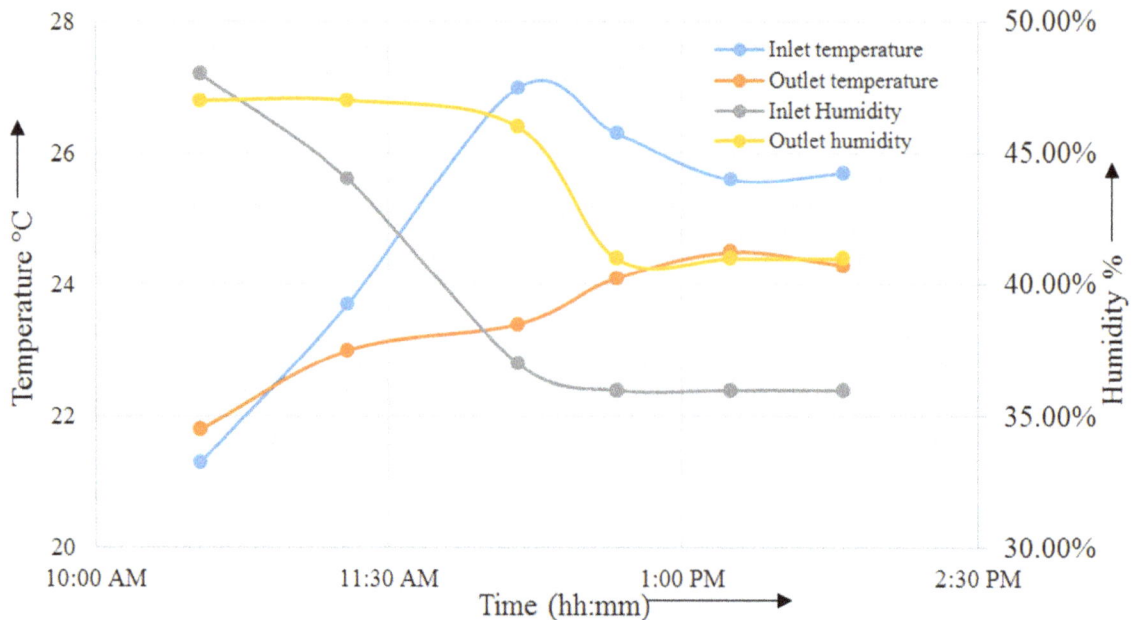

**Figure 15.** *Variation of temp. and RH for MS with time under inlet and exit condition (day).*

However, during the day time, MS pipe has shown scattered readings. Sometimes it has cooled the air and sometimes it has heated the air. It has been found that the humidity remains in the region of 40%.

### 4.6. Variation of Temp. and RH for PVC Filtered Pipe with Time Under Inlet and Exit Condition (Night & Day)

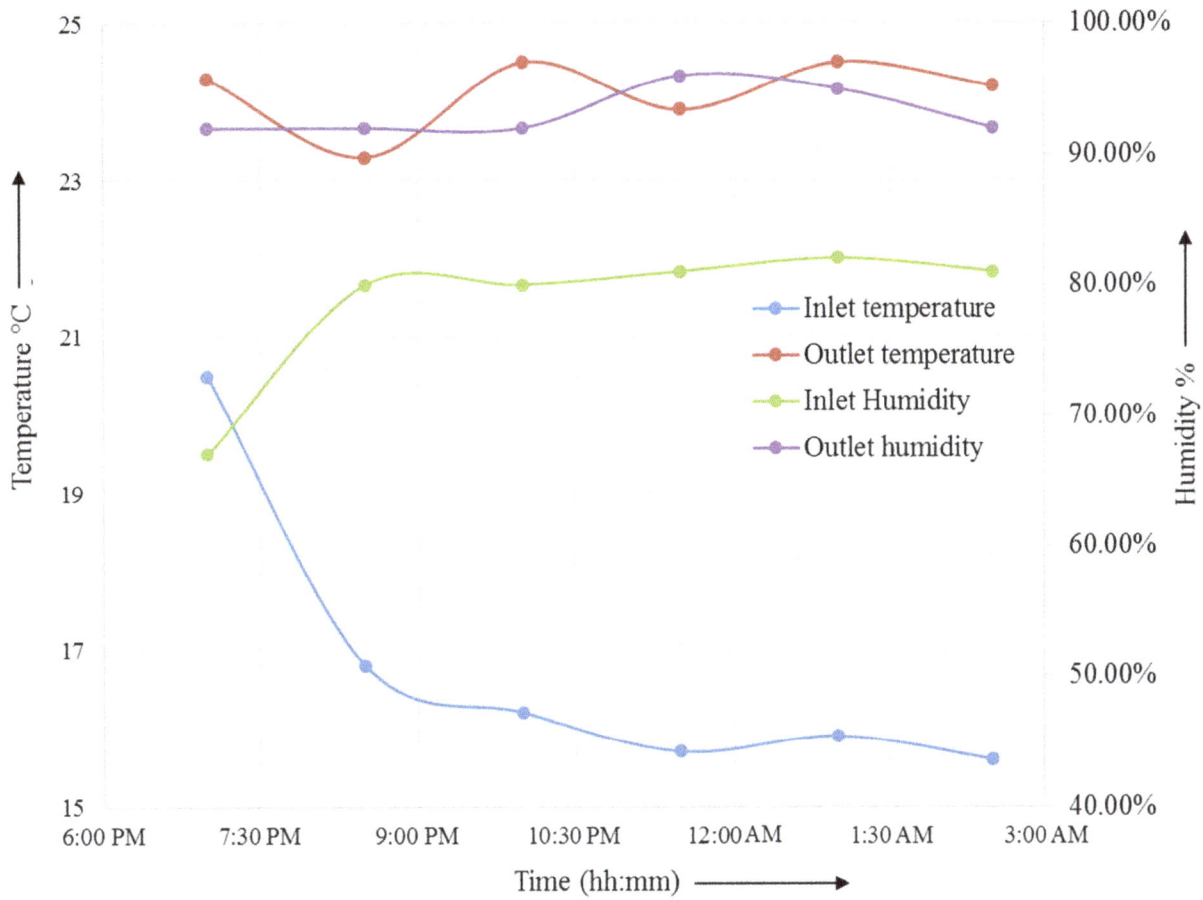

**Figure 16.** *Variation of temp. and RH for PVC filtered pipe with time under inlet and exit condition (night).*

PVC filtered pipe has shown the best heating properties of all the pipes during the night. But, it has increased the humidity of the air to a very high level. The highest outlet temperature recorded was 24.5°C.

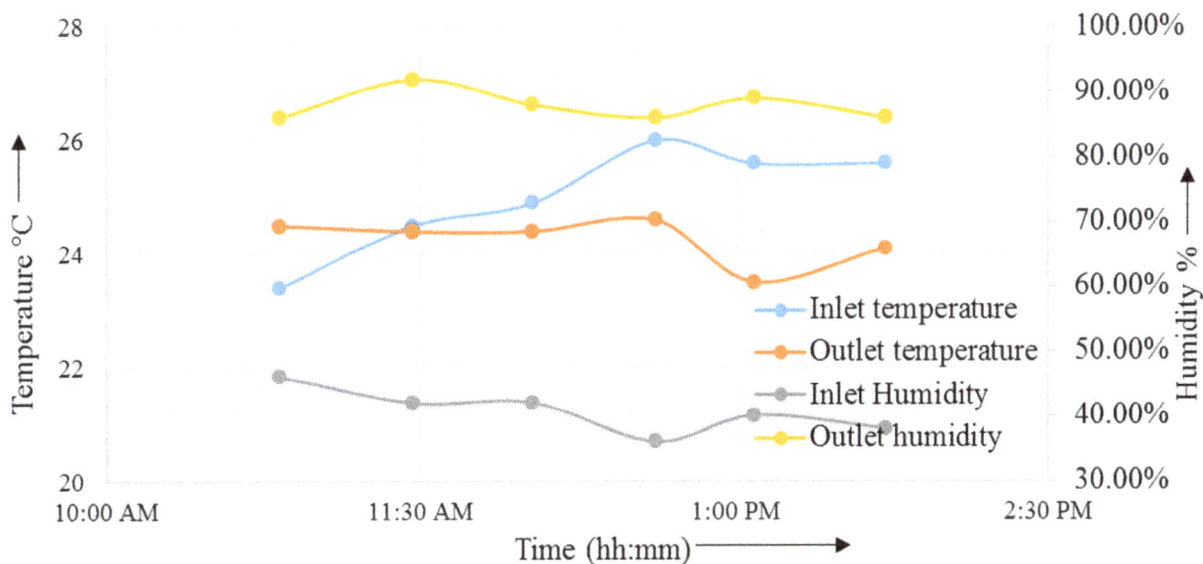

**Figure 17.** *Variation of temp. and RH for PVC filtered pipe with time under inlet and exit condition (day).*

PVC filtered pipe has shown more or less similar characters during the day and night time. It has increased the temperature and humidity of the air. The lowest outlet temperature recorded was 23.5°C.

### 4.7. Variation of Temp. and RH for Bamboo with Time Under Inlet and Exit Condition (Night & Day)

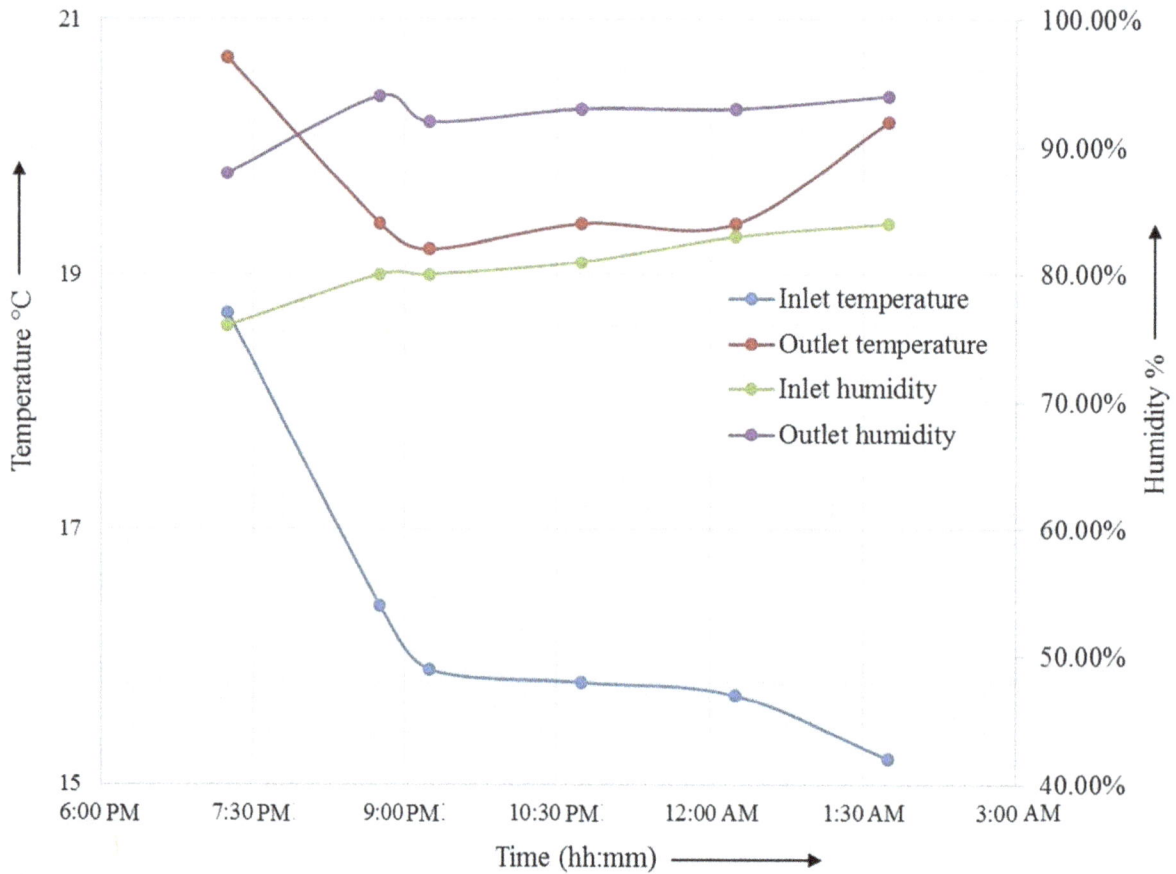

**Figure 18.** *Variation of temp. and RH for Bamboo with time under inlet and exit condition (night).*

Bamboo has increased the temperature of the air during night. It has also resulted in a high humidity. The highest outlet temperature recorded was 20.7°C.

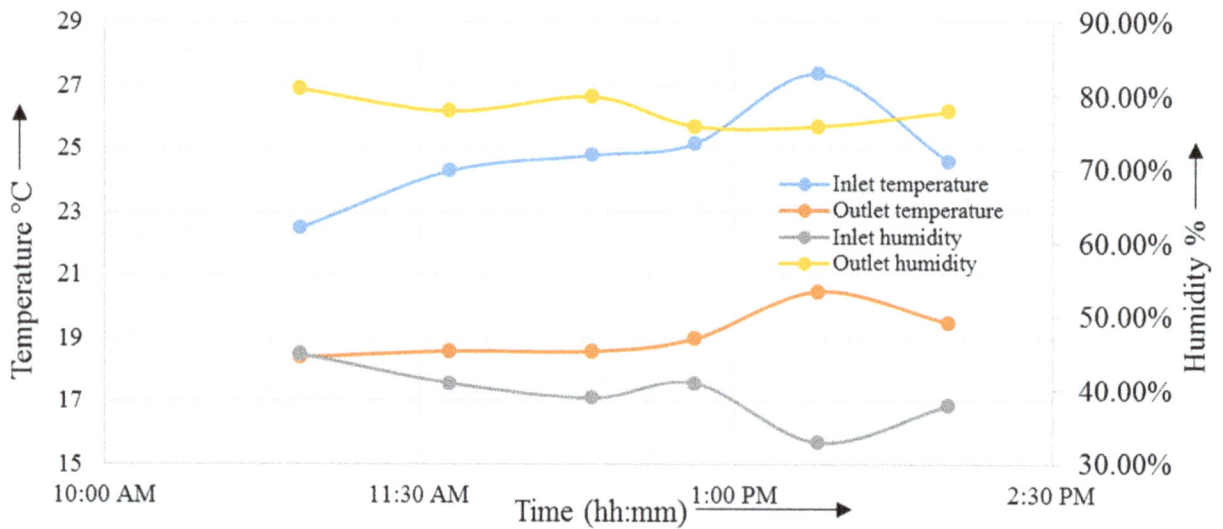

**Figure 19.** *Variation of temp. and RH for bamboo with time under inlet and exit condition (day).*

Figure 19 shows the variation of temperature and relative humidity for bamboo with respect to time. The data was taken by digital hygro-thermometer. The bamboo has shown great cooling performance in the day time. It had the highest cooling rate among all the pipes despite giving the lowest performance during the night time. The lowest outlet temperature recorded was 18.4°C.

## 4.8. Variation of COP with Time (Night & Day)

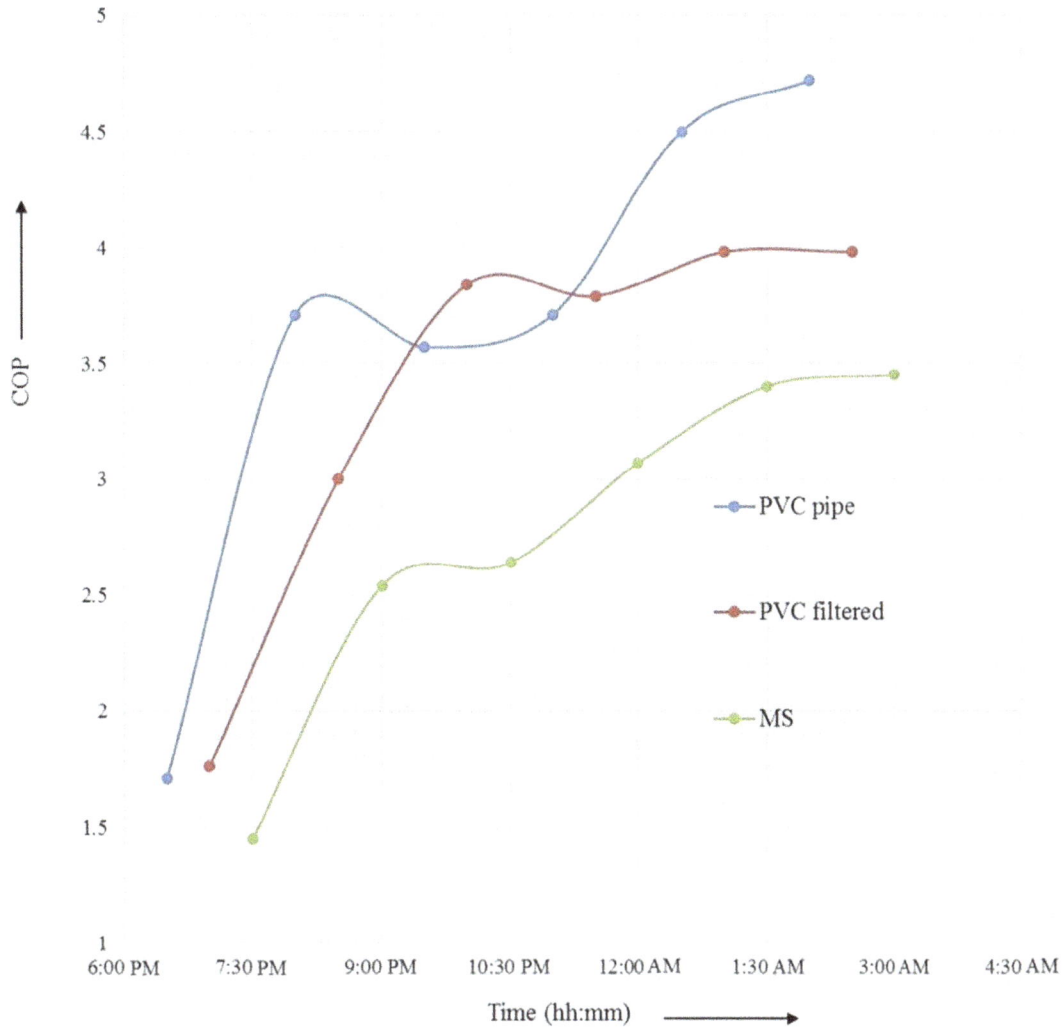

**Figure 20.** *Variation of COP with time (night).*

As the time passed, the performance of all the pipes grew higher. The lowest COP was 1.4 found for MS pipe and the highest COP was 4.7 found for PVC pipe.

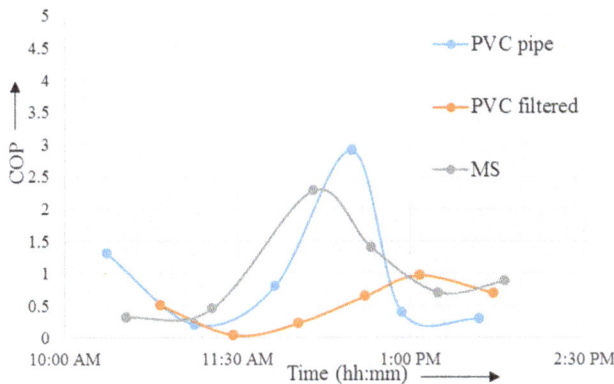

**Figure 21.** *Variation of COP with time (day).*

Figure 21 shows the variation of the co-efficient of performance of various pipes with time during the day time. During day time, the COP has varied very randomly with

time. The lowest COP was 0.05 found for PVC filtered pipe and the highest COP was 2.91 found for PVC pipe.

## 4.9. Maximum COP of Various Pipes During Day and Night

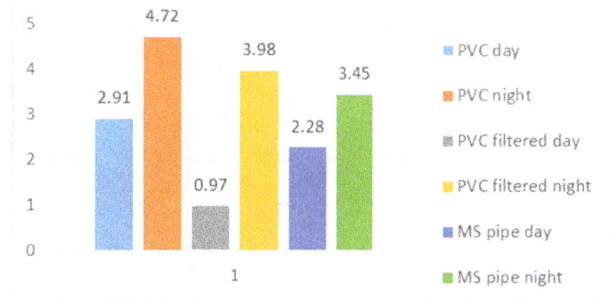

**Figure 22.** *Maximum COP of various pipes during day and night.*

Figure 22 shows the maximum COP of various pipes at day and night time. The highest COP was found for PVC pipe at night and it was 4.72.

## 4.10. Cost Estimation

The overall cost of the system was 6540 taka (approximately $82). The most cost effective system can be designed by PVC pipe. The labor cost or excavation cost can be reduced by proper planning. However while installing the system in household purpose, we would use only one pipe. So, the overall cost would not contain the cost of other pipes.

*Table 1. Cost estimation.*

| Item | Description | Amount (Taka) |
|------|-------------|---------------|
| PVC Pipe | 1.5 inch diameter, 50 ft. long | 750 |
| MS Pipe | 1.5 inch diameter, 50 ft. long | 1250 |
| PVC Filtered Pipe | 1.5 inch diameter, 50 ft. long | 1000 |
| Bamboo | 1.5 inch diameter, 50 ft. long | 1100 |
| Elbow | 4 PVC, 2 MS Pipe | 140 |
| Excavation Cost | Labor cost for excavation | 2300 |
| Total | | = 6540 |

## 4.11. Quantity of Materials

*Table 2. Quantity of materials.*

| Material | Amount (kg) |
|----------|-------------|
| PVC Pipe | 6 |
| PVC Filtered Pipe | 6.75 |
| MS Pipe | 9.1 |
| Bamboo | 54 |

PVC pipe is also the lightest material. The weight of the PVC pipe for the whole system was 6 kg. While the heaviest pipe was bamboo.

# 5. Conclusion

The design and construction of the system was a simple one. Though it takes huge space, there are very little moving parts. So the system has a very simplified design. The performance study was a lengthy one. To obtain precise characteristics of underground heating and cooling system for various materials, thorough observation and accurate data collection at regular interval was ensured. Our objective was to provide simultaneously heating action in the winter season and cooling action in the summer. A considerable amount of heating action was obtained from PVC, MS and PVC filtered pipe. Bamboo provided a reduced heating performance. Thus, with the use of PVC/ MS/ PVC filtered pipe, an economical and energy saving heating action can be obtained. On the other hand, with bamboo a considerable amount of cooling action can be achieved. Whereas the other pipes had less cooling properties than bamboo.

- The average COP of PVC pipe at night is 3.65, during day time it is 0.99.
- The average COP of MS pipe at night is 2.76, during day time it is 1.003.
- The average COP of PVC filtered pipe at night is 3.39, during day time it is 0.518.
- The COP of bamboo could not be measured as the exit velocity was very little.

The set-up of PVC and MS pipe was very much secure and sealed. It was ensured there was a minimum of leakage. So, the exit condition of these two pipes has shown a constant and stable nature. In the case of PVC filtered pipe, due to the perforations in the pipe, the air passing through the pipe came in direct contact with the moisture content of the soil. That is why the humidity content was found to be very high. In case of bamboo, to ensure continuous flow of air, the bamboo was slit longitudinally into two parts. Then the joints were separated and the two split parts were joined again in order to give it a pipe shape. The whole process kept a huge amount of leakage due to unavoidable misalignment. This allowed the air to come in ample contact of soil and take moisture content of soil with it.

# Recommendation

At first, we considered Bamboo as our prime target for cost efficient material. But, practically it was found that bamboo is neither cheap nor its set up and processing is simple. It takes a lot of work to split the bamboo into half, remove the joint and rejoin the two parts without any leakage with proper alignment. Thus the performance obtained from Bamboo was not satisfactory for the night time. But it gave a very good cooling performance during day time.

On the other hand, PVC filtered pipe gave a very good heating property during night. But it did not give any mentionable cooling property. So, for the cold weathers, use of PVC filder pipe would give very good performance.

# Night Condition

In the case of PVC and MS pipe, it was found that the temperature increases and the humidity decreases. In the case of PVC filtered pipe, the humidity increases to a very high level. However, the humidity level of the air obtained from PVC and MS pipe was significantly low. Hence, in both cases, the air may be uncomfortable for the user. This is a huge limitation of the system. A device can be used to control or regulate the moisture content of the air.

# Day Condition

In the case of PVC and MS pipe, it was found that the temperature and the humidity remains more or less identical. As the day passes from morning to noon, a considerable amount of heat transfer occurs and cooling action is obtained. In the case of PVC filtered pipe, the humidity increases to a very high level. However, the humidity level of the air obtained from PVC and MS pipe was significantly low. Hence, in both cases, the air may be uncomfortable for the user for day time too.

We had to keep the pipe length to 30 feet for scarcity of available space and cost considerations for the material and digging. The system is likely to give better performance if the length of the pipe is increased. The longer the air remains in contact with the soil, the more is the rate of heat transfer.

A reduced pipe diameter gives better cooling and heating

performance. But it also gives a very low exit velocity due to high frictional loss. Again, a pipe of larger diameter would take a fan of higher power which would increase the energy requirement of the system. Hence, a pipe of moderate diameter is suggested.

Data collection over a vast period of time is suggested. It would be more convenient if data is taken at different regions depending upon geographical and climatic conditions. The soil composition of different places may vary resulting a different heat transfer rate.

# References

[1]   World Meteorological Organization, Press Release, 2nd July, 2003.

[2]   Hollmuller, P., Lachal, B., Cooling and preheating with buried pipe systems: monitoring, simulation and economic aspects, Energy and Buildings 1295 (2000), 1-10.

[3]   Sharan, G., Jadhav, R., 2003, Performance of Single Pass earth-Tube Heat Exchanger: An Experimental Study, Ahmedabad, Indian institute of management.

[4]   Lee, K. H., Strand, R. K., The cooling and heating potential of an earth tube system in buildings, Energy and Buildings 40 (2008), 486-494.

[5]   Thevenard, D., September 6, 2007, Bibliographic search on potential of earth tubes, Waterloo, Canada, Numerical Logics Inc.

[6]   Mirianhosseinabadi, S., Cho, S., Kang, E. C., Lee, E. J., Simulation modeling of earth-to-air heat exchanger (Buried pipe) for the performance analysis of a school building in mid-atlantic region, ASRAE/IBPSA-USA, 2014, 378-384.

[7]   Holman, J. P., Bhattacharyya, S., 2011, Heat Transfer, New Delhi, Tata Mcgraw Hill Education Private Limited.

[8]   https://en.wikipedia.org/wiki/Electricity_sector_in_Bangladesh

[9]   http://www.michell.com/uk/documents/HumidityDewPointCatalogueUK-3.pdf

[10]  https://en.wikipedia.org/wiki/Anemometer

# 4

# Forest bioenergy or forest carbon

**Keith Openshaw**

Retired forest/energy economist. Formerly with the International Resource Group (IRG ENGILITY) of Washington D.C.

**Email address:**

openshaw.keith@gmail.com

**Abstract:** An article by Jon McKechnie *et al* entitled Forest Biomass or Forest Carbon purports to use an integrated life cycle assessment (LCA) and forest carbon analysis when examining the use of wood for electrical generation. Most publications assume that the $CO_2$ emitted is carbon neutral because plants will re-absorb the $CO_2$ through photosynthesis. However, the article challenges this hypothesis and states that incomplete LCAs are undertaken. The article demonstrates that it will take many years to recapture the $CO_2$ when the wood is used for bioenergy. But when analyzing the capture of $CO_2$, only regrowth is considered and not the tree growth of the whole forest. If in the example given, a full account is taken of the above-ground yield, it is shown that the annual increment from the management units is nearly double the potential removals for wood products, including bioenergy. Thus, rather than a decrease in forest capital there is an increase. Proper and full LCAs must be undertaken of the whole forest, rather than partial analysis: the latter results in erroneous accounting. It is very misleading and should not be used or cited. This same error has been made by a number of other quoted publications.

**Keywords:** Carbon Accounting, Carbon Sequestration and Use, Bioenergy, Life Cycle Assessment

## 1. Introduction

An article entitled Forest Bioenergy or Forest Carbon: Assessing Trade-offs in Greenhouse Gas Mitigation with Wood-based Fuels, by Jon McKechnie *et al*,[1] which was published in the Journal of Environmental Science and Technology # 45, 2011, uses integrated life cycle assessment (LCA) and forest carbon analysis when examining the use of wood for electrical generation or ethanol production compared to leaving the trees in the forest to sequester carbon. It concludes that "Application of the method to case studies of wood pellet and ethanol production from forest biomass reveals a substantial reduction of forest carbon due to bioenergy production. For all cases, harvest-related forest carbon reductions and associated GHG emissions initially exceeded avoided fossil fuel-related emissions, temporarily increasing overall emissions. In the long term, electricity generation from pellets reduces overall emissions relative to coal, although forest carbon losses delay net GHG mitigation by 16-38 years depending on biomass source ---. Forest carbon more significantly affects bioenergy emissions when biomass is sourced from standing trees compared to residues and when less GHG-intensive fuels are displaced. In all cases, forest carbon dynamics are significant. Although study

results are not generalizable to all forests, we suggest the integrated LCA/forest carbon approach be undertaken for bioenergy studies".

The article is based on the supply of biomass from standing trees and harvest residues from the Great Lakes-St. Lawrence (GLSL) forest region of Ontario. The supply of hardwood and softwood species comes from 10 forest management units covering 5.25 million hectares in the GLSL forests. Historically, significantly less than the allowable harvest has been harvested. Therefore, the 'excess' wood could be used for bioenergy. In addition, residues from tree tops and large branches etc. are a second source for bioenergy. Biomass availability is derived from actual forest management plan simulations undertaken by the Strategic Forest Management Model (SFMM), predicated on forest inventories formulated through Ontario's forest management process. While the article should be based on the sustainable use of wood for all purposes from the 5.25 million hectares, it concentrates on 'forest regrowth' rather than the annual increment of the trees in the 10 management units.

## 2. Methods Used in the McKechnie Article

Under methods, the article states "We develop a

---

[1] Henceforth referred to as the McKechnie article

framework integrating two analysis tools: life cycle inventory (LCI) analysis and forest carbon modeling. ---. The LCI is based on the assumption of immediate biomass carbon neutrality, as is common practice, and is therefore employed to quantify the impact of all emissions on atmospheric GHGs with the exception of biomass-based $CO_2$". It further states: "Forest carbon modeling quantifies the impact of biomass harvest on forest carbon dynamics, permitting an evaluation of the validity of the immediate carbon neutrality assumption. If biomass-based $CO_2$ is fully compensated for by forest regrowth, biomass harvest will have no impact on forest carbon stocks. Reduced forest carbon indicates that a portion of biomass-based $CO_2$ emissions contribute to increased atmospheric GHGs and should be attributed to the bioenergy pathways". And, "The total emissions associated with a bioenergy system are the sum of the two sets of GHG flows (those resulting from the LCI and those from the forest carbon analysis). GHGTot(t) = $\Delta$FC(t) + GHGBio(t) where GHGTot(t) is the total emissions associated with bioenergy, $\Delta$FC(t) is the change in forest carbon due to biomass harvest for bioenergy, and GHGBio(t) is the GHG emissions associated with bioenergy substitution for a fossil fuel alternative [all reported in metric tonne $CO_2$ equivalent (t $CO_2$equiv)] at time t." Furthermore, "The change in forest carbon, $\Delta$FC (t) is the difference in forest carbon stocks between harvest scenarios: those 'with' and 'without' bioenergy production. ---. Carbon in biomass harvested for bioenergy is assumed to be immediately released to the atmosphere. However, forest regrowth will capture and store atmospheric $CO_2$ over time".

The article stresses that many previous papers and assumptions have not employed proper LCAs, yet this article only uses partial analysis by limiting the analysis to forest regrowth, not total tree growth, that will capture and store $CO_2$ over time. Naturally, in most temperate countries, it will take decades of regrowth in newly felled areas to sequester the carbon given off when used for energy. Therefore, there will be an automatic $CO_2$ deficit, when emissions are compared to regrowth until equilibrium is reached.

# 3. Bioenergy for Electricity Generation

The article examines the saving of fossil fuel using a 20% mix of pellets with coal to generate electricity in a conventional power station. First it estimates the quantity of wood that could be used sustainably for bioenergy production from the ten management units in the GLSL forest region. Table 1 gives a breakdown of potential wood supply. It is assumed that 4.46 million ha (85%) out of 5.25 million ha is managed for wood products. At present, the wood comes from clear felling (50%), shelterwood over-storey removal (25%) and selection harvesting (25%), [KBM Forestry Consultants Inc. 2008]. Little, if any thinning is undertaken, but in a fully managed forest, about 50% of the total removals could be from thinning, - a management tool that could be used to increase the sustainable off-take of wood energy etc., while increasing the value of the remaining stock.

The McKechnie article accounts for the use of fossil fuels for felling, extraction, pelletization and transport to the power plant. In addition, 15% of the wood (338,000 odt) is used in the conversion process of roundwood to pellets; therefore, available wood energy will be 1.914 Modt. An input/output factor of wood and coal for electrical generation of 1: 0.33 is assumed with the energy value for standard coal of 33.0 GJ per odt (82% carbon) and that of wood of 18.7 GJ per odt (50% carbon). The bottom line is that 1.914 Modt of pelletized wood could save 1.167 Modt of coal annually, emitting about 3.509 Mt $CO_2$.[2] If the wood is assumed to be carbon neutral then the 4.129 Mt $CO_2$ emitted from 2.252 Mt of bioenergy wood, including 338,000 odt for pelletization, would be reabsorbed quickly, resulting in no 'carbon debt'.

**Table 1.** *Estimated annual wood available for traditional and bioenergy use in the GLSL forest region.*

|  | Felling/(thinning) | Potential residues | Total |
|---|---|---|---|
|  | Modt[1] | Modt | Modt |
| Existing area | 2.836 | 0.383 | 3.219 |
| Available for bioenergy | 1.869[2] | (0.300)[3] | 1,869 |
| **Total** | **4.705** | **0.683** | **5.088** |
| Traditional use | 2.836 | 0 | 2.836 |
| Bioenergy potential | 1.869 | 0.383 | 2.252 |
| **Total** | **4.705** | **0.383** | **5.088** |
| Total m³ | 9.175[4] | 0.747[4] | 9.922[4] |

*Note.* 1. odt = oven dry metric tonne.
2. This has been adjusted from 1.811 Modt to correct an addition error.
3. This is the collection of residues from standing trees available for bioenergy. It was not considered in the original study nor has it been included here. It is assumed to rot on the forest floor. If used, it would save another 0.144 Mt coal each year, which would emit 0.433 Mt $CO_2$equivalent.
4. A conversion factor of 1.95 m³ per odt has been used. (KPMG 2008). This conversion factor may be too high. A figure of 1.80 m³ per odt may be more accurate, although in Table 4 below a factor of 1.67 is used. This would reduce the total estimated volume to 9.158 or 8.497 Mm³.
*Source.* McKechnie *et al*, 2011. Table S-1 and text on page S-5. KPMG 2008.

# 4. Forest Carbon Accounting

Because the 'Forest Bioenergy or Forest Carbon' article assumes that the annual emitted carbon dioxide from bioenergy comes from tree regrowth, – 4.129 Mt $CO_2$ from 2.252 Modt wood (excluding 300,000 odt of potential residues), it will take many years to recapture this $CO_2$. It is assumed that if the 1.869 Modt of stemwood are not felled annually for bioenergy, they never will be and so will be left to grow until they reach a point where either the current annual increment is zero and they eventually die, or through competition and age, some of the trees and/or branches will die and decompose slowly. Regarding the 383,000 odt of residues (Table 1), if they are not used they will be left on the forest floor and will gradually rot. Table S-7 of the supporting information on page S-14 gives decomposition

---

[2] If natural gas is used (energy value 38.1 GJ/t [35.2 MJ per m³]) then the annual saving, assuming an efficiency of 60% for a gas-fired boiler, would be about 586 Mt of natural gas; this would emit about 1.612 MtCO₂ equiv't

rates (k) for hardwoods and softwoods divided into pulpwood and sawntimber harvest, ranging from 0.042 to 0.083. This information is used to calculate the forest carbon stock change for residues and standing trees from continuous harvesting for bioenergy: it is given as Table 2 of the article. One-hundred years is the time frame chosen to track changes of continuous and constant harvesting. After 10 years, the forest carbon stock changes result in emissions, in Mt $CO_2$equivalent, of 8.2 for residues and 43.6 for standing trees. These negative values gradually rise to 15.2 for residues after 90 years. For standing trees the maximum emissions value of 150.8 is reached after 90 years and then starts to decline, reaching minus 150.7 at year 100. Thus, the combined total at 100 years is nearly -166 Mt $CO_2$equivalent. This is a stock decrease of 90 Modt wood. However, over the 100-year period, 225.2 Modt of wood will have been used to generate electricity[3] and save 116.7 Modt of 'standard' coal containing 95.7 MtC, which will emit 350.9 $MtCO_2$ equivalent, thus more than twice as much $CO_2$ will be saved by using wood in place of coal.

The article does not explain how the numbers in Table 2 have been derived. It appears that the 'residue' calculation is based only on the annual use of 383,000 odt of wood for electrical generation and does not include the 300,000 odt left to decay. My estimate for the former is a maximum cumulative emission of 13.4 Mt $CO_2$ after 100 years, and including the 300,000 odt (annual total of 683,000 odt), the cumulative sequestered total is 30.1 Mt $CO_2$, equal to 16.4 Modt of wood. Likewise, no rotation age is given for estimating the 'carbon balance' from using 1.869 Modt of wood annually from felling 19,700 ha equivalent. It seems that a 50-year rotation has been chosen, but a carbon balance is derived for both two fifty year rotations and one 100 year rotation. The bottom line is with the former, the net sequestration after 100 years is over 44 Mt $CO_2$ and with the latter it is over 59 Mt $CO_2$. This is entirely different from the emission figure given in Table 2 of 151 $MtCO_2$. Taking into account the saving of coal emissions, the cumulative sequestration totals range from 315 to 330 Mt $CO_2$ after 100 years. (See Annex 1).

Although the article states that the actual and potential wood use is (or will be) sustainably harvested, the method chosen to quantify forest carbon stock actually results in an annual reduction of forest stock of up to 90 million odt of wood after 100 years due to bioenergy use. As pointed out, this is because only regrowth of trees are considered. Generally, a forest comprises a range of age classes and species; this is the case for the GLSL forest areas. Therefore, it is pertinent to examine the growing stock and annual yield on the 4.46 Mha of 'managed forests', assuming that the other 790,000 ha (15%) out of a total of 5.25 Mha have been

set aside for protection, water catchment and leisure etc[4]. The article states that "The forest carbon dynamics relating to biomass harvest are evaluated using FORCARB-ON, an Ontario-specific adaption of the FORCARB2 model (Chen J et al 2008)". However, the Chen article states that while most of the carbon storage for the next 100 years will be in timber products from the forests, there will be a modest increase in carbon storage in the forests themselves over a hundred year period. Even the FORCARB2 model for the USA predicts a C storage increase in their forests, despite a decline in the forest area[5]. There was an estimated 6.1% gain in forest tree carbon between 1990 and 2010 and a loss of 2.1% in forest soils, for an overall carbon gain of 2.6% in the example given (USDA 2009).

Table 2 gives a 2011 estimate of forest carbon in trees, litter and forest soils for the Algoma Forest Management Unit (Chen J et al 2010) in terms of total and per-hectare stored carbon and the tree volume equivalent.

The estimated carbon in wood, including the under-story is 42% of the total of which 84% is above ground. The volume in above-ground trees is estimated at 264 $m^3$/ha,[6] (244 $m^3$/ha with a conversion factor of 1.8 $m^3$ per t of wood), 98% of which is in trees with a top diameter of 78 mm, (3 inches). The McKechnie article and the accompanying notes, give no information on volume or age-class distribution in the 10 management units of the GLSL forest area included in the assessment. Other publications were examined to try and assess the growing stock and yield for the 4.46 Mha under review.

*Tree growth.* To illustrate the growth of trees over time, information from the British Forestry Commission Management Tables for Oak (*Quercus robur, Q. petraea*) Yield Class (YC) 4 is used (HMSO 1971). This is the average stem yield in $m^3$/ha at the age when the current annual increment (CAI) dissects the mean annual increment (MAI). For YC 4, this age is 90 years. At this age, the main crop volume before thinning is 207$m^3$/ha and the cumulative volume from thinning/attrition is 153$m^3$/ha, for a combined total of 360$m^3$/ha. The yield table represents either a natural succession of a single crop over time, or different age classes in a population. In Ontario, very little thinning is undertaken, thus, it is assumed that through competition, the 'thinning volume' will not be removed, but left to die and rot on the forest floor. The plantation starts with 5,000 trees per hectare and by the time it is 90 years old, only 300 trees remain. If the area is left to grow, there may be only 108 trees

---

[3] The use of wood for ethanol production, by first breaking it down into simple sugars, is not considered here as the cost is prohibitive. Methanol (wood alcohol) produced from the dry distillation of wood or other biomass is cheaper and may be viable. Methanol can be used directly or as an input to make other fuels/organic chemicals or serve as a hydrogen carrier ($CH_3OH$)

[4] KBM Forestry Consultants Inc 2008 estimate that 91% of the Algoma forest is managed and the KPMG 2007 study on 'wood pellets' of the Algoma and Martel forest estimate that 93% of the area is managed. Therefore, the assumption that only 85% of Ontario's Crown Forests are managed is conservative.

[5] It was estimated that the overall increase in carbon stored in woody biomass was somewhat offset by a loss of carbon in forest soils due to a switch of forest areas to other uses. The forest carbon loss from forests resulted in a gain by other sectors, although some decrease would occur.

[6] For the forests in Ontario's GLSL forest region, the (latest) 2001 Canadian Forest Inventory (Canadian NFI 2001) gives an average per-ha volume of live trees of 155$m^3$/ha. This is equivalent to 186$m^3$/ha with branches, which is 89% of 209$m^3$/ha

remaining by 180 years with the main crop volume 211m³/ha, the CAI is zero and the MAI is 2.9m³/ha. (See Annex 2). Although the volume for YC 4 oak is rather conservative, it is used to estimate the standing volume/mass and yield of the main crop over time.

*Table 2. An estimate of forest carbon in the Algoma Forest Management Unit 2011.*

| | Total | Above ground[1] | Below ground[1] | Total[1] | Above ground[2] | Below ground[2] | Total[2] |
|---|---|---|---|---|---|---|---|
| | MtC | | tC per ha | | | m³ per ha | |
| Live trees | 30.1 | 53.7 | 10.6 | 64.3 | 209 | 41 | 250 |
| Dead trees | 3.5 | 6.3 | 1.2 | 7.5 | 25 | 5 | 30 |
| Downed wood | 3.5 | 6.3 | 1.2 | 7.5 | 25 | 5 | 30 |
| Understory | 0.8 | 1.4 | 0.3 | 1.7 | 5 | 1 | 6 |
| *Sub-total* | *37.9* | *67.7* | *13.3* | *81.0* | *264* | *52* | *316* |
| Forest floor[3] | 12.8 | 27.3 | - | 27.3 | | | |
| Soil[4] | 37.9 | - | 81.0 | 81.0 | | | |
| **Total** | **88.6** | **95.0** | **94.3** | **189.3** | | | |
| Area (ha) | 468,000 | | | | | | |
| Per ha | 189.3 | 95.0 | 94.3 | | 264 | 52 | 316 |

*Note.* 1. The above and below ground totals have been estimated from the FORCARB2 example (Table 1 - USDA 2009).

2. The volume estimates assume 50% carbon in oven dry wood and 1.95 m3 per odt wood.

3. This remains constant over time as new additions are cancelled by decomposition and incorporation into the soil.

4. The forest soil C may be underestimated. In the USDA example it was nearly 60%, increasing the per-ha value to over 200 tC.

*Source.* Chen J et al 2010. USDA 2009. Author's estimates.

# 5. Age Class Distribution of Trees in the 10 Management Units of the GLSL Ontario Forests

KBM Forestry Consultants Inc. undertook an audit of the Algoma and Martel forests covering 639,310 ha of managed forests. The age-class distribution, in 20-year steps, was given in Table 5 of their Annex on pages A-51 to A-54 for 16 working groups. The totals for the 10 age classes, up to 200 years were then tallied. Figure 5 on page 7 of the main report gives the age-class distribution in 10-year steps. The area of each age-class, depicted in a bar diagram, was estimated and compared to the area as given in Table 5 of the Annex. There is a good agreement between the two sources, except for the years 141-150, which seem to be out by a factor of 10: this was adjusted to agree with Table 5. The areas were then divided into 10-year age groups and the percentage by age-groups calculated as shown in Column 2 of Table 3 below. These percentages were then applied to the total area of 4.46 Mha, assuming that the Algoma and Martel forests have a similar age-class structure to the whole area.

*Table 3. Model for the managed forest area (4.46 Mha).*

| Age class | Area in each class | Area by class (rounded) | Stem volume, to 7cm t.d. before thin | Volume by age-class | | CAI | CAI by age-class to 7cm top diameter (t.d.) | |
|---|---|---|---|---|---|---|---|---|
| | | | | Stem (S) | S & branch | | Stem | S & branch |
| Years | Percent-age | 1000 ha. | m³/ha - midpoint | Mm³ | | m³/ha midpoint | Line 3 (area) x CAI 1000 m³ | |
| 1-10 | 0.6 | 27 | 5 | 0.14 | 0.17 | 0.4 | 10.8 | 13.0 |
| 11-20 | 1.7 | 76 | 10 | 0.76 | 0.91 | 1.1 | 83.6 | 100.3 |
| 21-30 | 1.2 | 54 | 30 | 1.62 | 1.94 | 4.4 | 237.6 | 285.1 |
| 31-40 | 1.6 | 71 | 79 | 5.61 | 6.73 | 5.3 | 376.3 | 451.6 |
| 41-50 | 3.0 | 134 | 107 | 14.34 | 17.21 | 5.6 | 750.4 | 900.5 |
| 51-60 | 7.1 | 317 | 135 | 42.80 | 51.36 | 5.6 | 1775.2 | 2130.2 |
| 61-70 | 10.6 | 473 | 161 | 76.15 | 91.38 | 5.2 | 2459.6 | 2951.5 |
| 71-80 | 12.2 | 544 | 183 | 99.55 | 119.46 | 4.8 | 2611.2 | 3133.4 |
| 81-90 | 20.4 | 910 | 200 | 182.00 | 218.40 | 4.3 | 3913.0 | 4695.6 |
| 91-100 | 11.8 | 526 | 212 | 111.51 | 133.81 | 3.8 | 1998.8 | 2398.6 |
| 101-110 | 7.7 | 343 | 218 | 74.77 | 89.73 | 3.2 | 1097.6 | 1317.1 |
| 111-120 | 6.8 | 303 | 221 | 66.96 | 80.35 | 2.7 | 818.1 | 981.7 |
| 121-130 | 7.0 | 312 | 220 | 68.64 | 82.37 | 2.2 | 686.4 | 823.7 |
| 131-140 | 3.5 | 156 | 219 | 34.16 | 40.99 | 1.8 | 280.8 | 337.0 |
| 141-150 | 2.2 | 98 | 218 | 21.36 | 25.63 | 1.4 | 137.2 | 164.6 |
| 151-160 | 1.6 | 71 | 216 | 15.34 | 18.41 | 1.0 | 71.0 | 85.2 |
| 161-170 | 0.4 | 18 | 214 | 3.85 | 4.62 | 0.6 | 10.8 | 13.0 |
| 171-180 | 0.3 | 13 | 212 | 2.76 | 3.31 | 0.2 | 2.6 | 3.1 |
| 181-190 | 0.2 | 9 | 210 | 1.89 | 2.27 | 0.0 | 0 | 0 |
| 191-200 | 0.1 | 5 | 208 | 1.04 | 1.25 | 0.0 | 0 | 0 |
| Total | 100 | 4,460 | | 825.25 | 990.30 | | 17,321 | 20,786 |
| *dry wt. 10⁶ t.* | | | | *423.21* | *507.85* | | *8.88* | *10.66* |

*Note.* Stem volume has been taken from the British Forestry Commission Management Tables (FCMT) for Oak, Yield Class 4. It has been extrapolated to year 200. It is assumed that stem and branch volume/increment is 1.2 times stem volume/increment. The thinning volume has been neglected. For the sake of simplicity, it is assumed that little, if any, thinning is undertaken and trees die through competition and are not removed. The Current Annual Increment has been taken from Oak YC4 for years 5, 15, 25, 35 etc. The values for years 5, 15, 155, 165, 175, 185, & 195 are estimated. The FCMT have a built-in reduction of 15% to allow for gaps etc. The increment is 2.1% of the growing stock.
See supporting information concerning Table 2 in McKechnie article.
*Source.* KBM Forestry Consultants Inc. 2008. HMSO 1971. Author's estimates.

## 6. Forest Volume and Annual Yield: GLSL Ontario Forests

The division of the area by age-classes is given in Column 3 of Table 3. Column 4 shows the estimated mid-point volume for the 20 age classes. This was taken from Oak YC 4 (HMSO 1971). The stem volume in each age class was then calculated by multiplying the area by the mid-point volume and then totaled, giving a standing stem volume of 825 $Mm^3$ (equal to 185 $m^3$ per ha). The volume of stem plus branches was then estimated by adding 20% to the stem volume. This gave a figure of 990 $Mm^3$ (222 $m^3$ per ha). Excluded from these estimates are dead and downed trees and the understory, which in Table 2 are 26% of the stem volume. If included, this would bring the total above-ground volume with branches to 280 $m^3$/ha and the stem volume to 233 $m^3$/ha. If the age-class structure is a good approximation for the managed forest in GLSL forest area, then it is very unbalanced. Only 8% of the area is less than 51 years old, with 62% being from 51 to 100 years old and 30% of the forest area is greater than 100 years old. Because the 'optimum' rotation age is generally less than 100 years, it will take many decades to adjust the age-class structure - 65% of the volume is below 101 years with 35% of it greater than 100 years. The annual increment by age class was then calculated by multiplying the area by the mid-year CIA for each age class. The stem volume increment is given in Column 8 of Table 3. The total for the 4.46 million hectares come to 17.3 $Mm^3$ (nearly 9 Modt). The largest share of increment is in the age classes 51 to 100 years. They occupy 62% of the area, but produce 73% of the annual increment. The older age classes from 150 years onwards, while occupying 3% of the area only provide 0.5% of the annual increment, indicating that they add very little new sequestered carbon in biomass: a case could be made for felling these old trees in order to increase the annual increment and wood raw material. If branches are included, then the estimated annual increment is nearly 21 $Mm^3$ – 10.7 Modt.

## 7. Faulty Accounting Leads to False Figures

Because accounting was done by only considering regrowth of felled trees, the McKechnie article estimates that in $CO_2$ equivalent terms, 15 Mt will be emitted from forest residues and 150 Mt from stemwood used for bioenergy over a 100 year period (Table 2 in the article). *But this is a partial*

*analysis.* A full analysis should examine the annual increment of the whole management area and then compare it to removals for all uses not just for energy. If the annual increment is less than annual consumption of wood products from the GLSL forest management areas, then the forest capital is being reduced and there will be a net emission of $CO_2$. On the other hand, if the annual increment is more than the annual removals, then there will be no 'carbon debt': in fact there should be an increase in the store of forest carbon, as stated in the article 'Carbon budget of Ontario's managed forests and harvested wood products' (Chen J *et al* 2010).

Table 1 above estimated that the managed forests in the GLSL area of Ontario supplied 2.836 Modt to existing industries in the region (5.5 $Mm^3$) in 2010 and could supply a further 2.552 Modt for bioenergy (5.0 $Mm^3$) for a total of 5.388 Modt (10.5 $Mm^3$). The estimated annual increment of stemwood from the 4.46 Mha in the GLSL forest areas is 8.88 Modt (17.3 $Mm^3$) and 10.66 Modt of stemwood and branches (20.8 $Mm^3$) disregarding dead and fallen trees (Table 3). Therefore, there is a surplus of 5.27 Modt of wood from the annual increment of the forest: this total is nearly double the estimated demand requirements. Thus, considerably more wood could be extracted for all purposes without reducing the forest capital. Contrary to what the McKechnie article claims, there is no carbon deficit and all the $CO_2$ emitted for bioenergy uses is more than reabsorbed by the annual growth of trees in the GLSL forests. Naturally, there will be an emission of $CO_2$ from bioenergy if only regrowth is considered: *this is partial analysis and creative accounting.*

The McKechnie article quotes other articles to back up the emissions thesis, namely 'Biomass sustainability and carbon policy study' (Manomet Center 2010) and Searchinger T *et al,* (2009). Both of these articles confine the accounting methodology to regrowth of trees and not to the total forest area growth. A detailed assessment of the Manomet project was undertaken and the flaw was pointed out, but as yet no reply has been received. Recent articles and papers have been published by Searchinger T (2012), 'Sound principles and important inconsistency in the 2012 UK bioenergy strategy', the Institute for European Environmental Policy (IEEP) (2012) 'Does biomass have a role to play in reducing Europe's GHG emissions' and the sensation-seeking article by the (UK) Royal Society for the Protection of Bird/Friends of the Earth/Greenpeace (2012) 'Dirtier than coal? Why [UK] government plans to subsidise burning trees are bad news for the planet'. All these writings use the partial analysis of only considering tree regrowth. *If a falsehood is repeated often enough it may be believed!*

# 8. Benefits of Sequestration and Use

Rather than calling the McKechnie article 'Forest bioenergy or forest carbon?' a more accurate title should be 'Forest bioenergy and forest carbon'. Both the articles by Chen J et al (2008; 2010) demonstrate that rather than just leaving the forests to sequestering carbon in trees there will be more carbon sequestered if the forests are used both for storage and use. In addition, if wood used for energy purposes substitutes for fossil fuels, then $CO_2$ emissions are saved, provided that on average the harvested wood is equal to or less than the annual increment of the whole forest area. This is illustrated with an example given in the IEEP (2012) report.

*Table 4. Stock and yield on a 50 ha forest under three management regimes over 200 years[1] units tC.*

| Carbon storage and yield | Sequestration only (SO) | | Semi-management | | Commercial forest (CF) | |
|---|---|---|---|---|---|---|
| | per-ha | 50 ha | per-ha | 50 ha | per-ha | 50 ha |
| Average above-ground stock | 220[2] | 11,000 | 166 | 8,300 | 70 | 3,500 |
| Average below- ground stock[3] | 55 | 2,750 | 41.5 | 2,075 | 17.5 | 875 |
| Sub-total | 275 | 13,750 | 207.5 | 10,375 | 87.5 | 4,375 |
| Soil carbon[4] | 250 | 12,500 | 188.6 | 9,430 | 79.5 | 3975 |
| **Total carbon stock** | **525** | **26,250** | **396.1** | **19,805** | **167** | **8,350** |
| Additional C stock compared to CF | | 17,900 | | 11.455 | | 0 |
| Annual above-ground yield for bio-energy per 50 ha | zero | zero | 100 for 100 years | 10,000 | 180 for 150 years | 27,000[5] |
| Additional yield compared to SO | | 0 | | 10,000 | | 27,000 |
| **Total stock &yield** | | **26,250** | | **29,805** | | **35,350** |
| Additional Stock &Yield compared to SO | | 0 | | 3,555 | | 9,100 |

*Note.* 1. Based on YC 12 for Sitka spruce (*Picea sitchensis*) from the British Forestry Commission Management Tables.
2. Reaches full stocking at age 142 years assuming area planted over a 50 year period.
3. Below-ground stock assumed to be 25% of above-ground stock, or about 20% of total stock.
4. Additional soil carbon assumed to be in proportion to woody biomass stock. For the semi-managed and commercial forests, this may be an underestimate compared to the non-commercial managed forest.
5. This 27,000 tC in wood, if used for electrical generation, would save approximately 29,000 t of standard coal that would emit 86,000 t $CO_2$, assuming the same end-use efficiency and 15% of the wood energy used to prepare wood pellets.
*Source.* Figure 2, IEEP 2012. (UK) Forestry Commission 2003. HMSO 1971, Author's estimates.

Figure 2 on page 32 of the report is a comparison of (above-ground) carbon stock between a forest under a) non-commercial management and b) a high-yielding commercial forest. It is based on Sitka spruce YC 12 from the British Forestry Commission Management Tables (HMSO 1971). This was taken from a UK Forestry Report on 'Forests, carbon and climate change': the UK contribution to climate change, (Forestry Commission 2003) submitted to the Intergovernmental Panel on Climate Change (IPCC). For illustrative purposes, it is assumed that both forests have an area of 50 ha, with one ha in each age class from 1 to 50. The two forest types are compared over a period of 200 years. Diagram 2a in the IEEP (2012) report illustrates what would happen if the newly created forest is left to grow to maturity under non-commercial management. It will reach maturity at year 92 and is then left to stand. The above-ground carbon stock on this 50 ha area will be 11,000 tC. (Table 4 above). The same diagram shows a stock fluctuation ranging from a maximum of 220 tC/ha to a minimum of 120 tC. This is because of the introduction of some management into the 'mature forest' with the off-take being used for bio-energy: it is an alternative for this forest. The estimated above-ground carbon stock on this 50 ha option is 8,300 tC and the annual off-take will be 100 tC. For the high-yielding commercial forest (diagram 2b in the IEEP report), the stock on the 50 ha will be 3,500 tC and the annual yield will be 180 tC. This can be removed yearly without reducing the forest capital. This is what Matthews R et al (2012) demonstrates in their paper 'Carbon impacts of using biomass in bio-energy and other sectors: forests', a report published in support of the UK bio-energy strategy. Figure 2 in the IEEP report only shows the above-ground biomass, but there will be carbon sequestration in the below-ground biomass and in the forest soil. Table 4 above examines these three alternatives.

While the un-managed forest has about three-times the amount of carbon in the wood and forest soil as does the commercial forest - the counterfactual assessment - when the total stock and yield are taken into account over the 200 year period, the total carbon store and production is 35% more in the commercial forest compared to the unmanaged forest: this gap increases with time. *This is a full LCA!*

In most, if not all cases, the management of forests, for stock and yield will generate more carbon production than just leaving forests unmanaged. Carbon will still be stored in the trees and in non-energy wood products and if used for bioenergy it will substitute for fossil fuels. Generally, this latter does not result in a reduction of forest carbon, unless the annual harvest is more than the annual growth of the trees. By only considering the growth of the present and future fellings, it will lead to underestimating carbon capture and to false and misleading accounting as in the McKechnie and other quoted articles. It is important to correct this error and set the record straight. A full knowledge of perennial crops and their growth is important when analyzing carbon sequestration and use, otherwise bogus conclusions will be drawn.

## Annex 1: McKechnie *et al* Article: Analysis of Table 2.

Table 2 of the article is given below for stock chances for residues and standing trees together with the recalculated figures.

*Table 2.* Forest carbon stock change (Mt $CO_2$ equivalent)

| Age (yr) | 10 | 20 | 30 | 40 | 50 | 60 | 70 | 80 | 90 | 100 |
|---|---|---|---|---|---|---|---|---|---|---|
| Residues | -8.2 | -11.8 | -13.0 | -13.5 | -13.9 | -14.3 | -14.7 | -15.0 | -15.2 | -15.2 |
| Standing trees (ST) | -43.6 | -80.9 | -106.3 | -112.5 | -113.4 | -112.7 | -132.8 | -143.6 | -150.8 | -150.7 |
| *Recalculated cumulative estimates* | | | | | | | | | | |
| Residues | -6.6 | -12.1 | -13.8 | -12.4 | -8.4 | -2.2 | 5.2 | 13.4 | 21.8 | 30.1 |
| ST 100 yr rotation | -43.6 | -82.1 | -100.7 | -102.8 | -90.3 | -67.4 | -39.0 | -6.5 | 27.0 | 59.5 |
| ST 50 yr rotation | -43.6 | -67.7 | -69.5 | -35.4 | 22.2 | -21.4 | -45.5 | -47.3 | -13.2 | 44.4 |
| *Recalculated cumulative estimates with 'coal savings'* | | | | | | | | | | |
| Residues | -0.8 | -0.6 | 3.5 | 10.7 | 20.5 | 32.5 | 45.6 | 59.6 | 73.8 | 87.9 |
| ST 100 yr rotation | -16.5 | -28.0 | -19.5 | 5.44 | 45.0 | 95.0 | 150.4 | 209.9 | 270.4 | 330.0 |
| ST 50 yr rotation | -16.5 | -13.5 | 11.7 | 72.8 | 157.5 | 141.0 | 144.0 | 169.2 | 230.3 | 315.0 |

*Note.* Negative values indicate a GHG emission source indicating that forest carbon stocks are reduced due to biomass harvest. For the 100 year rotation oak yield class 4 is used and for the 50 year rotation, oak yield class 8 is used based on the UK's F.C. Management Tables (HMSO 1971). For residues, the formula given in the notes is used. These carbon stock changes are just for the present and future fellings, (19,700 ha /yr) and not for the whole area.
*Source.* McKechnie J. *et al* 2011. HMSO 1971. Author's calculations.

No explanation is given as to how the standing tree stock changes are calculated. For residues a formula is given to calculate the carbon stored in uncollected residues, based on the decay rates for hardwood and softwoods and pulpwood and sawlog harvest. An average decay factor (k) of 0.05 has been used, which I consider to be on the low side. However, using this figure, the residue total becomes positive in years 61-70. This is much different from the figures given in Table 2.

*Table A.* Production of stem wood from year 1 to year 100 in 10 year intervals.

| Age group | CAI/ha mid-point | Re-growth on 19,700 ha to 7 cm top diameter | | Multiplier for 10 years | Total for 10 year intervals | |
|---|---|---|---|---|---|---|
| years | m$^3$ | m$^3$ | odt | | million odt | mill t $CO_2$ equiv |
| 1-10 | 0.4 | 7,880 | 4,041 | 55 | 0.22 | 0.41 |
| 11-20 | 1.1 | 21,670 | 11,113 | 155 | 1.72 | 3.16 |
| 21-30 | 4.4 | 86,680 | 44,451 | 255 | 11.34 | 20.80 |
| 31-40 | 5.3 | 104,410 | 53,544 | 355 | 19.01 | 34.88 |
| 41-50 | 5.6 | 110,320 | 56,574 | 455 | 25.74 | 47.24 |
| 51-60 | 5.6 | 110,320 | 56,574 | 555 | 31.40 | 57.62 |
| 61-70 | 5.2 | 102,440 | 52,533 | 655 | 34.41 | 63.14 |
| 71-80 | 4.8 | 94,560 | 48,492 | 755 | 36.61 | 67.18 |
| 81-90 | 4.3 | 84,710 | 43,441 | 855 | 37.14 | 68.16 |
| 91-100 | 3.8 | 74,860 | 38,390 | 955 | 36.66 | 67.28 |

*Note.* The 10 year multiplier is the number of individual 19,700 ha in each 10 year age group. For example in the first 10 years it is 1+2+3+4+5+6+7+8+9+10 = 55. The current annual increment (CAI) for years 5 and 15 has been estimated from FC Yield Class 4 Table for Oak.
*Source.* HMSO 1971. Author's calculations.

For standing trees the nominal rotation age is not stated. It may be 50 years or it may be 100 years. I have calculated the changes in forest carbon based on 100 year and 50 year rotations. My results for standing trees are shown on lines 6 and 7 in the above table and for residue on line 5. Detailed work sheets are given below. As can be seen for the 100 year calculation for standing trees, the cumulative stock change is positive by year 90 and for the 50 year rotation it is positive by year 50. Again for residues, it is positive by year 70. The following tables show how my figures are derived.

Based on the Model for the managed forest area (4.46 million ha), the average stemwood standing stock is 185 m$^3$/ha. Felling the equivalent of 19,700 ha per year would give 3,644,500m$^3$ or 1.869Modt. Table A gives the estimated production of wood in 10-year intervals based on YC 4 CAI.

It is proposed to use 1.869 Modt of stemwood for bio-energy each year. This will give off 3.430 Mt$CO_2$ equivalent or 34.3 Mt$CO_2$ equiv. for a 10-year period. Table 2 has the equivalent of 43.6 Mt$CO_2$equiv. for the first 10-year period. This is 9.3 Mt $CO_2$equiv. more than from the bio-energy wood (34.3 Mt $CO_2$). It is assumed that this additional total is given off by carbon in roots and soil. In the first 10 years, 0.41 Mt $CO_2$ equiv. are captured by tree re-growth. So in year zero, the stock of $CO_2$ will be 44.01 Mt $CO_2$ equiv. It is also assumed that with the re-growth, the emissions of $CO_2$ from old root and soil will be countered by root re-growth and soil carbon accumulation, so by year 50 there will be no net-emissions from roots and soil. Table B gives the estimate of the carbon dioxide balance for the 100 year period in 10 year intervals, by comparing emissions to stem re-growth.

*Table B. Estimate of the carbon balance from using 1.869 Modt stemwood for bioenergy and tree re-growth on 19,700 ha.*

Units: Mt CO₂ equivalent

| Period: yrs | 1-10 | 11-20 | 21-30 | 41-40 | 41-50 | 51-60 | 61-70 | 71-80 | 81-90 | 91-100 |
|---|---|---|---|---|---|---|---|---|---|---|
| Emissions | -44.01 | -41.68 | -39.36 | -37.03 | -34.71 | -34.71 | -34.71 | -34.71 | -34.71 | -34.71 |
| Capture | 0.41 | 3.16 | 20.80 | 34.88 | 47.24 | 57.62 | 63.14 | 67.18 | 68.16 | 67.28 |
| Net em. | -43.60 | -38.52 | -18.56 | -2.15 | 12.53 | 22.91 | 28.43 | 32.47 | 33.45 | 32.57 |
| Culm em. | -43.60 | -82.12 | -100.68 | -102.83 | -90.30 | -67.39 | -38.96 | -6.49 | 26.96 | 59.53 |

Note. Culm em = cumulative emissions (em.). Negative values indicate a GHG emission source specifying that forest carbon stocks are reduced due to biomass harvest. Positive values indicate that carbon is being accumulated.
Source. McKechnie J. *et al* 2011. Author's calculations.

Compared to Table 2, assuming a 100-year rotation for the felled area for bio-energy, by year 50, there is a net accumulation of carbon stock and from then onwards it increases and these increases negate the cumulative emissions by year 81. By the end of the 100 years, the net accumulation of wood is over 32 million t.

Alternatively, Table 2 may be based on a 50-year rotation. If that is the case, I assume that the annual growth of trees in the felled area must be better than yield-class 4. I have assumed that it will be yield class 8 for oak from the F.C's management tables. This will give an annual yield by year 50 slightly in excess of 1.869 million t of wood from 19,700 ha. Thus, in theory, no new areas of forests need to be cleared or if they are, the output of wood for bio-energy could be doubled. Table C gives the estimated production of wood in 10-year intervals based on YC 8 CAI for oak assuming a 50 year rotation.

*Table C. Production of stem wood from year 1 to year 50 in 10 year intervals.*

| Age group | CAI/ha mid-point | Re-growth on 19,700 ha. | To 7 cm top diameter | Multiplier for 10 years | Total for 10 year intervals | |
|---|---|---|---|---|---|---|
| years | m³ | m³ | odt | | Modt | Mt CO₂ equiv |
| 1-10 | 1.9 | 37,430 | 19,195 | 55 | 1.06 | 1.94 |
| 11-20 | 6.8 | 133,960 | 68,697 | 155 | 10.65 | 19.54 |
| 21-30 | 8.2 | 161,540 | 82,841 | 255 | 21.12 | 38.76 |
| 31-40 | 11.0 | 216,700 | 111,128 | 355 | 39.45 | 72.39 |
| 41-50 | 11.1 | 218,670 | 112,138 | 455 | 51.02 | 93.63 |

*Note.* The 10 year multiplier is the number of individual 19,700 ha in each 10 year age group. For example in the first 10 years it is 1+2+3+4+5+6+7+8+9+10 = 55. The current annual increment (CAI) for years 5 and 15 has been estimated from FC Yield Class 8 Table for Oak.
*Source.* HMSO 1971. Author's calculations.

*Table D. Estimate of the carbon balance from using 1.869 Modt stemwood for bio-energy and tree re-growth on 19,700 ha for two 50 year rotations.*

Units: Mt CO₂ equivalent

| Period | 1-10 | 11-20 | 21-30 | 31-40 | 41-50 | 1-10 | 11-20 | 21-30 | 31-40 | 41-50 |
|---|---|---|---|---|---|---|---|---|---|---|
| Emissions | -45.54 | -43.61 | -40.61 | -38.29 | -36.01 | -45.54 | -43.61 | -40.61 | -38.29 | -36.01 |
| Capture | 1.94 | 19.54 | 38.76 | 72.39 | 93.63 | 1.94 | 19.54 | 38.76 | 72.39 | 93.63 |
| Net em. | -43.60 | -24.07 | -1.85 | 34.10 | 57.62 | -43.60 | -24.07 | -1.85 | 34.10 | 57.62 |
| Culm em. | -43.60 | -67.67 | -69.52 | -35.42 | 22.20 | -21.40 | -45.47 | -47.32 | -13.22 | 44.40 |

*Note.* Culm em = cumulative emissions (em.). Negative values indicate a GHG emission source specifying that forest carbon stocks are reduced due to biomass harvest. Positive values indicate that carbon is being accumulated.
*Source.* Author's calculations.

Table D give the estimated CO₂ balance for two 50 year periods in 10 year intervals, by comparing emissions to stem re-growth.

By year 41, there is an accumulation of forest carbon and by the end of the first rotation there is a positive accumulation, equivalent to 12 million t of wood and by the end of the second rotation it is the equivalent of 24 million t of wood.

Turning to residues, an average decomposition factor of 0.05 has been used in the calculations. This has to be done in two parts. The first part is to compare the negative emissions from burning 383,000 t of wood each year with the positive emissions if the wood is left to decompose. The second part is to compare the emissions from the annual decomposition of 300,000 t of wood with the growth of branch wood in the replanted areas.

Table E looks at the use of 383,000 t of residues for bio-energy (0.702 Mt CO₂/year) or letting 383,000 t decompose on the forest floor. Table F compares the growth of branch wood against the emissions from branch wood on the forest floor.

*Table E.* Estimate of the carbon balance from using 0.383 Modt branch wood for energy or letting the same amount decompose on the forest over 100 years.

Units: Mt $CO_2$ equivalent

| Period: years | 1-10 | 11-20 | 21-30 | 31-40 | 41-50 | 51-60 | 61-70 | 71-80 | 81-90 | 91-100 |
|---|---|---|---|---|---|---|---|---|---|---|
| Emissions | -7.02 | -7.02 | -7.02 | -7.02 | -7.02 | -7.02 | -7.02 | -7.02 | -7.02 | -7.02 |
| Decomposition | 1.67 | 3.82 | 5.10 | 5.87 | 6.33 | 6.61 | 6.78 | 6.87 | 6.93 | 6.97 |
| Net em. | -5.35 | -3.20 | -1.92 | -1.15 | -0.69 | -0.41 | -0.24 | -0.15 | -0.09 | -0.05 |
| Culm em. minus coal | -5.35 | -8.55 | -10.47 | -11.62 | -12.31 | -12.72 | -12.96 | -13.11 | -13.20 | -13.25 |
| Coal savings | 5.77 | 5.77 | 5.77 | 5.77 | 5.77 | 5.77 | 5.77 | 5.77 | 5.77 | 5.77 |
| Net em. | 0.42 | 2.57 | 3.85 | 4.62 | 5.08 | 5.36 | 5.53 | 5.62 | 5.68 | 5.72 |
| Culm savings with coal | 0.42 | 2.99 | 6.84 | 11.46 | 16.54 | 21.90 | 27.43 | 33.05 | 38.73 | 44.45 |

*Note.* Culm em = cumulative emissions. Negative values indicate a GHG emission source specifying that forest carbon stocks are reduced due to biomass harvest. The decomposition is treated as a positive value.
An estimated 185,000 t of coal containing 0.577 Mt $CO_2$ will be saved annually from using 383,000 of residues, or 5.77 Mt $CO_2$ every 10 years.
*Source.* McKechnie J. *et al* 2011. Author's calculations.

It seems that only emissions from the use of residues for fuel had been considered in Table 2, because there is a net accumulation of carbon from year 61 onwards (Table G)! Also, if the saving of emissions from coal is considered then there is a saving of emissions from year 1. After 100 years the cumulative saving is over 12 million t C.

Table F compares the decomposition of 300,000 t of wood residue on the forest floor from the felling of 19,700 ha per year to the annual growth of the branch wood on the 19,700 ha annual planting/regeneration. Because the annual growth of branch wood is on average more than 300,000 t, a reduction factor of 0.80 has been applied to the gross growth figure. YC 4 has been assumed. Again a decomposition factor of 0.05 has been used in the formula given on page S-14 in the articles' accompanying notes. Table G combines Tables E and F to examine total emissions.

*Table F.* Estimate of the carbon balance from letting 300,000 odt branch wood decompose on the forest floor compared to the growth of branch wood on newly planted/regenerated area over 100 year period.

Units: Mt $CO_2$ equivalent

| Period: years | 1-10 | 11-20 | 21-30 | 31-40 | 41-50 | 51-60 | 61-70 | 71-80 | 81-90 | 91-100 |
|---|---|---|---|---|---|---|---|---|---|---|
| Decomposition | -1.31 | -2.99 | -4.00 | -4.60 | -4.96 | -5.18 | -5.31 | -5.38 | -5.43 | -5.46 |
| Capture | 0.08 | 0.65 | 4.26 | 7.14 | 9.68 | 11.80 | 12.93 | 13.76 | 13.96 | 13.78 |
| Net em. | -1.23 | -2.34 | 0.26 | 2.54 | 4.72 | 6.62 | 7.62 | 8.38 | 8.53 | 8.32 |
| Culm em. | -1.23 | -3.57 | -3.31 | -0.77 | 3.95 | 10.58 | 18.20 | 26.58 | 35.11 | 43.43 |

*Note.* Culm em = cumulative emissions (em.). Negative values indicate a GHG emission source specifying that forest carbon stocks are reduced due to biomass harvest.
*Source.* McKechnie J. *et al* 2011. Author's calculations.

*Table G.* Estimated total emissions from residues over a 100-year period.

Units: Mt $CO_2$ equivalent

| Period: years | 1-10 | 11-20 | 21-30 | 31-40 | 41-50 | 51-60 | 61-70 | 71-80 | 81-90 | 91-100 |
|---|---|---|---|---|---|---|---|---|---|---|
| Culm em. without coal savings | -6.58 | -12.12 | -13.78 | -12.39 | -8.36 | -2.14 | 5.24 | 13.47 | 21.91 | 30.18 |
| Culm em. With coal savings | -0.81 | -0.58 | 3.53 | 10.69 | 20.49 | 32.48 | 45.63 | 59.63 | 73.84 | 87.88 |
| Table 2 figures | -8.2 | -11.8 | -13.0 | -13.5 | -13.9 | -14.3 | -14.7 | -15.0 | -15.2 | -15.2 |

*Note.* Culm em = cumulative emissions. Negative values indicate a GHG emission source specifying that forest carbon stocks are reduced due to biomass harvest. Table 2 figures are from the McKechnie article.
*Source.* McKechnie J. *et al* 2011. Author's calculations.

Comparing the second line in Table G, with the article's Table 2 figures, given on the bottom line of Table G, it will be seen that there is a considerable discrepancy from years 51-60 and onwards without coal savings and from year 1 with coal savings. The cumulative savings of carbon is nearly 24 million t.

On a related issue, I think that the decomposition rate is too low. The lifetime of the decaying wood with a decay factor of 0.05 is over 130 year. In my opinion, for tops and branches, the rate should be about 0.15. This would give the lifetime of about 30 years for decomposing small diameter wood.

Table H1&2 looks at the adjusted emission figures for stem wood taking into consideration the coal saving factor. The two rotations have been considered, namely 50 & 100 years. The estimated annual saving of coal from substituting 1.914 Mt of wood is 1.085 Mt coal which would give off 2.708 Mt $CO_2$ equivalent or 27.08 Mt every 10 years.

It seems there are large discrepancies between Table 2 and the above calculations. It is also certain that the article did not consider the annual growth of all the trees, just the areas to be used for bio-energy. If it had done, then it should have concluded that annual growth of wood exceeds annual removals and there are no net carbon emissions. Rather there is carbon capture.

*Table H1.* Estimate of the carbon balance from using 1.869 Modt stemwood each year for bio-energy and tree re-growth on 19,700 ha for two 50 year rotations.

Units: Mt CO₂ equivalent

| Period: years | 1-10 | 11-20 | 21-30 | 31-40 | 41-50 | 1-10 | 11-20 | 21-30 | 31-40 | 41-50 |
|---|---|---|---|---|---|---|---|---|---|---|
| Emissions | -45.54 | -43.61 | -40.61 | -38.29 | -36.01 | -45.54 | -43.61 | -40.61 | -38.29 | -36.01 |
| Capture | 1.94 | 19.54 | 38.76 | 72.39 | 93.63 | 1.94 | 19.54 | 38.76 | 72.39 | 93.63 |
| Net em. | -43.60 | -24.07 | -1.85 | 34.10 | 57.62 | -43.60 | -24.07 | -1.85 | 34.10 | 57.62 |
| Culm em. | -43.60 | -67.67 | -69.52 | -35.42 | 22.20 | -21.40 | -45.47 | -47.32 | -13.22 | 44.40 |
| Coal savings | 27.08 | 27.08 | 27.08 | 27.08 | 27.08 | 27.08 | 27.08 | 27.08 | 27.08 | 27.08 |
| Net em. with coal | -16.52 | 3.01 | 25.23 | 61.08 | 84.70 | -16.52 | 3.01 | 25.23 | 61.08 | 84.70 |
| Culm em. with coal | -16.52 | -13.51 | 11.72 | 72.80 | 157.50 | 140.98 | 143.99 | 169.22 | 230.30 | 315.00 |
| | | | | | **Table 2 figures. McKechnie article** | | | | | |
| Standing trees | -43.6 | -80.9 | -106.3 | -112.5 | -113.4 | -112.7 | -132.8 | -143.6 | -150.8 | -150.7 |

*Note.* Culm em. = cumulative emissions (em.). Negative values indicate a GHG emission source specifying that forest carbon stocks are reduced due to biomass harvest. Positive values indicate that carbon is accumulating. Carbon saving including coal by year 100 = 86 Mt and without coal 12 Mt. Table 2 of the McKechnie article shows a carbon debt of 41Mt.
*Source.* McKechnie J. *et al* 2011. Author's calculations.

*Table H2.* Estimate of the carbon balance from using 1.869 Modt stemwood each year for bio-energy and tree re-growth on 19,700 ha assuming a rotation of 100 years.

Units: Mt CO₂ equivalent

| Period: years | 1-10 | 11-20 | 21-30 | 41-40 | 41-50 | 51-60 | 61-70 | 71-80 | 81-90 | 91-100 |
|---|---|---|---|---|---|---|---|---|---|---|
| Emissions | -44.01 | -41.68 | -39.36 | -37.03 | -34.71 | -34.71 | -34.71 | -34.71 | -34.71 | -34.71 |
| Capture | 0.41 | 3.16 | 20.78 | 34.85 | 47.20 | 57.57 | 63.09 | 67.13 | 68.10 | 67.22 |
| Net em. | -43.60 | -38.52 | -18.58 | -2.18 | 12.49 | 22.86 | 28.38 | 32.42 | 33.39 | 32.51 |
| Culm em. | -43.60 | -82.12 | -100.68 | -102.83 | -90.30 | -67.39 | -38.96 | -6.49 | 26.96 | 59.53 |
| Coal savings | 27.08 | 27.08 | 27.08 | 27.08 | 27.08 | 27.08 | 27.08 | 27.08 | 27.08 | 27.08 |
| Net em. with coal | -16.52 | -11.44 | 8.50 | 24.90 | 39.57 | 49.94 | 55.46 | 59.50 | 60.47 | 59.59 |
| Culm em. with coal | -16.52 | -27.96 | -19.46 | 5.44 | 45.01 | 94.95 | 150.41 | 209.91 | 270.38 | 329.97 |
| | | | | | **Table 2 figures. McKechnie article** | | | | | |
| Standing trees (ST) | -43.6 | -80.9 | -106.3 | -112.5 | -113.4 | -112.7 | -132.8 | -143.6 | -150.8 | -150.7 |

*Note.* Culm em. = cumulative emissions (em.). Negative values indicate a GHG emission source specifying that forest carbon stocks are reduced due to biomass harvest. Positive values indicate that carbon is accumulating. Carbon saving including coal by year 100 = 90 Mt. and without coal 16 Mt. Table 2 of the McKechnie article shows a carbon debt of 41 Mt.
*Source.* McKechnie J. *et al* 2011. J. Author's calculations.

# Annex 2.

*Table 1.* UK Forestry Commission: Management Table for Oak Yield Class 4

| Age years | Plants left per-ha number | Plants thinned/ died/ha number | Main crop Before Thinning: m³/ha To 7cm top diam. | After | Thinning/death Per 5 yr m³/ha To 7cm top diam. | Cumulative | Cumulative total m³/ha 7cm t.d. | CAI m³/ha To 7cm top diam. | MAI m³/ha |
|---|---|---|---|---|---|---|---|---|---|
| 0 | 5000 | 0 | 0 | 0 | 0 | 0 | 0 | 0 | 0 |
| 25 | 4200 | 800 | 30 | 30 | 0 | 0 | 30 | 4.4 | 1.2 |
| 30 | 3750 | 450 | 54 | 54 | 0 | 0 | 54 | 4.9 | 1.8 |
| 35 | 2363 | 1387 | 79 | 66 | 13 | 13 | 79 | 5.3 | 2.3 |
| 40 | 1702 | 661 | 93 | 79 | 14 | 27 | 106 | 5.5 | 2.7 |
| 45 | 1285 | 417 | 107 | 93 | 14 | 41 | 134 | 5.6 | 3.0 |
| 50 | 1006 | 277 | 121 | 107 | 14 | 55 | 162 | 5.6 | 3.2 |
| 55 | 822 | 184 | 135 | 121 | 14 | 69 | 190 | 5.6 | 3.5 |
| 60 | 681 | 141 | 148 | 134 | 14 | 83 | 217 | 5.4 | 3.6 |
| 65 | 573 | 108 | 161 | 147 | 14 | 97 | 244 | 5.2 | 3.8 |
| 70 | 492 | 81 | 172 | 158 | 14 | 111 | 269 | 5.0 | 3.9 |
| 75 | 428 | 64 | 183 | 169 | 14 | 125 | 294 | 4.8 | 3.9 |
| 80 | 376 | 52 | 192 | 178 | 14 | 139 | 317 | 4.5 | 4.0 |
| 85 | 335 | 41 | 200 | 186 | 14 | 153 | 339 | 4.3 | 4.0 |
| 90 | 300 | 35 | 207 | 193 | 14 | 167 | 360 | 4.0 | 4.0 |
| 95 | 270 | 30 | 212 | 198 | 14 | 181 | 379 | 3.8 | 4.0 |
| 100 | 244 | 26 | 216 | 202 | 14 | 195 | 397 | 3.5 | 4.0 |
| 105 | 221 | 23 | 218 | 204 | 14 | 209 | 413 | 3.2 | 3.9 |

| Age years | Plants left per-ha number | Plants thinned/ died/ha number | Main crop Before After Thinning: m³/ha To 7cm top diam. | | Thinning/death Per 5 yr Cumulative m³/ha To 7cm top diam. | | Cumulative total m³/ha 7cm t.d. | CAI m³/ha To 7cm top diam. | MAI m³/ha |
|---|---|---|---|---|---|---|---|---|---|
| 110 | 201 | 20 | 219 | 206 | 13 | 222 | 428 | 2.9 | 3.9 |
| 115 | 185 | 16 | 221 | 208 | 13 | 235 | 443 | 2.7 | 3.9 |
| 120 | 171 | 14 | 221 | 209 | 12 | 247 | 456 | 2.4 | 3.8 |
| 125 | 159 | 12 | 220 | 209 | 11 | 258 | 467 | 2.2 | 3.7 |
| 130 | 149 | 10 | 220 | 210 | 10 | 268 | 478 | 2.0 | 3.7 |
| 135 | 141 | 8 | 219 | 210 | 9 | 277 | 487 | 1.8 | 3.6 |
| 140 | 134 | 7 | 218 | 210 | 8 | 285 | 495 | 1.6 | 3.5 |
| 145 | 128 | 6 | 218 | 211 | 7 | 292 | 503 | 1.4 | 3.5 |
| 150 | 123 | 5 | 217 | 211 | 6 | 298 | 509 | 1.2 | 3.4 |
| 155 | 118 | 5 | 216 | 211 | 5 | 303 | 514 | 1.0 | 3.3 |
| 160 | 114 | 4 | 215 | 211 | 4 | 307 | 518 | 0.8 | 3.2 |
| 165 | 111 | 3 | 214 | 211 | 3 | 310 | 521 | 0.6 | 3.2 |
| 170 | 109 | 2 | 213 | 211 | 2 | 312 | 523 | 0.4 | 3.1 |
| 185 | 108 | 1 | 212 | 211 | 1 | 313 | 524 | 0.2 | 3.0 |
| 180 | 108 | 0 | 211 | 211 | 0 | 313 | 524 | 0.0 | 2.9 |

*Note.* The yield table has been extrapolated from 150 to 180 years. At this point the current annual increment (CAI) = zero. MAI = mean annual increment. The yield table represents either a natural succession of a single crop over time or different age classes in a population. At an early age, many plants that are thinned or die have a top diameter less than 7cm. The MAI at year 180 is 2.9 m³/ha; this is only 72.5% of the maximum MAI – 4.0 m³/ha. The table has been modified to give main crop after thinning and the cumulative thinning totals every five years. Excluded are top height, mean diameter, basal area, volumes to 18cm and 24cm of main crop and thinnings, cumulative basal area and basal area increment.
*Source.* British Forestry Commission Booklet No. 34. Her Majesty's Stationery Office (HMSO) 1971. London, UK.

# References

[1] Canadian NFI 2001. https://nfi.nfis.org/standardreports.php?lang=en

[2] Chen J *et al* 2010. Carbon budget of Ontario's managed forests and harvested wood products 2001-2010, Forest Ecology and Management 259 (2010) 1385-1398.

[3] Chen J *et al* 2008. Future carbon storage in harvested wood products from Ontario's crown forests, Can. J. For. Res. 38 1947-1958.

[4] HMSO 1971. British Forestry Commission Booklet No. 34 (Metric). Her Majesty's Stationery Office. London.

[5] Institute for European Environmental Policy 2012. The GHG intensity of biomass: does biomass have a role to play in reducing Europe's GHG emissions. IEEP, 15 Queen Anne's Gate, London, SW1H 9BU. UK.

[6] KBM Forestry Consultants Inc. 2008. Algoma Forest – Independent Forest Audit Report to Ontario Forest Service. KBM, 349 Mooney Av, Thunder Bay, ON P7B 5L5 Canada.

[7] KPMG LLP 2008. Wood pellet plant cost study for the Algoma and Martel forests in the western portion of the Great Lakes/St. Lawrence forests. KPMG LLP, 4600-333 Bay St. Bay Adeland Centre, Toronto, Ontario M5H 2S5 Canada.

[8] Manomet Center 2010. Biomass sustainability and carbon policy study. Manomet Center for Conservation Sciences. 81 Stage Point Road, Plymouth, MA 02360, USA.

[9] Matthews R *et al* 2012. 'Carbon impacts of using biomass in

bio-energy and other sectors: forest'. A report for DECC – project TRN 242/08/2011 published in support of the UK bio-energy strategy. Forest Research and North Energy UK.

[10] McKechnie J *et al* 2011. Forest bioenergy or forest carbon? Assessing trade-offs in greenhouse gas mitigation with wood-based fuels, (with supporting information). Environ. Sci. Technol. 45 789-795.

[11] Searchinger T 2012. 'Sound principles and important inconsistency in the 2012 UK bioenergy strategy', with appendix. Woodrow Wilson School of Public and International Affairs, Princeton University. Princeton, NJ, USA.

[12] Searchinger T *et al* 2009. Fixing a critical climate accounting error. Science 2009, 326 (5952) 527-528.

[13] The (UK) Royal Society for the Protection of Birds/Friends of the Earth/Greenpeace 2012. 'Dirtier than coal? Why [UK] government plans to subsidise burning trees are bad news for the planet'. Royal Society for the Protection of Birds, the Lodge, Potton Road, Sandy, Bedfordshire, SG 19 2DL, UK.

[14] UK Forestry Commission 2003. Forests, carbon and climate change: the UK contribution. Report to the IPCC. F.C. Environmental Research Branch, Alice Holt Lodge, Wrecclesham, Farnham, Surry, GU10 2LH, England.

[15] USDA 2009. FORECARB2: an updated version of the U.S. forest carbon budget model, Linda S. Heath et al. Forest Service, Northern Research Station, General Technical Report NRS-67. U.S. Forest Service, 11 Campus Blvd, Suite 200, Newtown Square, PA 19073, USA.

[16] Zang Y *et al* 2010. Life cycle emissions and cost of primary electricity from coal, natural gas and wood pellets in Ontario Canada, (with supporting information). Environ. Sci. and Technol. 44 538-544.

# 5

# Performance analysis of power generation by producer gas from refuse derived fuel-5 (RDF-5)

**Natthawut Dutsadee[1, *], Nigran Homdoung[1], Rameshprabu Ramaraj[1], Khamatanh Santisouk[2], Shangphuerk Inthavideth[2]**

[1]School of Renewable Energy, Maejo University, Sansai, Chiang Mai-50290, Thailand
[2]Department of Mechanical Engineering, Faculty of Engineering, National University of Laos, Vientiane, Laos

**Email address:**
natthawu@yahoo.com (Dussadee N.), natthawu@mju.ac.th (Dussadee N.)

**Abstract:** At present, municipal and city corporation governments throughout the world are facing choices about how to manage the unending stream of waste generated by their residents and businesses. In many places landfills and dumpsites are filling up, and all landfills and dumpsites leak into the environment; due to increasing populations, the issue of waste becomes more urgent and more complicated. Many regions are already facing a waste crisis, and drastic measures are needed. In the past, the main approach to waste management operations is the landfill which is causes many environmental pollutions and health hazards. Furthermore, extending the land for land filling is the one of best solutions. This paper demonstrated the performance analysis of power generation producer gas from RDF-5 in Chiang Mai University, Thailand. The efficiency of different ratio waste composition and of RDF-5 was revealed. In addition, the humidity, density and heat capacity of RDF-5 are also focused. In order to analyze the compositions, heat capacity of producer gas, fuel consumption, efficiency of producer gas system, waste water and quantity of ash; RDF-5 have been tested by using producer gas in different ratio of oxygen and fuel. In term of automobile application, the performance of RDF-5 and Diesel-RDF-5 are compared; and the specific factors such as power, specific fuel consumption rate, carbon dioxide, sound level and fuel feeding were included that comparison. Consequently, this paper mainly focused and concerned with the production and properties of refuse derived fuel-5 for use in energy from waste technologies.

**Keywords:** Municipal Solid Waste, Combustion Engine, Renewable Energy

## 1. Introduction

Waste management is collection of waste which generated by every person in the society. The majority of the waste consists of household waste, agricultural waste, industrial waste and bio medical waste. They consist of recyclable waste and biodegradable waste. Some of them are hazardous waste and toxic waste which cause health hazardous and environmental pollution. One of the best solutions for waste management is converted waste into renewable energy in the form of thermal energy as well as electrical energy. Refuse Derived Fuel (RDF) is combustible or, in other word, high calorific fraction recovered from Municipal solid waste (MSW). There are other terms used for MSW derived fuel such as Recovered Fuel (REF), Packaging Derived Fuel (PDF), Paper and Plastic Fraction (PPF) and Process Engineered Fuel (PEF) [1, 2].

The refuse-derived fuel (RDF-5) is convenient for transportation and storage, high in calorific value, and low in pollution. The waste composition and physical and chemical characteristics of each waste fraction were determined to evaluate the suitability of the waste for recycling and reuse as RDF-5 [3]. In 2005, the production of RDF-5 waste reached 13 million tons in EU, indicating the rapid development of RDF-5 waste as an alternative energy source. The use of RDF-5 in Asia is gradually becoming popular, e.g., in Japan and Taiwan [4].

RDF-5 is the first alternative fuel which is obtained from municipal solid waste [5]. However, producing RDF-5 does not make household waste and industrial waste disappear.

Presently, in the development of RDF-5 in the form of thermal and electrical energy is only 5 MW [6]. This is due to restriction of laws and regulations and lack of information for decision makers and investors to introduce RDF plant.

Moreover, the communities cannot managed the waste as well as it should be and most people do not believe that the technology can be overcome those problems. In 2007 the wastes increased 14.72 million tons or 40332 tons/day which is 14,432 tons/day or 35.8% is treated in the right way [5], show in Table 1. In 2006, the northern of Thailand, the wastes are 2,195 tons/day [6, 8]. It would be useful if this amount of waste is converted into renewable energy, such as thermal and electrical energy as supplementary fuel for industries. Therefore, it could be reduced the amount of import fuel from oversea as well as reduced environment pollution.

**Table 1.** *Solid waste management in 2007.*

| Region | Solid waste (ton/day) | | |
|---|---|---|---|
| | Generated | Eliminated | Proportion (%) |
| Bangkok | 8532 | 8532 | 21.2 |
| Municipal and Pattaya (1,277 places) | 13600 | 4810 | 11.9 |
| Outside of Municipal 6,500 places) | 18200 | 1090 | 2.7 |
| Total | 40332 | 14432 | 35.8 |

Currently, there are several technologies to eliminate the municipal solid waste; anaerobic digestion, biogas from waste landfill, producer gas RDF-5, manufacturing of fuel, plasma arc, converted plastic waste into fuel and incinerator. In this paper, producer gas RDF-5 is chosen as eliminate municipal solid waste technology. The steps and advantages of producer gas process are shown in the list below:

1. Preliminary liberation: this step involves separating municipal solid waste into bio-degradable, glass, rags, paper, plastic, leather and rubber, metals and other domestic hazardous, inert.
2. Size screening: size screening involves separating the municipal waste based on the size and shape of the particle. It helps in material handling comfortably.
3. Shredding: destructing the large amount of solid waste into smaller pieces by crushing and cutting for easy handling and transporting.
4. Magnetic separation: separating the metal particles from the crushed particles.

According to the ASTM E-75 standard, producer gas divided into 7 types [4]: RDF (Municipal Solid Waste (MSW)), RDF-2 (Coarse RDF), RDF-3 (Fluff RDF), RDF-4 (Dust RDF), RDF-5 (Densified RDF), RDF-6 (RDF Slurry) and RDF-7 (RDF Syn-gas). Producer gas technologies are included mechanical system and thermal system. In this paper, producer gas Down-Daft Gasified type is used. The system divided into three zones: Distillation zone, Combustion zone and Reduction zone. The chemistry reaction of each zones divided into combustion zone, reduction zone, distillation or pyrolysis zone and drying zone [5,9].

# 2. Experimental Design

## 2.1. Experiment Apparatus and Measurement System

Producer gas RDF-5 shown in Figure 1 and Figure 2 illustrated the producer gas system Down Draft type which is used in this research. The system consists of producer gas system, cleaning system and heat exchanger (producer gas cooling), piping and its associated devices. The producer gas has a neck diameter of 16.6 cm. The location of nozzle is 13 cm along the height of producer gas from the neck diameter plane of producer gas. There are 5 nozzles in this system which is 15 cm diameter for each. The gasifier reactor capacity is 35-40 kg for charcoal and 100-110 kg for RDF-5. The top of the reactor is the feedstock hopper which has the capacity 10 kg and 30 kg of charcoal and RDE-5, respectively. The charcoal (biomass) consumption rate is about 5-20 kg/hr. To eliminate the ashes, the filter is installed and there is storage for ash which is contained 70 liters of water. The air input and output controlled by 220 V, 185.5 W fan.

**Figure 1.** *Producer gas RDF-5 [7].*

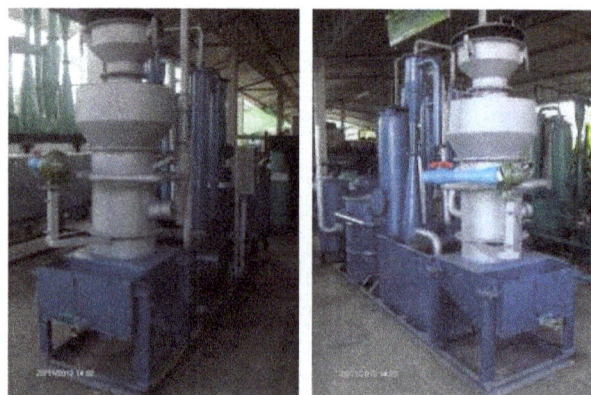

**Figure 2.** *Producer gas Down Draft type.*

The cleaning system was demonstrated in Figure. 3. The system consists of cyclone, wet collectors, filters, and trap tank. For the heat exchange is exhibited in Figure 4. The producer gas is transmitted by the pipe which has 5.08 cm diameter and welded with the cyclone. The 90 liters square steel tank contains the cool water to drop the temperature of producer gas out from 400-500 °C to 200-250 °C. The gas transmitter and air controller are used the 3 phase, 400 W, 50 Hz, 380 V fans as show in Figure 5.

Figure 3. Cleaning system.

Figure 4. Heat exchanger and producer gas transmitter system.

Figure 5. producer gas engine and load (heater).

Table 2. Environmental and algal biomass measurements

| Gasifier System Specification | |
| --- | --- |
| Type of Gasifier | Downdraft |
| Feeding | Manual/batch feeding |
| Biomass consumption | 5-20 kg/h |
| Hopper capacity | Charcoal: 35-40 kg, RDF-5: 100-110 kg |
| Producer gas cooling | Water |
| Producer gas flow rate | 80 m³/h (maximum) |
| Biomass size | 10 mm (minimum)-50 mm(maximum) |
| Producer gas cleaning | Ventury Scrubber/Pack bed Scrubber Biomass filter/Fubrics filter/Paper filter |

For the producer gas engine, the ignition engine NISSAN

Z16, 4 cylinders, 4 strokes, 1595 cc, maximum power 61.74 kW is used, as shown in Figure 5. The compression ratio have modified from 8.8:1 to 10.5:1 as more details refer to table 2 and 3.

Table 3. Performance testing parameters of producer gas engine.

| SI producer gas engine specification | |
| --- | --- |
| Engine make & model | NISSAN/Z16 |
| Cylinder/alignment/cycle | 4/Inline/4stroke |
| Compression ratio | 8.8:1 (10.5) |
| Bore x Stroke (mm) | 83 x 73.7 |
| Cylinder capacity | 1595 cc |
| Gross engine power | 61.74 kW |
| Ignition system | Battery-based distributor type with ignition advance |
| Total oil capacity | ELECKING 1 phases 50 Hertz 220 V 2kVA |
| Load | Electrical heater |

### 2.2. Experimentation and Implementation

The experiment setup and measurement system is illustrated in Figure 6. The angle velocity of the engine is measured by a Tacho meter at the pulley of the engine while exhaust emission analysis is installed at the exhaust emission.

Performance testing of producer gas engine is started by measured the temperature, pressure at the atmosphere pressure. The average pressure and temperature are 0.92 kPa and 32 ± 3°C, respectively, air density is 1.1 kg/m3. In order before run the test, RDF-5 should be weigh first to obtain feeding capacity. After that, the gas generator is started to ignition about 20 min, the gas quality is checked by ignition at the gas check point (1) as shown in Figure 7; when observe that the gas is balance, the valve is opened to allow the producer gas flow into the cleaning system and flow throughout the heat exchanger and then check the quality of producer gas again at check point (2). If the producer gas is balance, the valve is opened to allow the producer gas flow and mix with the air in the appropriated composition and then enter to the producer gas engine. In the case of the amount of gas in the storage is exceed the other valve will open to release it to the environment.

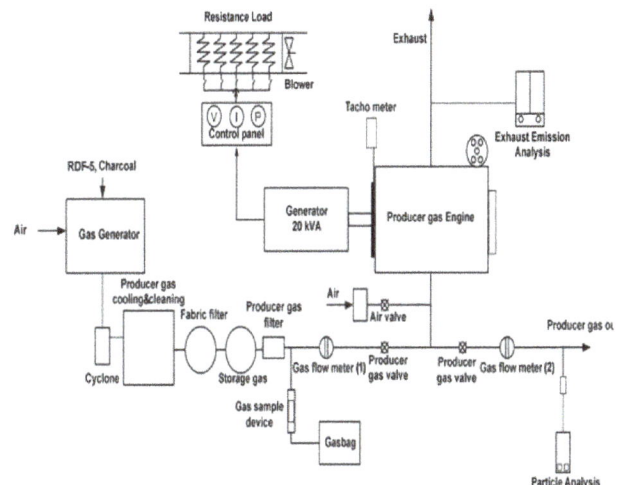
Figure 6. Overall schematic diagram of experiment apparatus and instruments setup.

*Figure 7. Checking quality of producer gas and running generator.*

The test are divided into 6 cases of load; 1.5 kW, 3 kW, 4.5 kW, 6 kW, 7.5 kW and 9 kW and in the case of unloading. The consumption of RDF-5, producer gas flow rate, angle velocity of engine, ignition angle, emissions of engine and sound of engine are recorded 3 times for each case.

## 3. Results and Discussion

The result illustrates that the maximum efficiency of producer gas is 62.4%. The substances obtained from producer gas RDF-5 are 8.35% CO, 10.23% $H_2$, 3.52% CH4, 17.70% $CO_2$, 8.11% $O_2$ and 52.10% $N_2$, respectively. The average of heat capacity of producer gas is about 3,434.45 $kJ/Nm^3$. For the engine power systems, producer gas RDF-5 produced maximum electric power 9 kW at 1500 rpm, combustion located at 25°C, minimum specific fuel consumption 1.53kg/kWh at the load of 7.5 kW and obtained maximum 3.21% of CO and 74.5 ppm of hydrocarbon. The minimum of CO and hydrocarbon occurred at the load 1.5 kW are 0.26% and 42 ppm, respectively.

The power generated by producer gas engine is increased gradually by increased energy or producer gas flow rate as show in Figure 8. The producer gas engine generated maximum electrical power 9 kW and minimum 1.5 kW at producer gas flow rate 63.35 m3/hr and 17.7 m3/hr, respectively. The producer gas RDF-5 is better than charcoal 4-14 % in terms of energy input for the engine as show in Figure 9. Advantage is the only technology that can dispose waste properly, by reducing mass and volume, and at the same time, can generate the green energy in the form of electricity with no harmful to environment. Experimental study has been carried out to study the possibility of using RDF-Gasification technology for power generation by using waste generated within University as Model. Waste has been sorted out of non-combustible materials, as well as recyclable material from source separation. The rest paper and plastic are used to prepare RDF. The specific fuel consumption rate of producer gas engine is the rate of solid wastes or charcoal input compare with electrical power output.

The results show that at the load 2-6 kW the specific fuel consumption rate of producer gas engine trended reduce and increase at the load 8-9 kW. These are occurred both RDF-5 and charcoal as illustrate in Figure 10. The minimum of specific fuel consumption rate of producer gas engine occurred at 7.5 kW of load, the rate of fuel consumption is 1.53 kg/kWh for RDF-5 and 1.2 kg/kWh for charcoal. The maximum of specific fuel consumption rate of producer gas engine occurred at 1.5 kW of load, the rate of fuel consumption is 2.4 kg/kWh for RDF-5 and 1.93 kg/kWh for charcoal.

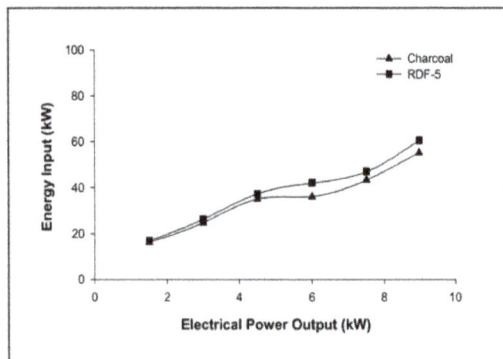

*Figure 8. Power of producer gas engine form RDF-5.*

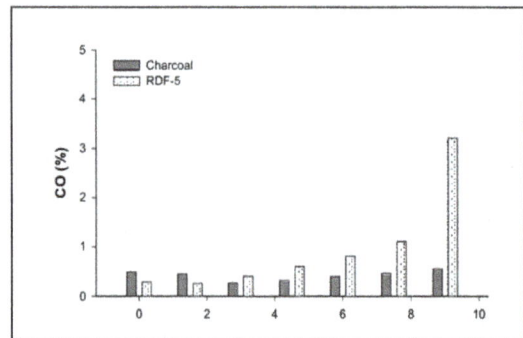

*Figure 10. Carbon monoxide emission of producer gas engine.*

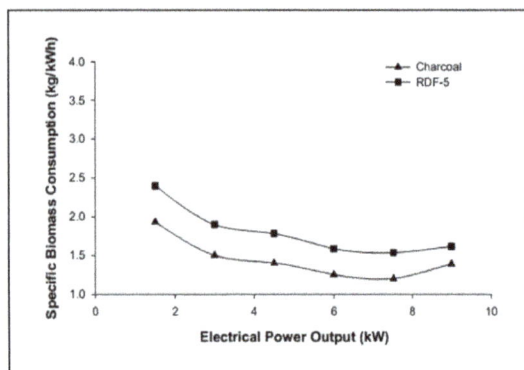

*Figure 9. Specific fuel consumption rate of producer gas engine*

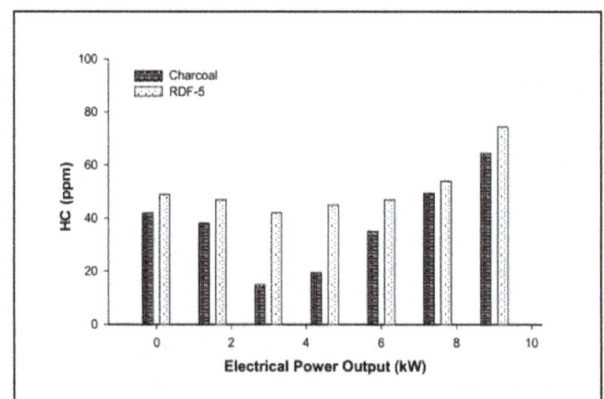

*Figure 11. Hydrocarbon emissions of producer gas engine.*

The quantity of carbon monoxide (CO) from RDF-5 and charcoal increased when the load is lightly and heavily as show in Fig. 10. At the middle load to maximum load CO from RDF-5 is 89% greater than charcoal because of the combustion of the engine trend to be in-combustion [5]. The maximum and minimum of CO from RDF-5 is 3.21 % and 0.26 % at 9 kW and 1.5 kW, respectively. The maximum and minimum of CO from charcoal is 0.55 % and 0.27 % at 9 kW and 3 kW, respectively. At the load of 7.5 kW CO from RDF-5 is lower than carbon monoxide emission standard 26%.

Similar to carbon monoxide case, the quantity of hydrocarbon from RDF-5 and charcoal increased when the load is lightly and heavily as show in Figure 11. The maximum and minimum of hydrocarbon from RDF-5 is 74.5 ppm 42 ppm at 9 kW and 3 kW, respectively. The maximum and minimum of hydrocarbon from charcoal is 64.5 ppm and 15 ppm at 9 kW and 3 kW, respectively. The results also illustrated that hydrocarbon from RDF-5 is lower than hydrocarbon emissions standard 62.7%.

The maximum sound intensity level from RDF-5 of producer gas engine is 96.4 dB at the maximum load while the maximum sound intensity level from charcoal is 94.4 dB at the same condition of load. The sound intensity is increased gradually along with increasing the loads as show in Figure 12. Fortunately, the sound intensity level from RDF-5 and charcoal are lower than the standard sound intensity level 100 dB. The exhaust temperature of the producer gas engine depends on the load, if the load is increased the exhaust temperature also increased as show in Figure 13. The exhaust temperature from RDF-5 of producer gas engine is lower than the exhaust temperature from charcoal. At the maximum load the exhaust temperature of RDF-5 and charcoal is 339°C and 342°C, respectively. At present, bio-energy refers to useable energy (such as electricity and heat) that is converted from biomass. Similar to wind and solar energy, bio-energy is an excellent energy recycling technology, but it has the most prospects, given its ability to turn refuse into bio-energy [9]. Several technological processes are available to convert this waste into usable energy resources and products, such as ethanol, biodiesel, electric power, and plastics. For example, biomass can be converted to provide an electric power source for automobiles [10]. As such, refuse-derived fuel (RDF) technology has in recent years become the refuse processing technology adopted by advanced nations [11,12].

The American Society for Testing and Materials (ASTM, 2006) divides RDF into seven main categories. The most common type of refuse-derived fuel comes from sewage sludge (RDF-5), where the sewage sludge is made into solid pellet-shaped fuel after going through various processes [13]. Sewage sludge contains a significant amount of organic matter which is predominantly proteins and carbohydrates. These account for approximately 90% of the volatile suspended solids after the sludge is concentrated by settling at 4 °C for 24 h [14, 15, 16]. The main characteristics of RDF-5 lie in its size, its high and constant heating value (3000–6000 kcal/kg), the differing amounts due to the source of waste material, but is usually two-thirds that of coal), its low level of pollution, and

that it does not emit a foul smell. In addition, as it is reduced to one-tenth of its original size after processing, it can be easily transported or stored. It is also convenient in use as it can be stored under normal temperatures for 6–12 months without decaying. The fuel can also be used directly in the fired boiler as the main combustible or when mixed with other fuels [15–17]. (Alter, 1996, Raili and Marttl, 1996 and Weber et al., 2009).

Accordingly, the main advantage of RDF-5 is conversion of waste into energy which helps in effective handling of Municipal Waste. The other advantage includes the problem associated with solid waste management and fossil fuels are eliminated. RDF-5 is one of the alternative and renewable resources of fuelwhich is derived from municipal waste. Production and utilization of RDF leads to green environment.

## 4. Conclusions

From this study, it can be conclude that the waste generated within the university campus has the potential to be converted into high heating value gaseous fuel via gasification process. The optimum operating condition for the gasification process of RDF-5 collected from Chiang Mai University, Thailand. The essential characteristics of a refuse-derived fuel-5 (RDF-5) and the combustion behaviors were performed in this study. The test data demonstrate good results for the development of energy recovery technology of organic sludge or waste. The ash deposit formation propensity has been based on pretreatment, temperature and the ratio of organic sludge to sawdust. The usage of organic sludge and waste as an alternative fuel is cost effective and has environmental benefits.

## References

[1] United Nations Environment Programme (UNEP), "Solid Waste Management", Volume I, 2005.

[2] A. Gendebien, A. Leavens, K. Blackmore, A. Godley, K. Lewin, K.J. Whiting, R. Davis, J. Giegrich, H. Fehrenbach and U. Gromke, N. del Bufalo, D. Hogg, "Refuse Derived Fuel, Current Practice and Perspectives- Final Report, European Commission, Report No: CO 5087-4, 2003.

[3] W. Punin, S. Maneewan, C. Punlek, "The feasibility of converting solid waste into refuse-derived fuel 5 via mechanical biological treatment process", Journal of Material Cycles and Waste Management, 2014, 16: 753–762.

[4] F. C. Wu, P. H. Wu, R. L. Tseng, R. S. Juang, "Use of refuse-derived fuel waste for the adsorption of 4-chlorophenol and dyes from aqueous solution: Equilibrium and kinetics", Journal of the Taiwan Institute of Chemical Engineers, 2014, 4: 2628–2639.[

[5] Ministry of Energy, "Renewable Energy Development Plan (REDP)", Available from <http://www.eppo.go.th/ccep/energy_3-5.html> [Accessed 1 August 2014], 2014.

[6] MNRE, "Ministry of Natural Resources and Environment Digital Library", Available from <http://lib.mnre.go.th/index.php/> [Accessed 1 August 2014], 2014.

[7]  K. Sombatsompop, "Wastewater treatment by sequencing batch reactor system", The Journal of KMUTNB, 2010, 18: 96–103.

[8]  Pollution Control Department, "Reports of pollution in Thailand in 2012", http://infofile.pcd. go.th/mgt/DraftPol, 2013.

[9]  P. Nutongkaew, J. Waewsak, T. Chaichana, Y. Gagnon, "Greenhouse gases emission of refuse derived fuel-5 production from municipal waste and palm kernel" Energy Procedia, 2014, 52: 362– 370.

[10] Report2555_25560214.pdf? CFID =13119758& CFTOKEN=22373626. Retrieved 1 Apr 2013. N. B. Chang, Y. H. Chang, W. C. Chen, "Evaluation of heat value and its prediction for refuse-derived fuel", Science of the Total Environment, 1997, 197:139–148.

[11] W. S. Chen, F. C. Chang, Y. H. Shen, M. S. Tsai, "The characteristics of organic sludge/sawdust derived fuel", Bioresource Technology, 2011, 102: 5406–5410.

[12] Z. M. Fu, X. R. Li, H. Koseki, "Heat generation of refuse derived fuel with water", Journal of Loss Prevention in the Process Industries, 2005, 18: 27–33.

[13] T. Kupka, M. Mancini, M. Irmer, R. Weber, "Investigation of ash deposit formation during co-firing of coal with sewage sludge, saw-dust and refuse derived fuel", Fuel, 2008, 87: 2824–2837.

[14] American Society for Testing and Materials (ASTM), "Standard Definitions of Terms and Abbreviations Relating to Physical and Chemical Characteristics of Refuse Derived Fuel", Volume 11.04 Waste Management, Annual Book of ASTM Standards 2006. ASTM International, West Conshohocken. 2006.

[15] C. Ryu, "Potential of Municipal Solid Waste for Renewable Energy Production and Reduction of Greenhouse Gas Emissions in South Korea", Journal of Air & Waste Management Association, 2010, 60: 176–183.

[16] J. Nithikul, "Potential of Refuse Derived Fuel Production from Bangkok Municipal Solid Waste". A thesis for the degree of Master of Engineering in Environmental Engineering and Management, Asian Institute of Technology School of Environment, Resources and Development, Thailand, 2007.

[17] H. Yuan, Y. Chen, H. Zhang, J. Su, Q. Zhou, G. Gu, "Improved bioproduction of short-chain fatty acids (SCFAs) from excess sludge under alkaline conditions", Environment Science Technology, 2006, 40: 2025–2029.

[18] H. Alter, "The recycling of densified refuse-derived fuel", Waste Management and Research, 1996, 14: 311–317.

[19] V. Raili, F. Marttl, "Organic emissions from co-combustion of RDF with wood chips and milled peat in a bubbling fluidized bed boiler", Chemosphere, 1996, 32: 681–689.

[20] R. Weber, T. Kupka, K. Zajac, "Jet flames of a refuse derived fuel", Combustion and Flame, 2009, 156: 922–927.

6

# Renewable Energy Application for Organic Agriculture

**Rameshprabu Ramaraj, Natthawud Dussadee**[*]

School of Renewable Energy, Maejo University, Sansai, Chiang Mai-50290, Thailand

**Email address:**
rrameshprabu@gmail.com (Ramaraj R.), rameshprabu@mju.ac.th (Ramaraj R.), natthawu@yahoo.com (Dussadee N.), natthawu@mju.ac.th (Dussadee N.)

**Abstract:** Agriculture is still the occupation of the majority of Thai people, despite the share of industry and services rising constantly. In terms of agricultural lands, Thailand is also one of the largest countries in the world, especially in Asia. Fruits and field crops make up for the most of vegetable products, rice being the leading crop. Currently, the market demand for organic food is increasing mainly due to consumer perceptions of quality and safety of these products. The primary goal of organic agriculture is to optimize the health and productivity of interdependent communities of soil life, plants, animals and people. Organic agriculture is expected that this requirement will continue to grow in the near future. On the other hand, energy is used in many organic agricultural inputs, including irrigation, mechanization and fertilizer. Both organic and conventional farming systems are mainly dependent on fossil energy, which is particularly crucial given rapidly growing energy costs. However, renewable resources are abundant. Many commercial technologies are available to connect these resources, and with suitable support, additional technologies could be brought to market. The aim of this research review is to investigate the utility of renewable energies for organic agricultural activities. In this concept, solar energy, biomass energy, wind energy, geothermal energy and hydropower are discussed by application. There is significant potential for organic agricultural involvement in the production and consumption of solar, wind, geothermal and biomass energy.

**Keywords:** Renewable Energy, Organic Agriculture, Renewable Sources, Energy System

## 1. Introduction

Organic agriculture refers to a farming system that enhances soil fertility through maximizing the efficient use of local resources. It relies on a number of farming practices based on ecological cycles, and aims at minimizing the environmental impact of the food industry, preserving the long term sustainability. Organic agriculture is often assumed that organic farming is synonymous with sustainable agriculture. The broad goals of sustainable agriculture include economic profitability, environmental stewardship, and community vitality [1]. The principle aims of organic production and processing as following [2,3]:

- To produce food of high quality in sufficient quantity.
- To interact in a constructive and life-enhancing way with natural systems and cycles.
- To consider the wider social and ecological impact of the organic production and processing system.
- To encourage and enhance biological cycles within the farming system, involving micro-organisms, soil flora and fauna, plants and animals.
- To develop a valuable and sustainable aquatic ecosystem.
- To maintain and increase long term fertility of soils.
- To maintain the genetic diversity of the production system and its surroundings, including the protection of plant and wildlife habitats.
- To promote the healthy use and proper care of water, water resources and all life therein.
- To use, as far as possible, renewable resources in locally organized production systems.
- To create a harmonious balance between crop production and animal husbandry.
- To give all livestock conditions of life with due consideration for the basic aspects of their innate behavior.
- To minimize all forms of pollution.

- To process organic products using renewable resources.
- To produce fully biodegradable organic products.
- To produce textiles which are long-lasting and of good quality.
- To allow everyone involved in organic production and processing a quality of life which meets their basic needs and allows an adequate return and satisfaction from their work, including a safe working environment.
- To progress toward an entire production, processing and distribution chain which is both socially just and ecologically responsible.

In addition, organic agriculture offers alternatives to energy-intensive production inputs such as synthetic fertilizers which are likely to be further limited for poor rural populations by rising energy prices. In developing countries, organic agricultural systems achieve equal or even higher yields, as compared to the current conventional practices, which translate into a potentially important option for food security and sustainable livelihoods for the rural poor in times of climate change [4].

Agriculture is a major contributor to global warming through greenhouse gases (GHG) from activities such as deforestation, soil treatment and methane emissions from livestock. It is also one of the main users of fossil fuels, thus also contributing further to GHG emissions. The cost of agricultural products is highly dependent on, and vulnerable to, fuel prices [5]. It is therefore appropriate to assess alternative sources of energy for the future of agriculture. However organic agricultural activities, such as irrigation, could be powered by renewable sources. Organic farm machinery could also be renewably powered, but the machinery would need to be adapted to use renewable electricity, instead of liquid fuel [6]. Renewable technologies are now supplying or supplementing many on-farm energy requirements, from water pumping to space heating [7]. Increasingly, farmers and ranchers are selling energy (e.g., electricity generated from wind turbines, biofuels, and products from biomass). This is contributing to greater energy security in normal and organic agriculture through increased diversity of energy sources, more self-supply of energy, and reduced environmental impact. Main renewable energy sources for applicable in organic agriculture and their usage forms presented in Table 1 [8,9].

*Table 1. Main renewable energy sources and their usage form*

| Energy source | Energy conversion and usage options |
|---|---|
| Hydropower | Power generation |
| Modern biomass | Heat and power generation, pyrolysis, gasification, digestion |
| Geothermal | Urban heating, power generation, hydrothermal, hot dry rock |
| Solar | Solar home system, solar dryers, solar cookers |
| Direct solar | Photovoltaic, thermal power generation, water heaters |
| Wind | Power generation, wind generators, windmills, water pumps |
| Wave | Numerous designs |
| Tidal | Barrage, tidal stream |

The Thailand faces a choice of energy futures. Fossil energy for mechanized agriculture has been an important driver of the "Green Revolution" of increasing farm productivity. Energy is one of the major parameters for establishing growth and progress of a country, rather than the standard of living, which depends directly upon the per capita energy consumption [10].

Renewable energy can address many concerns related to fossil energy use. It produces little or no environmental emissions and does not rely on imported fuels. Renewable resources are not finite and many are available throughout the country [11]. The sustainable energy approach promotes renewable energy in the organic agriculture sector, especially in remote or rural areas all over the world where solar energy is available in abundance. The purpose of the current paper is to present an integrated review of the methodologies, approaches and the related activities of the renewable energy application for organic agriculture. Furthermore, this article examines the domestic status and opportunities for a number of renewable energy technologies - solar, wind, geothermal and biomass.

## 2. Solar Energy Systems for Organic Agriculture

Solar energy is the energy derived directly from the Sun. The fastest growing type of alternative energy, increasing at 50 percent a year, is the photovoltaic cell, which converts sunlight directly into electricity. The Sun delivers yearly more than 10,000 times the energy that humans currently use [10]. Solar technologies produce electrical or thermal energy. Photovoltaic cells (or "solar cells") that convert sunlight directly into electricity are made of semiconductors such as crystalline silicon or various thin-film materials. Solar thermal technologies collect heat from the sun and then use it directly for space and water heating or convert it to electricity through conventional steam cycles, heat engines, or other generating technologies [6].

In organic agriculture, solar cells can economically provide electricity where the distance is too great to justify new power lines. Solar electric systems are used to provide electricity for lighting, battery charging, small motors, water pumping, and electric fences. The number of solar energy applications is expected to grow as new technologies increase solar cell efficiency and reduce costs. New "quantum dot" materials could theoretically more than double efficiency, converting 65 percent of the sun's energy into electricity, as compared to the best commercially available solar cells today, which have conversion efficiencies of up to 30 percent.

The first generation solar cells are based on Si wafers, beginning with Si-single crystals and the use of bulk polycrystalline Si wafers. These cells are now marketed and produce solar conversion efficiencies between 12% and 16% according to the manufacturing procedures and wafer quality. In Figure 1, one of the collections of solar modules that were used for the production of electricity in separate areas is presented. The energy storage was based on lead–acid batteries [12]. Therefore, energy storage system is an appropriate process for organic agriculture.

*Figure 1.* Solar system based on Si-single crystals [12].

# 3. Wind Energy for Organic Agriculture

Wind power differs from solar power in that it is available, in principle, for 24 h per day. Wind technologies provide mechanical and electrical energy for organic agriculture. Wind power grew fastest at 52% and will multiply by seven times to 2010, overtaking biopower. The reduction in greenhouse gas emissions can be achieved by production of environmentally-friendly power generation technologies (e.g. wind, solar, fuel cells, etc.). The challenge is to match leadership in greenhouse gas reduction and production of power from renewable energy resources by developing a major research and manufacturing capacity of environmentally-friendly technologies [10,13].

Wind turbines operate on a simple principle: Wind turns rotor blades, which drive an electric generator, turning the kinetic energy of the wind into electrical energy. The typical wind turbines were demonstrated in Figure 2. The wind is a renewable energy source, and windmills do not produce harmful environmental emissions. Utility-scale turbines range in size from 750 kilowatts (kW) to 5 megawatts (MW), with most turbines exceeding 1 MW. Turbines are often grouped into wind farms, which provide bulk power to the electrical grid. Small wind turbines range in size from 0.4 to 1.5 kW generators for small loads, such as battery charging for sailboats and small cabins, to 3 to 15 kW systems for a home, to those that generate up to 100 kW of electricity for larger loads, such as small commercial operations. Wind power technology is already in widespread use due to substantial progress in reducing costs for areas with consistently high wind speeds [14]. Recent' state-of-the-art wind turbines, operating in high-wind areas, can produce electricity for a few cents per kilowatt-hour (kWh), which is competitive with the cost of fossil fuel-fired plants. Accordingly, small wind systems can serve organic agriculture in traditional ways, such as using mechanical energy to pump water or grind grain. As costs decrease, small systems used to generate electricity may also become economically efficient by avoiding the expense of installing transmission wires, especially in more remote applications [15]. As technological improvements continue to increase the economic efficiency of wind energy, organic agricultural producers are likely to increase their use of wind power to lower energy costs and become more energy self-sufficient.

*Figure 2.* The typical wind turbines.

# 4. Geothermal for Organic Agriculture

Direct-use of geothermal energy is one of the oldest, most versatile and a common form of utilization of geothermal energy in agriculture [16]. Recently geothermal energy is utilizing in organic agriculture. There are three types of geothermal power plants are operating today: dry steam plants, flash steam plants, and binary-cycle plants. High-temperature geothermal resources (greater than 149°C) are used for power generation in organic agriculture. Utilization of geothermal energy in general can be divided into 2 types, namely utilization (indirect use) and use (direct use). Utilization of the indirect use of geothermal energy for power generation, while the direct use of the direct use of heat contained in the geothermal fluid to various fields such as conventional and organic agriculture / agro-industry, fisheries, etc [17].

Geothermal energy has many agricultural applications. Vegetables, flowers, ornamentals, and tree seedlings are raised in 43 greenhouse operations heated by geothermal energy. Forty-nine geothermal aquaculture operations raise catfish, tilapia, shrimp, alligators, tropical fish, and other aquatic species. Organic agri-industrial applications include food dehydration, grain drying, and mushroom culture. The drying of onions and garlic is the largest industrial use of geothermal energy [17].

Individual power plants can be as small as 100 kW or as large as 100 MW. The technology is suitable for rural electric mini-grids, as well as national grid applications. The heat from geothermal energy can also be utilized directly. Geothermal fluids can be used for such purposes as heating buildings, growing plants in greenhouses, dehydrating onions and garlic, heating water for fish farming, and pasteurizing milk. Generally, low-to-medium temperature resources (between 21°C and 149°C) are used. Another technology, geothermal heat pumps, can provide space heating and cooling. This technology does not require a hydrothermal (hot water) resource, but instead uses the near-surface ground as a heat source during the heating season and as a heat sink during the cooling season.

Direct or non-electric generation provides over 10,000 thermal megawatts, including geothermal heat pumps. The power from direct use systems is measured in megawatts of heat as opposed to power plants that measure power in megawatts of electricity [17]. Some geothermal projects "cascade" geothermal energy by using the same resource for different purposes simultaneously, such as heating and power. Cascading uses the resource more efficiently and improves economics. The geothermal resource base for low-to-medium temperatures is much more plentiful and widespread than the high-temperature resource base. Hence, geothermal approaches are reliable resource for organic agriculture.

## 5. Biorefineries for Organic Agriculture

Thailand has a great potential for use renewable raw materials in biorefineries and applicable for organic agriculture. It is one of the largest producers of agricultural and animal commodities, which produce large amounts of residues and wastes [28]. Discussion of renewable energy from biomass centers on the concept of the "biorefinery," where new technologies are being used to extract energy and other valuable products from biomass resources. Like oil refineries, biorefineries are envisioned as industrial facilities that convert a stream of raw material into a varied slate of products, maximizing value by shifting the mix of output to match dynamic market conditions. Potential biorefinery products include liquid fuels, such as ethanol and biodiesel, electricity, steam, and high-value chemicals and materials [18]. Many of these products have the potential to replace petroleum, either as a vehicle fuel or as a chemical feedstock, resulting in increased energy security and reduced environmental emissions.

They process corn into ethanol, corn syrup, animal feed, and other products, or transform trees into a variety of wood products, electricity, and heat. A thermochemical process (the syngas platform) involves heating biomass to turn it into a gas composed of a few basic molecules, then processing this raw material into fuels and products through chemical or biological techniques. Using biomass in a biorefinery concept instead of oil for producing energy and chemicals via organic and conventional agriculture in Figure 3.

Biorefineries use completely renewable organic agricultural products to produce a range of products for uses as wide ranging as fuel, high protein animal feed, gluten and electricity. They also have a vital role to play in reducing transport greenhouse gas emissions; from their own operations and the renewable ethanol from transport fuels they produce.

## 6. Renewable Energy Source Water Pumping Systems for Organic Agriculture

To meet the energy demands and reduce the environmental impact, the idea of integrating RESs such as solar photovoltaic, solar thermal, wind, biomass and hybrid forms of energy with water pumps has been proposed by many researchers around the world [19]. Comparing with conventional fuel, solar water pumping system has numerous advantages, for instance, besides of no cost for fuel and maintenance, the system it has no noise and pollution for the environment [20, 21]. Although there are solar water pumps with high capacity (10 of kW can be used), usually the pumps that are used in remote areas are small scale one (usually less than 1500 W). The main issue regarding to these systems is maintenance while at the same time 24 h electrical service is not demanded [7, 22]. Solar water pump system is generally divided into two groups; solar photovoltaic and solar thermal water pumping systems. The research developments with renewable energy source water pumping systems are presented in this section. Renewable energy source water pumping systems are classified into five major groups as follows: (1) solar photovoltaic water pumping systems, (2) solar thermal water pumping systems, (3) wind energy water pumping systems, (4) biomass water pumping systems and (5) hybrid renewable energy water pumping systems. The typical solar water pump system was demonstrated in the Figure 4.

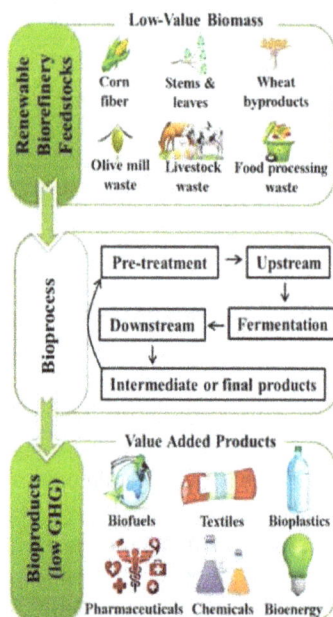

*Figure 3.* Using biomass in a biorefinery concept instead of oil for producing energy and chemicals [18] via organic and conventional agriculture.

*Figure 4.* Solar water pump system, Chiang Mai Province, Thailand.

There is a significant potential for solar water pumping in countries like Thailand which receive abundant amounts of solar radiation. Solar powered water pumps offer several advantages over their diesel and petrol counterparts. For the user, the life cycle costs are significantly lower. There is no fuel to be purchased, maintenance is greatly reduced and the system lifetime is longer. There are also significant environmental advantages. There are no pollutants, they are silent, require no fuel and contamination from spills, etc are eliminated [23]. They also contribute to energy self-sufficiency and reduce dependence on imported oil. Solar water pumping has been implemented around the globe as an alternative electric energy source for remote locations. The systems are cost effective in many remote applications especially in organic agricultural field. And this system is a mature technology to convert sunlight into electricity. Consequently, this review report concludes that renewable energy sources play a vital role in reducing the consumption of conventional energy sources and its environmental impacts for water pumping applications.

# 7. Renewable Energy Source as Fertilizer for Organic Agriculture

The standard biogas plant was illustrated in Figure 5. Anaerobic digestion is an optimal way to treat organic waste matter, resulting in biogas and residue. Utilization of the residue as a crop fertilizer should enhance crop yield and soil fertility, promoting closure of the global energy and nutrient cycles [24]. The digestate byproduct is the result of a mineralization process which, in general, can be used as fertilizer since it is rich in nutrients (high nitrogen-to-carbon ratio), enhances nutrient-penetration into the soil and reduces up to 80% of odours from the feedstock [25, 26]. This high content is advantageous to the organic agricultural crops as they are primarily capable of utilizing ammonium nitrogen. The slurry from the digester is rich in ammonium and other nutrients used as an organic fertilizer [24]. It is often possible to replace nitrogen from commercial fertilizer by digested slurry and thus save money [27].

*Figure 5. Biogas plant, Mae Taeng. Chiang Mai Province, Thailand [28].*

The biogas produced from swine manure and urine was used for cooking, lighting, or to maintain the temperature inside the greenhouse for optimum vegetable growth and the digestate were used as a fertilizer to replace chemical fertilizers. During winter, the low temperature and sunlight levels increases the application of chemical fertilizers. This frequent use of chemical fertilizer not only increased the cost of expenses but also decreased the vegetable quality during the winter. However, the substitution of chemical fertilizer with digestate increased the vegetable yield by 18.4% and 17.8% for cucumber and tomato respectively [29]. Consequently, Biogas is a promising renewable energy source and biogas final digestive byproduct applied to croplands as fertilizer for the organic agriculture farm.

# 8. Conclusion

Agriculture is the sole provider of human food. Organic agriculture is one sustainable alternative to avoid the negative environmental effects often caused by conventional agriculture. Most farm machines are driven by fossil fuels, which contribute to greenhouse gas emissions and, in turn, accelerate climate change. Such environmental damage can be mitigated by the promotion of renewable resources such as solar, wind, biomass, tidal, geo-thermal, small-scale hydro, biofuels and wave-generated power. These renewable resources have a huge potential for the agriculture industry. The farmers should be encouraged by subsidies to use renewable energy technology. Hence, there is a need for promoting use of renewable energy systems for organic agriculture, e.g. solar photovoltaic water pumps and electricity, greenhouse technologies, solar dryers for post-harvest processing, and solar hot water heaters. Clean development provides industrialized countries with an incentive to invest in emission reduction projects in developing countries to achieve a reduction in $CO_2$ emissions at the lowest cost. The mechanism of clean development is discussed in brief for the use of renewable systems for sustainable agricultural development specific to solar water pumps in Thailand and the world. Therefore, renewable energy technologies are being used in a variety of applications on farms, ranches and particularly in organic agriculture filed.

# References

[1] J. R. Goldberger, "Conventionalization, civic engagement, and the sustainability of organic agriculture", Journal of Rural Studies, 2011, 27: 288–296.

[2] IFOAM, "Basic Standards for Organic Production and Processing", IFOAM Tholey-Theley, Germany, 1998.

[3] D. Rigbya, D. Cáceres, "Organic farming and the sustainability of agricultural systems", Agricultural Systems, 2001, 68: 21–40.

[4] N. El-Hage Scialabbaa, M. Müller-Lindenlaufa, "Organic agriculture and climate change", Renewable Agriculture and Food Systems, 2010, 25: 158–169.

[5]   U. Bardi, T. El Asmar, A. Lavacchi, "Turning electricity into food: the role of renewable energy in the future of agriculture", Journal of Cleaner Production, 2013, 53: 224–231.

[6]   Fischer JR, Finnell JA, Lavoie BD. (2006) Renewable energy in agriculture: Back to the future. Choices - 1st Quarter, 21: 27–31.

[7]   K. Meah, S. Flecher, S. Ula. "Solar photovoltaic water pumping for remote locations", Renewable and Sustainable Energy Reviews, (2008) 12: 472–87.

[8]   A. Demirbas, "Global renewable energy resources", Energy Sources, Part A: Recovery, Utilization, and Environmental Effects, 2006, 28:779–792.

[9]   N. L. Panwar, S. C. Kaushik, S. Kotharia, "Role of renewable energy sources in environmental protection: A review", Renewable and Sustainable Energy Reviews, 2011, 15: 1513–1524.

[10]  A. Chel, G. Kaushik, "Renewable energy for sustainable agriculture", Agronomy for Sustainable Development, 2011, 31: 91–118.

[11]  R. Baños, F. Manzano-Agugliaro, F. G. Montoya, C. Gila, A. Alcayde, J. Góme, "Optimization methods applied to renewable and sustainable energy: A review. Renewable and Sustainable Energy Reviews, 2011, 15: 1753–1766.

[12]  W. A. Badawy, "A review on solar cells from Si-single crystals to porous materials and quantum dots", Journal of Advanced Research, 2013. DOI: 10.1016/j.jare.2013.10.001

[13]  A. M. Omer, "Clean energies for sustainable development for built environment", Journal of Civil Engineering and Construction Technology, 2012, 3: 1–16.

[14]  American Wind Energy Association, "U.S. wind industry ends most productive year, sustained growth expected for at least next two years, 2006", Available online: http://www.awea.org/news/ US_Wind_Industry_Ends_Most_Productive_Year_012406.ht ml.

[15]  M. Bergey, "Small wind systems for rural energy supply. Presentation from Village Power 2000, Washington, DC", December 4-8. 2000. Available online: http://www.rsvp.nrel.gov/vpconference/vp2000/ vp2000_conference/technology_mike_bergey.pdf.

[16]  M. H. Dickson, M. Fanelli (editors), "Geothermal Energy: Utilization and Technology", UNESCO Renewable Energy Series, 2003.

[17]  J. W. Lund, D. H. Freeston, T. L. Boyd, "Direct utilization of geothermal energy 2010 worldwide review", Geothermics, 2011, 40: 159–180.

[18]  A. ElMekawy, L. Diels, H. D. Wever, D. Pant, "Valorization of Cereal Based Biorefinery Byproducts: Reality and Expectations", Environmental Science & Technology, (2013) 47: 9014–9027.

[19]  C. Gobal, M. Mohanraj, P. Chandramohan, P. Chandrasekar, "Renewable energy source water pumping systems – A literature review",Renewable and Sustainable Energy Reviews, 2013, 25: 351–370.

[20]  U. Çakır, K. Çomaklı, Ö. Çomakl, S. Karsl, An experimental exergetic comparison of four different heat pump systems working at same conditions: As air to air, air to water, water to water and water to air, Energy, 2013, 58: 210–219.

[21]  S. Mekhilef, S. Z. Faramarzi, R. Saidur, Z. Salam, "The application of solar technologies for sustainable development of agricultural sector", Renewable and Sustainable Energy Reviews, 2013, 18: 583–594.

[22]  T. D. Short, P. Thompson, "Breaking the mold: solar water pumping– the challenges and the reality", Solar Energy, 2003, 75: 1–9.

[23]  M. Jafar, "A model for small-scale photovoltaic solar water pumping" Renewable Energy, 2000, 19: 85–89.

[24]  V. Arthurson, "Closing the Global Energy and Nutrient Cycles through Application of Biogas Residue to Agricultural Land – Potential Benefits and Drawback", Energies, 2009, 2: 226–242.

[25]  P. Weiland, "Biogas production: current state and perspectives", Applied Microbiology and Biotechnology, 2010, 85: 849–860.

[26]  C. Rodriguez-Navas, E. Björklund, B. Halling-Sørensen, M. Hansen, "Biogas final digestive byproduct applied to croplands as fertilizer contains high levels of steroid hormones" Environmental Pollution, 2013,180: 368–371.

[27]  H. Ørtenblad, T. Birkmose, L. Knudsen, "Næringsstofudnyttelse af afgasset gylle", Landbrugets Rådgivningscenter (1995).

[28]  N. Dussadee, K. Reansuwan, R. Ramaraj, "Potential development of compressed bio-methane gas production from pig farms and elephant grass silage for transportation in Thailand", Bioresource Technology, 2014, 155: 438–441.

[29]  K. Rajendran, S. Aslanzadeh, M. J. Taherzadeh, "Household biogas digesters - a review", Energies, 2012, 5: 2911–2942.

# 7

# Energy Autonomy in Small Islands in the Frame of Their Sustainable Development Exploring Biomass Energy Potential in Samothrace (Greece)

O. Christopoulou[1], M. Fountoukidou[1], St. Sakellariou[1], St. Tampekis[1], F. Samara[1], A. Sfoungaris[1], A. Stergiadou[2], G. Tsantopoulos[3], K. Soutsas[3], I. Sfoungaris[1]

[1]University of Thessaly, School of Engineering, Department of Planning and Regional Development, Thessaly, Greece
[2]Aristotelian University of Thessaloniki, School of Agriculture, Forestry and Natural Environment, Department of Forestry and Natural Environment, Thessaloniki, Greece
[3]Democritus University of Thrace, Department of Forestry Management of the Environment and Natural Resources, Komotini, Greece

**Email address:**
ochris@prd.uth.gr (O. Christopoulou), nanty@for.auth.gr (A. Stergiadou), tsantopo@fmenr.duth.gr (G. Tsantopoulos)

**Abstract:** One of the major problems faced by small islands in the Aegean Sea, is meeting their energy needs and the increased costs these needs generate because of their distance from the mainland and the increasing price of oil. To the above, the dramatic reduction of fossil fuels and climate change should be added, which obligate Member States of the European Union to comply with Directive 2009/28 / EC, which aims at a contribution of RES of 20% (Greece 18%) in the total EU energy consumption. It is therefore desirable to explore the energy autonomy capabilities offered in each case. This paper explores the possibility of utilizing biomass and in addition the use of livestock and poultry waste for biogas production.

**Keywords:** Utilization of Biomass, Biogas, Small Islands

## 1. Introduction

Sustainable development has become a primary objective of strategic planning at a global, national, regional and local level. Sustainability, inspired by the rational management of forests since the late 19th century, is today leading the effort to achieve balance between economic growth and environmental protection. But, environmental and economic problems influence each other and are linked by social factors such as unemployment, poverty, social exclusion and ultimately social welfare.

Cornerstone of sustainable development is the rational management of natural resources, in a manner that, according to Brutland report, the ability of future generations to meet their needs is not undermined (Our Common Future, 1987). Therefore, the need for the conservation of the renewability of renewable resources and the conservation / non-exhaustion of the non-renewable ones becomes clear.

In Chapter 17 of Agenda 21, adopted at the Rio Conference, it is written that "islands are spatial units, of particular interest from both an environment and a development point of view; they are very fragile and vulnerable and in the context of sustainable development, energy is the cornerstone of their planning strategies". In the final document of the First European Conference on Sustainable Island Development, it is stated that: "Non-renewable energy sources must be considered as provisional solutions, unsuitable as a long-term solution to the energy problem in islands". [1]

Biomass, the wind, solar power, water power and geothermal potential are the renewable energy sources that can provide sustainable energy services, which rely on the use of domestic resources. [2].

In this study, the potential of biomass utilization for energy purposes is being examined, on a Greek island. Last decades, mainly the oil importing countries make efforts for the sustainable exploitation of biomass (wood residues, crop residues, fruit residues etc.) in order to substitute the use of oil [3]. Also, the Action Plan for the biomass produced by the European Commission, promotes the increase of the energy production from wood, waste and agricultural crops [4].

Bioenergy is considered (for many regions of the European Union) as a considerable factor of sustainable development due to the established forest plantations and the agricultural energy crops [4].

## 2. Biomass

Biomass comes from products or waste from various human activities and is divided into four categories according to Easterly and Burnham [5]:

a) energy crops, b) wood chips, c) agricultural residues, d) solid municipal waste.

a) Energy crops are cultivated species, traditional or new, which produce biomass for energy use, i.e. heat and electricity production, production of liquid biofuels etc. For the successful development of an energy crop in a region, certain conditions must be met, according to Venturi and Venturi [6], such as: appropriateness of soil and climatic conditions, ease of insertion of an energy crop in existing crop, continuous and uniform performance in terms of quality and the quantity, competitive income in comparison to traditional crops, positive energy balance resulting from production and consumption, cultivation techniques in harmony with sustainable development, resistance to major biotic and abiotic adversities, availability of biomass sources, such as seeds, appropriate roots for the region, necessary machinery particularly for harvesting operations, specific for the given crop, which can be used in other crops, with small changes.

Energy crops that have been investigated over the last fifteen years in Greece by the Centre for Renewable Energy Sources for the production of solid biofuels are: Reed, Cardoon, Switchgrass and Miscanthus (Perennial), Kenaf, Fibrus Sorghum (Annual), Eucalyptus and Black locust (Forest plantations) (see Table 1).

*Table 1. Energy produced by dry biomass, output in product and energy. Source: Centre for Renewable Energy Sources [7].*

| type | product | heat energy (µj/kg) | average yield in dry biomass (t/ha/year) | yield in energy (gj/ ha/year) |
|---|---|---|---|---|
| Perennial crops | Reed | 18,0 | 1,0 – 2,0 | 18,0 – 36,0 |
| | Cardoon | 18,0 | 1,0 – 1,5 | 18,0 – 27,0 |
| | Switchgrass | 16,0 | 1,0 – 2,0 | 18,0 – 36,0 |
| | Miscanthus | 18,0 | 1,0 – 1,5 | 18,0 – 27,0 |
| Annual crops | Kenaf | 18,6 | 0,8 – 1,8 | 14,9 – 33,4 |
| | Cellulosic Sorghum | 16,0 | 2,0 – 3,5 | 36,0 - 63,0 |
| Forest crops | Eucalyptus | 19,4 | 1,8 – 3,0 | 31,8 – 58,0 |
| | Black locust | 17,8 | 0,8 – 1,3 | 14,3 – 23,2 |

A wide range of energy crops has been tested in Europe as biomass crops in the last two decades in small or large scale trials in order to test the adaptability, performance and quality characteristics under different soil and climatic conditions. Energy crops were grouped according to their end use, in oil seed crops for biodiesel production, sugar and starch crops for the production of bio-ethanol and cellulosic plants, and woody crops for heat and electricity production. Some of these cultivations can produce more than one product, such as cardoon or hemp, which can be used for producing oil and solid biomass, and grains, which can be used for the production of bioethanol and solid biomass.

The choice of the crop in each region depends on its suitability in various climatic constraints (rainfall, maximum and minimum temperature of air and soil), access to irrigation water, if necessary, and soil conditions (quality of arable land). An important parameter to consider is the acceptance of the farmers, especially in the case of perennial crops, which have a lifespan of at least 15 years.

According to Alexopoulou and Skretschmer [8], a series of fifteen promising energy crops were distinguished according to the final products or the produced energy in crops for biodiesel production, fiber crops, crops for bioethanol and ligno-cellulosic crops.

In the northern parts of the Mediterranean, mainly in the mountains, poplar crops, miscanthus, reed, corn, sunflower, sorghum, flax, sugar beet, soybean, rapeseed and kenaf crops are favored, while in the southern parts of the Mediterranean, and not so mountainous, it is more common to cultivate reed, cardoon, eucalyptus, sorghum and flax.

b) Wood chips mainly originate from industries and forests. Large quantities of wood residues come from the wood industry and / or its products, such as paper mills, sawmills and furniture manufacturing plants. While wood industries use their waste to produce energy, there is a significant amount of residue left over. In forests, there is also available biomass usable for energy purposes. Logging residues such as barks, branches, leaves and needles of coniferous trees remain on the forest ground as well as trees that have been removed during the creation of firebreaks

c) Agricultural residues can be categorized into plant and animal origin. Agricultural residues of plant origin are essentially residues remaining in the field after harvest, such as leaves, stems, fruits, twigs and food wastes, and they can be used locally or in regional biomass power plants. Residues of animal origin include waste from poultry farms, dairy farms, pig farms and slaughterhouses, as well as animal faeces,

d) Municipal waste represent a significant source of biofuels. Municipal waste include solid municipal waste and sewage. Municipal solid wastes include domestic, craft, and commercial waste, waste water as well as sanitation wastes etc. The values of density and moisture contained in the various types of waste vary depending on the country, the time of year and the region, and are critical features regarding their burning.

### Biomass Energy

Biomass can be used to meet energy needs (heat, electricity) either by direct combustion, or through the production of solid, liquid and gaseous biofuels using thermochemical or biochemical methods. Since the utilization of biomass usually experiences problems regarding its large dispersal, high volume, and collection - transport - storage difficulties, it must be exploited as close as possible to the place of production. Thus, it can have many applications, according to the Center for Renewable Energy Sources [7]:

1. Cogeneration of electricity - heat to cover the needs in agricultural and other industries. I.e., through the combined production of heat and electricity, from the same source, most of the energy is recovered and profitably used. Thus, considerable energy saving is achieved, the degree of energy conversion of the fuel is increased and pollutant emissions get reduced. Indeed, conventional power plants have an efficiency of 15-40%, while that efficiency reaches up to 75-85% in co-generation systems.

2. Teleheating of residential areas, meaning safeguard of hot water for both space heating and for its direct use in a set of buildings, a settlement, a village or a city, originating from a central heat plant. The generated heat is transported by a pipeline network from the station to the heated buildings. Teleheating is growing strongly in many countries, as it presents significant advantages such as a higher degree of efficiency, reduction of pollution of the environment and the possibility of using non-conventional fuels, thus resulting in additional economic and environmental benefits.

3. Heating greenhouses. The utilization of biomass in heat production units to heat greenhouses constitutes an interesting and economical perspective for their owners. Already, in about 10% of the total area of heated greenhouses in the country, various types of biomass are utilized.

4. Biofuel production. Under Directive 2003/30 / EC, "Biofuels are liquid or gaseous fuel for transport, produced from biomass". Through anaerobic digestion of biomass, biogas is produced, which consists mainly of methane and carbon dioxide and can be used to generate heat and / or electricity. Biogas production is typically encountered on pig and poultry farms, dairy farms, as well as municipal waste landfill sites. Generally, 20 to 50% of municipal solid waste originates from urban fruit and vegetable residues [9]. Cow and pig manure, produced in villages around the cities, are included in the municipal solid waste, which is a raw biomass material. The organic matter of this biomass is converted into biogas through anaerobic digestion. Also, through the fermentation of biomass bioethanol is produced, which can be used as fuel in the transport sector. Through the process of the fast pyrolysis, bio-oil, a high energy density liquid biofuel, can be quickly produced, which can be used as an oil substitute in heating applications and in internal combustion engines for power generation.

5. Energy crops. The lingo-cellulosic type of biomass, derived mainly from forest energy crops, is exploited for the production of solid biofuel in the form of pellets or briquettes, producing heat and / or electricity, through their direct burning. Apart from solid biofuels, derived primarily from wood chips, there are plants that contain sugar (eg sugar beet) or starch crops (eg maize), which are primarily suitable for the producing of bioethanol, and sunflower and rapeseed for biodiesel production. Biodiesel presents no technical barriers and could be used as an additive to conventional diesel. More specifically, it can be blended with diesel fuel at a rate of volume up to 5% according to the EN 590: 2004 standard. Recently, research programs have been focusing on different engines (truck engines, etc.) on which the use of blends of biodiesel / diesel, ranging from 2% to 98% with the use of certain additives, has been tested [10].

First-generation biofuels (biodiesel, bioethanol, biogas), seem to occupy and create some concerns to the scientists. Their main disadvantage is the confrontation "food or biofuels?" that has begun, as well as the increase in food prices, allegedly due to the increased production of these biofuels [11]. These biofuels have been the subject of many lifecycle analyzes focusing on energy and the balance of greenhouse gases. Although reduction in the production of greenhouse gases has been achieved, greenhouse gas emissions from land use change are not considered. Recent studies show that the benefits of reducing greenhouse gas emissions locally can be offset by increased emissions elsewhere due to the intensification and deforestation [12].

Second-generation biofuels are not produced from products that can be used as food, but from lingo-cellulosic biomass, i.e. from wood biomass. These are advanced biofuels, since they are produced from plants that exist in abundance in nature, are not edible, and their production is in a pilot phase. [11]. According to Lakaniemi et al. [13], algae are the next generation of biofuels. They are photosynthetic microorganisms, which convert sunlight, water and carbon dioxide into lipids. They have high growth rates and tolerance to adverse environmental conditions. They survive and grow in low water and seawater quality. The last characteristic makes them useful in wastewater treatment plants. Their development requires smaller amounts of water than conventional agricultural crops, thus, saving water and not requiring the addition of pesticides or herbicides. Algae do not bring about changes in land use, since they can be grown in brackish water in non-arable land and thus not endanger food production, livestock and other products deriving from land crops. According to Piccolo [14], countries with a coastline on the Mediterranean between 45° and 30° north, are suitable for growing microalgae and particularly countries in the Northern Mediterranean, where temperatures throughout the year do not fall much below 15 ° C. In these areas, the growth of algae in a system of open or closed ponds is favored. According to the above, but also

Singh and Gu [15], a number of countries of the Mediterranean basin, like Israel, have great potential for microalgae cultivation.

## 3. Research Area

Samothrace is an island in the northeastern Aegean. Its surface is 178 km2, while the highest peak has an altitude of 1.611 m. It is the highest Greek island with the exception of the two large islands of Crete and Euboea. Its permanent population, according to the 2011 census, is 2,859 inhabitants, while the economy of Samothrace is based mainly on the tertiary sector (tourism) and the primary sector (44,73% of the working population). The main physical characteristic of the island is the steepness of the slopes of the mountain Saos. The image presented by the landscapes of the north-northeast and south-southwest of Samothrace is very different, as if they were two different islands. In the first case, the green slopes of Mt Saos, with oak forests mainly (Quercus frainetto, Q. pubescens and Q. dalechampuii), fruit crops, dense scrubland and the extremely impressive waterfalls of the streams that flow into the sea dominate. The dominant species of the riparian vegetation is oriental plane (Platanus orientalis). In the second, the landscape is hilly with sparse scrubland, farmland and olive groves, typical image of a North Aegean Sea island. The key environmental problems faced by the island are erosion due to overgrazing (more than 50 thousand sheep), given the strong slopes and torrential phenomena, a risk that exists throughout the Mediterranean [16], as well as inadequate waste management which leads either to pollution of streams, aquifers and seas from uncontrolled dropping, or excessive costs due to transport them off the island.

### 3.1. History of use: Energy Production in Samothrace

Before 2000, energy in Samothrace was produced in an engine room operating with oil. In 2000 Samothrace was connected to two underwater cables, 10MW of power each, with the mainland. The daily power in the winter months is approximately 2MW, while in the summer months, when maximum demand is observed, it reaches 4MW. Water heating takes place, in several cases, with the use of solar water heaters, while heating of premises, mainly, with the use of oil and, less often, of wood. In 1984, oil tanks were

manufactured outside of the main town of the island (Chora). They are private and serve the island's needs for oil. The pipeline that comes from the sea passes close by houses, which increases the need for maintenance, to avoid the risk of leakage. In 1992, they built four small, 55KWh power, wind turbines on the north coast of the island, however, the Public Power Corporation dismantled them. The initial design for their replacement by turbines of more power (900 KW) through reintegration into REPOWERIN program did not materialize, whereas, during the decade from 2000 to 2010, many new applications were filed regarding the creation of new wind parks, which were not materialized.

The island, being close to the mainland, is interconnected to the continental electric network, importing all required electricity via underwater cable. Also, large amounts of fossil fuels are imported to the island via maritime transport to meet, mainly, demands in transport and heating. In this way, it becomes apparent that the island relies entirely on imported energy.

According to the Sustainable Islands Network "Daphne" [17], the distribution of primary energy demand per energy carrier for Samothrace is formed as shown in the diagram below.

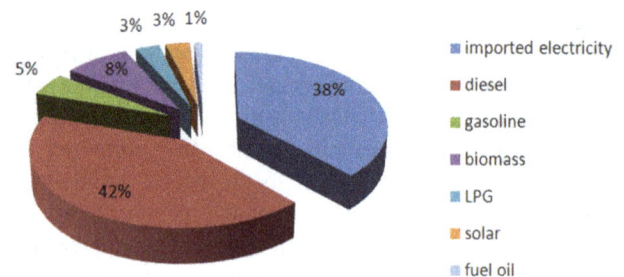

*Figure 1. Distribution of primary energy demand per energy carrier in Samothrace, Source: [17].*

According to the Network of Sustainable Islands "Daphne" [17], the final energy demand in the residential sector in the base year 2005 for Samothrace is apportioned according the following table:

*Table 2. Final residential sector energy demand in the base year 2005 [MWh], Source: [17].*

|  | Electricity | Diesel | LPG | Solar | Biomass | Total | Allocation of final energy demand in the domestic sector (%) |
|---|---|---|---|---|---|---|---|
| Residential sector | 4.921 | 9.312 | 1.035 | 758 | 2.419 | 18.444 | |
| Hot water | 899 | 733 | 0 | 758 | 228 | 2.618 | 14.2 |
| Heating and cooling | 1.365 | 8.579 | 725 | 0 | 1.933 | 12.602 | 68.3 |
| Lighting | 738 | 0 | 0 | 0 | 0 | 738 | 4.0 |
| Cooking | 516 | 0 | 310 | 0 | 258 | 1.083 | 5.9 |
| Refrigerators and freezers | 724 | 0 | 0 | 0 | 0 | 724 | 3.9 |
| Washers and dryers | 44 | 0 | 0 | 0 | 0 | 44 | 0.2 |
| Washing dishes | 62 | 0 | 0 | 0 | 0 | 62 | 0.4 |
| Televisions | 57 | 0 | 0 | 0 | 0 | 57 | 0.3 |
| Other electrical devices | 516 | 0 | 0 | 0 | 0 | 516 | 2.8 |

The most common energy carriers in the domestic sector are electricity and oil, with the latter mainly covering space heating needs of homes. LPG is mainly used for heating and cooking, as well as biomass which mainly concerns the consumption of firewood which in most cases are burned in open fire-places. Finally, solar energy is used exclusively for the production of domestic hot water through solar water heaters. The heating and cooling sectors are the most energy consuming sectors followed by hot water production.

Since the heating and cooling sector accounts for 68.3% of the final energy demand in the residential sector in Samothrace and mainly comes from diesel, a solution is necessary in order to reduce the consumption of diesel fuel and therefore emissions associated therewith.

In the primary sector, which reflects the agricultural and fishing activities of the island, the most used energy carriers are electricity, oil and biomass, covering for the most energy needs for irrigation, heating and cooling, lighting and the use of various equipment features. In the secondary sector, the subsectors of manufacturing and construction are the most energy consuming, and to cover the final energy demand electricity oil and fuel oil are used, while, in the tertiary sector, the most used energy carriers are electricity and oil. LPG and biomass (in the form of coal) are used mainly in restaurants for cooking, while solar energy is used only to meet the demand for hot water by hotels. Finally, in the transport sector, final energy demand is divided as follows: 59% gasoline and 41% diesel. The distribution of the final energy demand into individual subsectors is as follows: 62% private use transports, 35% road freight transport and removal services, 1% passenger road transports (means of public transport, taxi, tourism, etc.) and 2% for other public and private services [17].

Energy problems currently faced by all Greek islands are focusing on the following, according to the Energy Office Aegean [18]

- They have great energy dependence on oil and a high cost of conventional power generation
- They present a serious commercial power supply problem and an incapable of base charge due to the lack of large industrial units
- They present high rates of growth in energy demand per year and large seasonal fluctuations in charge demand, mainly due to increased residential development and growth of tourism

### 3.2. Biomass Potential in Samothrace

The largest part of the island is occupied by rangelands. Maquis vegetation, transitional shruby vegetation, broadleaf forests and conifer forests then follow.

According to information received from the Greek Statistical Authority [19], the main crops of the island (year 2008) are as follows:

**Table 3.** Area of tree crops, arable crops and vines in Samothraki [19].

| Crop Type | Surface in Ha |
|---|---|
| Tree Crops | |
| Olive trees for edible olives | 400 |
| Olive trees for olive oil extraction | 500 |
| Pear trees, apple trees, peach trees, cherry trees, fig trees, almond trees, walnut trees | 20 |
| Total tree crops | 920 |
| Arable Crops | |
| large crop plants and other crops | 1645.8 |
| Horticultural land, greenhouses, commercial flower gardens, seedbeds | 37.2 |
| Fallow lands for 1-5 years | 16.0 |
| Land, maintained in good agricultural and ENV. condition | 163.4 |
| Total arable crops | 1862.4 |
| Total Crops | |
| Total arable crops | 1862.4 (66.64% of total crops) |
| Tree crops | 920.0 (32.92% of total crops) |
| Vines, Raisinvines | 12.2 (0.4% of total crops) |
| Total crops | 2794.6 |

From the above table, it is concluded that 98% of tree crops are olive trees, covering an area of 9,000 Ha. Also, 88% of arable land is covered by field crops (large crop plants), such as wheat, barley, maize, fodder peas, vetch hay, perennial clover etc. These plants are used as animal feed, which are of great demand in Samothrace. The percentage distribution of crops is shown in the following figures.

Regarding available forest biomass that can be used to generate heat energy, the Greek Statistical Authority [19] gathered data on the annual production of firewood, collected in agricultural lands from 2000 to 2008. To calculate the power of the thermal energy power plant, its average value will be used for the years 2000-2008.

**Table 4.** Annual production in tones, of firewood collected in farmlands in Samothrace, from 2000 to 2008.

| YEAR | 2000 | 2001 | 2002 | 2003 | 2004 | 2005 | 2006 | 2007 | 2008 |
|---|---|---|---|---|---|---|---|---|---|
| Annual production of firewood in tones | 700 | 500 | 450 | 500 | 450 | 500 | 480 | 400 | 350 |

It should be noted that the Department of Crop and Animal Production of the Directorate of Rural Economy and Veterinary Medicine of the District of Evros was not able to provide data on the residues of plant origin, partly because the quantities of plant origin residues are within very extended limits, because a large portion of grain land is grazed without being harvested in the known manner of harvesting (combine harvester) due to the large number of animals present on the island and the high demand in animal feed, on the other hand very strong winds are blowing on the island, thus, luring straw stays that have remained in the field after harvesting. Approximately, according the above mentioned Department, the residues of intensive cereal crop are about 1400-1800 kgr / Ha) with a humidity of a class of 5%. The total area of cereal crops was in 2008, 526.4 Hectares, thus the residues are ranging from 736.960 to 947.520 kgr with humidity of a class of 5%.

Additionally, the Forest Authority of Alexandroupoli was unable to provide data on the forest biomass of Samothrace, because there are no forest plantations, no forests management and there were no new firebreak zones opened after 2000.

Finally, the Greek Statistical Authority [19], gathered data on the number of animals in livestock and poultry farms in the region.

*Table 5. Number of livestock animals in Samothrace.*

| Animal group | Animal NO.s |
|---|---|
| pigs | 1143 |
| Sheep / goats | 51900 |
| poultry | 8500 |
| horses | 92 |
| TOTAL | 61635 |

### 3.3. Utilization of Firewood from Farmland to Produce Thermal Energy

#### 3.3.1. Methodology

To calculate the heat content of firewood harvested from agricultural land, the average annual production of firewood from 2000 to date should initially be calculated [20]. The heat content of firewood in kWh can be calculated by the formula:

$$Eth = Mn * C_p$$

where

Mn = the average annual production of firewood from 2000 to date

$C_p$ = the calorific value of firewood, taken equal to: 3,833 kWh / kg, according to Katsoulakos and Kaliampakos [21].

The strength of the combustion plant of firewood harvested from the farmlands will also be calculated. This unit will aim to produce thermal energy for district heating of the Chora (town) of Samothrace and other neighboring regions, under certain conditions. To calculate the power of a combustion plant for the production of thermal energy, the available amount of firewood per hour (DP) in tones or kilograms should be calculated as follows:

DP = Mn / (328 days per year * 24 hours a day)

Admission: The unit will operate 328 days a year. We consider that 10% of the year, i.e. 37 days, the plant will remain closed for maintenance purposes [20], and will also operate 24 hours a day.

According to Poulis [20], for the operation of a 1MW combustion plant 240 kg of wood per hour would be required. Therefore, the power of the unit in MW which can be constructed to exploit the available hourly production of

firewood can be calculated using the formula:

$$P = DP / 240$$

#### 3.3.2. Results

From the data available [19], the average annual production of firewood, for the years 2000 – 2008, was estimated at 481.11 tones. Thus, the heat content of firewood is estimated at 480.000 * 3,833 = 1.839.840 KWh = 1.839,84 MWh. The available quantity of firewood per hour is 61 kg. Therefore, the unit that can be constructed to exploit the 480 tons per year lies in the class of 0.25 MW.

This unit, although of small capacity, is not intended to fully cover the energy requirements for heating of the whole island, but only those of the Chora and Kamariotissa, which is situated just 5 kilometers from the country. Moreover, between these two cities resides 60% of the entire population of the island. Note that the combustion unit can work complementary throughout the year in combination with the use of fossil fuels, offering savings in their use.

### 3.4. Utilization of Grain Residues for Heat Production

#### 3.4.1. Methodology

The amounts of residues of plant origin, according to an official answer received by ELSTAT, fluctuate around 140-180 kgr with a humidity of 5%. Thus, 736960-947520 pounds with a humidity of 5%. For safer results the lower price will be used. According to Papazoglou and Kyritsis [22], the calorific value of grains is 17,891 MJ / ton of dry weight. The concept of an availability factor is introduced to the calculations of the available energy from cereal residues, which is the percentage of the potential of biomass that is energetically exploitable. According to Tzinevrakis et al. [23], its value depends on various factors such as ease of collection of biomass (whether there is a great dispersion, gathering in inaccessible places) and its alternative uses (eg for food). The value obtained is 0.35 with a variation of 20%. Given the dispersion of waste due to strong winds and their use as animal feed, we will take the minimum value, ie 0.28. Therefore:

Thermal Energy = Quantity of grain residues * dry weight percentage * availability factor *

Calorific grain strength

#### 3.4.2. Results

The total energy that can be produced from burning cereal residues is 736.960 tones * 95% * 0.28 * 17891MJ / ton of dry weight = 3,507.197MJ = 974,221KWh = 974.221MWh. This energy is relatively small to be used on an industrial scale, but could cover the energy needs of small family livestock and poultry farms or even the thermal needs of the greenhouse of Samothrace.

### 3.5. Utilization of Livestock and Poultry Waste for Biogas Production

#### 3.5.1. Methodology

The production of biogas depends on the organic solid

content of the raw material, but a good approximation is the content of the livestock waste in dry solids [24]..

The estimated amount of biogas in m3 that can be produced from the available livestock waste is given by the following equation:

$$BP = POP * DS * BY * AF,$$

where

POP is the number of animals per group

DS is the annual quantity in tons of dry solid per animal of each group

BY the biogas yield per animal of each group in m3 / dry tn.

AF is the availability factor (0 <AF <1) of each group of animals

The total amount of biogas estimated (LBP) for all groups of animals in m3 is the sum of the estimated amount of biogas in all groups.

The density of the estimated biogas in an area in m3 / Km2 can be calculated by the formula:

$$DBP = LBP / A$$

where A is the surface area in Km2

The estimated available energy in TJ originating from the estimated biogas is given by the fomula:

$$E = 21.6 * 10\text{-}6 * LBP$$

where 21.6 MJ / m3 is approximately the calorific value of biogas

The availability factor, according to Batzias et al. [24], was estimated depending on the conditions of production for each group of animals. Due to alternative uses of animal waste, biomass availability for biogas production may differ significantly between groups of animals.

### 3.5.2. Results

Data supplied by the Greek Statistical Authority for 2008 regarding the number of animals (domestic or otherwise) that exist in Samothrace, and according to the above methodology, lead to the table below:

**Table 6.** *Total animal waste, dry solids per animal group, availability factor, biogas yield per animal group, density of estimated biogas, and bioenergy in Samothrace.*

| ANIMAL GROUP | Total manure in tons per animal per year | POP (no. animals per group) | DS (tn/ head) | AF | BY (m3/drytn) | BP (m3) | DBP (m3/Km2 ) | E (TJ) |
|---|---|---|---|---|---|---|---|---|
| Pigs | 1,89 | 1143 | 0,216 | 0,8 | 649 | 128184,2 | | 2,76878 |
| Sheep/ Goats | 0,64 | 51900 | 0,222 | 0,35 | 120 | 483915,6 | | 10,45258 |
| Poultry | 0,034 | 8500 | 0,01 | 0,7 | 359 | 21360,5 | | 0,461387 |
| Horses | 8,82 | 92 | 2,6 | 0,1 | 160 | 3827,2 | | 0,082668 |
| TOTAL | | | | | | 637287,5 | 3580,27 | 13,76541 |

Estimated available energy from the exploitation of livestock and poultry waste amounts to 13.76 TJ and estimated quantity of biogas to 637,287.5 m³. Thus, the available energy amounts to 3,822.22 MWh. The produced biogas can be used in a district heating plant in order to meet the heating needs of the entire Samothrace.

## 4. Location of Energy Utilization Plants from Biomass or Biogas

Under the Special Framework for Spatial Planning and Sustainable Development for Renewable Energy Sources and restrictions imposed through that, Natura 2000 sites in Samothrace are incompatibility zones. Thus, the installation of a biomass power plant in them, will be excluded. However, in the rest of Samothrace the necessary investigations can be made in order to find a suitable location of such a unit, provided that the other criteria set by the Special Framework are met.

An Environmental Impact Assessment (EIA) is considered necessary in order to ensure the minimum, and if possible zero effect of the power plant on the ecosystem of Samothrace. The necessity of this study lies in the fact that part of Samothrace that does not belong to the network Natura 2000 (Figure 2), occupies a very small area which is located nearby the protected territory. The area occupied by the Natura 2000 network is 37,458 Ha (21,021 Ha Dir. 79/409/EEC and 16,434 Ha Dir. 92/43 EEC).

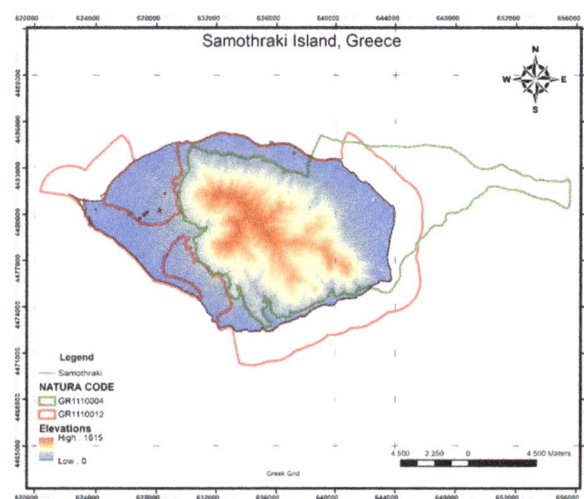

**Figure 2.** *Natura 2000 sites in Samothrace.*

## 5. Conclusions

Samothrace is an island of available bioenergy of 6,636.28 MWh, utilizing only the available firewood biomass, plant residues and livestock and poultry waste.

Calculations on the available energy content of firewood and produced biogas from livestock and poultry residues, lead to the following conclusions:

1. The total available energy from the burning of firewood and biogas is 3,822.22 + 1,839.84 = 5,662.06MWh. There are 2,859 inhabitants in Samothrace, therefore, 1.98 MWh correspond to each inhabitant. The annual final energy demand for heating from all energy consuming subsectors is 12,602 MWh, thus covering 45% of the annual final demand. This means that they can fully meet the need for heating the whole of the resident population of Samothrace for about 5 months.

2. Dividing the total available energy from the burning of firewood and biogas (5,662.06MWh) with the annual final energy demand for heating by diesel (8,579 MWh), it can be concluded that 66% of heating oil can be saved.

3. Taking into account that the lower calorific value of diesel is 10.70 KWh / l or 0.01070 MWh / l, according to the Centre for Renewable Energy, and the efficiency of a conventional oil boiler is 90%, it can be concluded that with 5,662.06 MWh, about 476,000 liters of oil can be saved, which, if burned would produce 1,266.160 kgr of $CO_2$. (One liter of heating oil produces 2.66 kg of $CO_2$). According to the Fuel Gas Prices Observatory of the Ministry of Development and Competitiveness, the average price for heating Diesel in October 2014 was 1.050 € / liter. So, the economic benefit of saving 476,000 liters of diesel is on average 499,800 €, ie about 700 € per household / year, a particularly important amount, given the financial crisis.

Bioenergy from grain residues is not available throughout the year, since the harvest of wheat is held in June and residues harvest is completed by July. Due to this seasonality, bioenergy can be used complementary during the summer months to meet the increased demand due to tourism. Also, the storage of grain residues for use in the time period October - March, would be more effective. During that period, the greenhouse and the mill of the island operate, whose thermal requirements could be met.

# 6. Suggestions

A turn towards biomass is necessary to produce cheap energy. If the use of available biomass is methodical and targeted, significant economic and environmental benefits can be yielded. Indeed Samothrace has outstanding prospects; however, these are not utilized extensively.

- The olive groves for oil production in Samothrace cover 54% of the total area of tree crops. That area of 5,000 acres is served by a mill with a capacity of 102.55 tons of oil produced annually. Unfortunately, there is no oil core plant on the island. An oil core plant could process the oil pomace of the olive mill and provide an energy-rich fuel, the olive pits, with a calorific value of 4,700 – 5,000 Kcal / Kgr. The utilization of olive pits could lead to realistic energy solutions for the island.
- The promotion of energy crops on the island will bring significant benefits not only to the farmers in the region, but generally throughout the local community. The penetration of energy crops on the domestic market can ensure adequate farm income compared to some conventional crops and enhance diversification of farmers' activities. Furthermore, the creation of a market for the production of biofuels, heat and electricity in the region, will contribute to the staying of the population in rural areas by creating new jobs and bringing additional income to the local community. And finally, the use of crops for energy reduces dependence on oil imports.
- Recycling of solid waste in conjunction with the installation of a landfill and the parallel biogas collection is proposed in the case of Samothrace, as, firstly, the quantity of waste would be reduced, and secondly, the energy potential of the island would increase. The problem of the management of solid waste on the island has not been solved as waste is transferred to the landfill of Alexandroupolis. Through proper public awareness and consensus about recycling, and a landfill establishment study on the island, in order not to affect protected ecosystems, there is an energy recovery potential from recyclable materials and produced from landfill biogas.
- The available forest biomass could increase if the authorities maintained firebreak zones in Samothrace. According to the Forestry of Alexandroupolis, there has been no opening of firebreaks since 2000. This is a troubling fact, since the maintenance of firebreaks is crucial in order to prevent the spread of fires and also, in the energy sector, forest biomass residues could be exploited for energy production. Note, also, that each year fires occur in Samothrace, which are strengthened by the strong winds that blow on the island.

The protection of the Natura 2000 network is an important criterion regarding the location of the biomass exploitation plant for energy production. Biodiversity conservation and the protection of flora and fauna should not be affected in any way by the installation of such a unit. EIAs are necessary, in order to ensure the balance of the ecosystems in the region.

# Acknowledgements

This research has been co-financed by the European Union (European Social Fund – ESF) and Greek national funds through the Operational Program "Education and Lifelong Learning" of the National Strategic Reference Framework (NSRF) - Research Funding Program: Thales. Investing in knowledge society through the European Social Fund.

# References

[1]  G. Notton, 2015. Importance of islands in renewable energy production and storage: The situation of the French islands. Renewable and Sustainable Energy Reviews 47: 260-269.

[2]  Antonia V. Herzog, Timothy E. Lipman, Daniel M. Kammen. The Encyclopedia of Life Support Systems (EOLSS) Forerunner, Volume "Perspectives and Overview of Life Support Systems and Sustainable Development," Part 4C. Energy Resource Science and Technology Issues in Sustainable Development, Renewable Energy Sources, and can be found at: http://www.eolss.com

[3]  E. Zafeiriou, G. Arabatzis, Th. Koutroumanidis, (2011). The fuelwood market in Greece: An empirical approach. Renewable and Sustainable Energy Reviews, 15:3008-3018.

[4]  G. Arabatzis, K. Kitikidou, St. Tampakis, K. Soutsas, (2012). The fuelwood consumption in a rural area of Greece. Renewable and Sustainable Energy Reviews, 16:6489-6496.

[5]  J. Easterly and M. Burnham, (1996), Overview of biomass and waste fuel resources for power production, Biomass and Bioenergy, 10: 79 – 92.

[6]  G. Venturi and P. Venturi, (2003), Analysis of energy comparison for crops in European agricultural systems, Biomass and Bioenergy, 25: 235 – 255

[7]  Centre for Renewable Energy Sources, http://www.cres.gr/kape/pdf/download/energy_crops_2006_L.pdf

[8]  E. Alexopoulou and B. Skretschmer (2011), Mapping biomass crop options for EU27: Biomass Futures Policy Briefings, available at: http://www.biomassfutures.eu/work_packages/WP6%20Policy/Biomass%20Futures%20D6.4_4Fcrops%20policy%20briefing.pdf

[9]  I. Boukis, N. Vassilakos, G. Kontopoulos, S. Karellas, (2009), Policy plan for the use of biomass and biofuels in Greece Part I: Available biomass and methodology, Renewable and Sustainable Energy Reviews, 13: 971 – 985.

[10]  W. Qiao, X. Yan, J. Ye, Y. Sun, W. Wang, Z. Zhang, (2011). Evaluation of biogas production from different biomass wastes with/without hydrothermal pretreatment, Renewable Energy, 36: 3313 - 3318

[11]  S. Naik, V. Goud, P. Rout, A. Dalai, (2010), Production of first and second generation biofuels: A comprehensive review, Renewable and Sustainable Energy Reviews, 14: 578–597

[12]  P. Havlik, U. Schneider, E. Schmid, H. Bottcher, S. Fritz, R. Skalsky, AokiK., S. DeCara, G. Kindermann, F. Kraxner, S. Leduc, I. McCallum, A. Mosnier, T. Sauer, M. Obersteiner, (2011). Global land-use implications offer stand second generation biofuel targets, Energy Policy, 39: 5690–5702

[13]  A. Lakaniemi, O. Tuovinen, J. Puhakka, (2013). Anaerobic conversion of microalgal biomass to sustainable energy carriers – A review. Bioresource Technology, 135: 222-231.

[14]  A. Piccolo, 2009. Algae oil production and its potential in the Mediterranean region 1st EMUNI Research Souk 2009 (EMUNI ReS 2009). The Euro-Mediterranean Student Research Multi-conference Unity and Diversity of Euro-Mediterranean, Identities, 9 June 2009

[15]  J. Singh, S. Gu, (2010). Commercialization potential of microalgae for biofuels production, Renewable and Sustainable Energy Reviews, 14: 2596–2610.

[16]  J. García-Ruiz, M. Nadal –Romero, E. Lana Renault, N., Beguería, S., (2013). Erosion in Mediterranean Landscapes: Changes and future challenges. Geomorphology, 198: 20-36.

[17]  Network for Sustainable Islands "Daphne", 2012, Island Action Plan for sustainable energy, available in: http://www.islepact.eu/userfiles/ISEAPs/Report/greece/ISEAP_Samothrace%20GR%20(draft_v2).pdf

[18]  Aegean Energy Office, 2009, Strategic Study on Energy Saving, promoting renewable energy sources and reducing emissions in the Aegean islands, available at :http://www.aegeanenergy.gr/gr/pdf/Strategic%20Energy%20Planning.pdf. (in Greek).

[19]  Greek Statistical Authority

[20]  A. Poulis A., (2012) Utilization of forest biomass for energy generation and district heating in the city of Konitsa, Workshop: The contribution of the National Technical University of Metsovo in the integrated development of the Municipality of Konitsa, Konitsa: Conference room of the City Hall of Konitsa. (in Greek).

[21]  N. Katsoulacos, D. Kaliampakos, (2010). Renewables and mountainous regions, 6th Interdisciplinary-Interuniversity Conference of NTUM "Integrated development of mountain areas", 16-19 September 2010, Metsovo.(in Greek).

[22]  E. Papazoglou., E., Kiritsis Sp., (2000). Environmental benefits from the disposal of agricultural residues in Greece for Energy Production. Agricultural University of Athens. (in Greek).

[23]  M. Tzinevrakis, I. Tzavara, Th. Tsoutsos, D. Vamvouka, K. Xifaras, (2006). Biomass potential for energy utilization in Crete. Proceedings of the 8th National Conference on Renewable Energy Sources, Thessaloniki, 29-31 March 2006, pp. 453-460. (in Greek).

[24]  F. Batzias, D. Sidiras, E. Spyrou, (2005). Evaluating livestock manures for biogas production: a GIS based method, Renewable Energy, 30: 1161–1176.

# 8

# Sustainable Proposal for Utilization of a University's Park and the Contribution of Different Types' of GPS for an Architecture Landscape Plan

**Stergiadou Anastasia[1], Kolkos Georgios[2]**

[1]Aristotle University of Thessaloniki, Faculty of Forestry and Natural Environment, Institute of Forest Engineering and Topography, Thessaloniki, Greece
[2]Forester-Enviromentalist AUTH, Ariadnis, Thessaloniki, Greece

**Email address:**
nanty@for.auth.gr (S. Anastasia), geo.kolkos@gmail.com (K. Georgios)

**Abstract:** Urbanization has brought city dwellers under stress; resulting in the renewal and development of open spaces to create urban green areas or remodeling existing ones; gives a breath of renewal in the wider area of an urban area. The aim of this paper is to demonstrate the appropriateness' of using handheld GPS for mapping urban green for developing regeneration proposals. Two different types of GPS and various topographic methods were used for comparing the results of field measurements with the "true value" - in our case was that of the KTIMATOLOGIO AE - because it gives acceptable by the State surveying quotes. The utilization of the park was based on the method of mild and low regeneration cost. The results of this research are given in two stages; the surveying and the planning. The surveying results from comparing measurements between two types of GPS and the Ktimatologio AE are within the industrial error of each instrument. As far as urban park planning and utilization in the campus of AUTH a reconstruction plan is proposed based on low cost in intervention on urban green.

**Keywords:** Sustainability, Utilization, GPS Types, Methods of Surveying Floss, Architecture Landscape Urban Plan

## 1. Introduction

The decade of 1960s, the development of the system Transit started, ancestor of the current GPS systems (Global Positioning System), by US government's organizations, including the military and NASA. In the beginning the reasons for the development of the system were for military applications (even today the operational control belongs to the USAF-United States Air Force). However in recent decades, political and geodetic applications have been developed. GPS is a global satellite system providing positioning in global coverage, all-weather and with high accuracy.

The purposes of this paper are: 1) to show how useful GPS surveying systems can be in topographical mapping of urban green 2) whether the differences in positioning are within the limit of topological errors and 3) if GPS systems can be used to plan reconstruction proposals of urban parks.

The creation of green areas and parks within urban spaces is necessary in our time, due to climate change, the greenhouse effect and the crowding of the population in urban centers. The reconstruction and development of these sites is very beneficial, because it gives a sense of renewal and alternation for both aesthetic and environmental reasons. This study aims to reflect the current situation of the park of the Observatory of AUTH and recommend gentle low-cost regeneration measures, because the rotation of planted links in urban green spaces, gives impetus to renewal within the urban centers.

## 2. Literature Review

The NAVSTAR / G.P.S. (Navigation Satellite Timing and Ranging - Global Positioning System) or simply GPS is a satellite-based global positioning system that uses coordinates of time and speed, anywhere on the Earth's surface or below it, in any time and regardless of the weather conditions. The system was designed in the 1970s, developed in the 1980s and is constantly under the control of the US Department of Defense (Department of Defense) (Fotiou and Pikridas, 2006).

The GPS belongs to the GNSS systems (Global Navigation Satellite Systems), such as the similar Russian system GLONASS (GLObal Navigation Satellite Systems) and promising purely political European GALILEO system.

Fixed points for creating topographic charts is a method used since ancient times. The use of geodetic, photogrammetric, satellite and other methods to capture and map an area are necessary to achieve the field measurements. (Papadimitriou and others, 2007, Doukas, 2002) makes reference to the necessity of land registration processes in forest cadastre. As "Urban Green" is mainly characterized the urban tissue space which is planned or is in development process of the city and has been evolved to remain without buildings and host some form of vegetation. The built-up areas have now flooded the surface of large cities, especially of megacities. The urban green is limited in parks or in rows on both sides of major roads, while the suburban green spaces have also declined dramatically with the continuous expansion of the building (mainly arbitrary) (Kassios, 2003). The importance of urban and suburban green is a climate forming factor of the city, which provides constant source of renewal of air, and a filter for large amounts of air pollutants (Dafis, 2001, Karameris 2009). The benefits of urban green range from physical and psychological in social cohesion and ecological balance, and conserve biodiversity (Fuller and Gaston, 2009).

The urban green, in the current times and in the conditions of big cities, should provide possibility of escape from the man densely built up environment in an area with other colors, shapes and sounds, cleaner air, milder noises, in a cool room, shady or simply sunny with blue sky or panoramic views. Stergiadou (2001) said that the development of space is a way of creative expression. It is a social necessity, economic and environmental, leading to the fulfillment of human needs. The combination of cartographic surveying of an urban green area using low-cost GPS and the redevelopment proposals of mild and low cost is required in our times, because in the present time of economic crisis it will also cover the socio-economic needs of the city inhabitants.

## 3. Research Area

As research area was chosen the Observatory Park on the campus of AUTH. The choice of the area was such because it is not only one of the busiest and popular areas of the University campus but also of the city of Thessaloniki generally. The site occupies an area of 19629, 29 m². The perimeter of the park of the Observatory, as it was measured by this study, is 513,1m. In the area stands the building of the Observatory, as well as the Jewish monument. There are two internal main streets, as well as two others that lead to the main entrance of the main building of the Observatory (fig. 1).

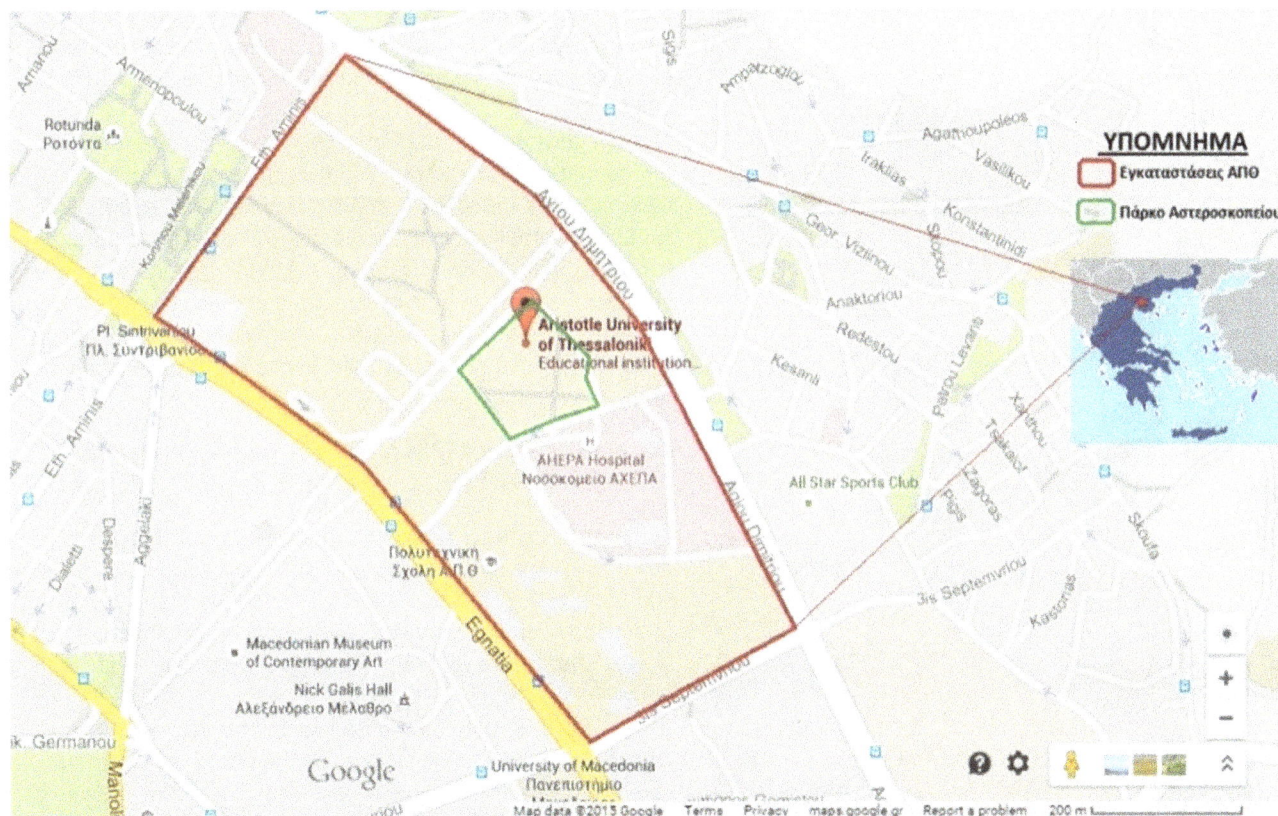

*Figure 1. Location of Observatory Park on campus of AUTH (source: Google Map).*

The vegetation consists of forest species (*Pinus nigra, Pinus Radiata, Cupressus Sempervirens* etc.) and rangeland vegetation (*Hordeum, Dactylis Glomerata, Lolium, Brumus,* *Poa,* etc.). The Park of Observatory represents the 13% of the total green campus of AUTH (Tab.1 & 2 and Fig.2 & 3).

**Table 1.** *Green Park areas in the campus of AUTH.*

| TOTAL AREA OF AUTH CAMPUS | 332388,94m² | 100% |
|---|---|---|
| GREEN AREA OF AUTH CAMPUS | 151799,50m² | 45,67% |

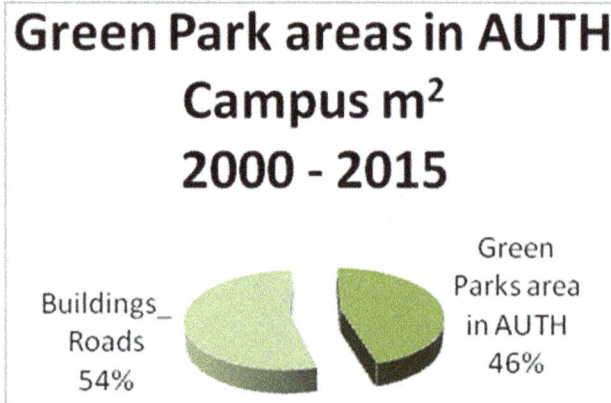

**Figure 2.** *Green Ratio relative to the total area of AUTH Campus.*

**Table 2.** *Green Parks in AUTH Campus.*

| PARK NAME | GREEN AREA (m²) |
|---|---|
| OBSERVATORY | 19629,3 |
| THEOLOGICAL SCHOOL | 8915,088257 |
| LIBRALY | 1611,563232 |
| METEOROSKOPIO | 26962,27954 |
| MEDICAL ACADEMY | 6191,908691 |
| PHILOSOPHY SCHOOL | 8596,597627 |
| UNIVERISTY GYM | 66630,39698 |
| FREE SPACES – SQUARE OF CHIMIO | 13262,37525 |
| TOTAL GREEN AREA | 151799,5096 |

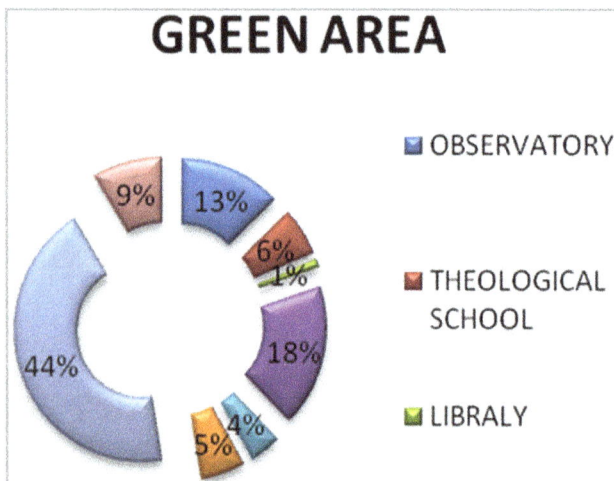

**Figure 3.** *Green area's in the AUTH Campus.*

# 4. Materials and Methods

Our research consists of two levels and therefore uses different materials and methods for completion of both studies, given in detail below:

## 4.1. Materials and Methods for Capturing the Park Observatory of AUTH Campus

The measurements were performed with: 1) GPS handheld Garmin eTrex Vista mapping and 2) with GPS for Android EGSA '87. Both two devices are in Class Handheld GPS, low cost, which is the issue in our times. The area was captured with the GPS perimeter initially and then inside with emphasis on key points (specific places-buildings) and internal roads. Also the area was measured using a measuring tape to calculate the length of the perimeter and internal roads.

Measurements took place at midday, with good weather, so we could avoid the error of multi-reflection, and in positions that do not impede the signal reception from the plants or by the existence of buildings in the area. To accept measurements within the factory errors of each instrument we have given each measure the necessary time to allow for the greatest possible number of satellites in each position. Thus higher accuracy was achieved, because of the resolution phase of satellites.

The measurement procedure was surveying each point as: X (East - C), Y (north - N) and Z (altitude - elevation). Also recorded the deviation of each position based on the measurements of each institution and the satellite geometry (Stergiadou et all, 2003).

The "true values" for comparison of differences between GPS handheld and preparing the topographic plan was received by KTIMATOLOGIO AE. All the measurements are in the coordinate system EGSA'87 (Greek Geodetic System 1987).

The purpose of the mapping was the cartographic representation of the terrain and the record of the accuracy of measurements based on Forest Cadastre (urban green) using suitable electro and satellite bodies (GPS). The method of comparison of measurement results with various handheld GPS was based on geodetic methods, particularly in the theory of errors by calculating the mean square error and the average squared error in order to find the deviations of measurements of the "true values" (Doukas K., 2004).

## 4.2. Materials and Methods for Proposing Redevelopment of Observatory Park

In urban centers there are three types of arrangement of green areas for optimum use of their potential: linear, radial and concentric. By Law no. 2508/1997 three basic types of urban renewal are provided depending on the intensity of urban and intervention: reconstruction, renewal and improvement.

Based on Legislation and also the type of users who visit the Observatory Park, we chose an intervention refreshing combination between linear and radial defining centered design the building of the Observatory. With this combination of landscape architecture we aim to achieve redevelopment of the park and the creation of recreational areas and relaxation for students, teachers and companions of hospitalized in adjoining AHEPAN's hospital.

# 5. Results

The results of this study are divided into two levels: 1) The reliability of handheld GPS measurements to capture urban green areas, and 2) The recreation landscape architectural plan of the Observatory Park.

## 5.1. Results for Imprinting with Two Types of Low-Cost Handheld GPS

*Table 3. Surveying perimeter points with GPS Garmin eTrex Vista mapping.*

| POINT | X | Y | Accuracy AC (m) | ELAVATION | Distance from previous point (m) |
|---|---|---|---|---|---|
| P1 | 411829 | 4498070 | 5 | 52 | 0 |
| P2 | 411817 | 4498057 | 6 | 51 | 15 |
| P3 | 411811 | 4498048 | 5 | 50 | 15 |
| P4 | 411800 | 4498034 | 5 | 50 | 15 |
| P5 | 411789 | 4498025 | 5 | 48 | 15 |
| P6 | 411785 | 4498016 | 5 | 47 | 7,7 |
| P7 | 411785 | 4498015 | 6 | 46 | 2,1 |
| P8 | 411771 | 4498002 | 6 | 46 | 15,5 |
| P9 | 411769 | 4498001 | 5 | 46 | 2,2 |
| P10 | 411762 | 4497996 | 5 | 45 | 15 |
| P11 | 411747 | 4497981 | 7 | 44 | 12,9 |
| P12 | 411747 | 4497983 | 6 | 43 | 3,1 |
| P13 | 411745 | 4497982 | 6 | 42 | 1,9 |
| P14 | 411757 | 4497967 | 6 | 44 | 15 |
| P15 | 411769 | 4497959 | 8 | 45 | 15 |
| P16 | 411775 | 4497954 | 7 | 44 | 8,1 |
| P17 | 411773 | 4497954 | 6 | 42 | 6,6 |
| P18 | 411786 | 4497939 | 6 | 43 | 15 |
| P19 | 411804 | 4497916 | 7 | 43 | 15 |
| P20 | 411810 | 4497907 | 5 | 44 | 15 |
| P21 | 411810 | 4497898 | 5 | 45 | 13,3 |
| P22 | 411810 | 4497905 | 8 | 44 | 2,4 |
| P23 | 411817 | 4497909 | 6 | 45 | 2,4 |
| P24 | 411834 | 4497902 | 6 | 46 | 15 |
| P25 | 411836 | 4497912 | 7 | 48 | 15 |
| P26 | 411850 | 4497915 | 7 | 49 | 15 |
| P27 | 411869 | 4497915 | 6 | 49 | 15 |
| P28 | 411884 | 4497923 | 5 | 50 | 15 |
| P29 | 411896 | 4497927 | 6 | 50 | 18 |
| P30 | 411914 | 4497935 | 6 | 52 | 15,4 |
| P31 | 411923 | 4497934 | 6 | 54 | 6,8 |
| P32 | 411922 | 4497943 | 7 | 49 | 6,3 |
| P33 | 411919 | 4497940 | 7 | 47 | 5,6 |
| P34 | 411922 | 4497951 | 7 | 43 | 15 |
| P35 | 411913 | 4497968 | 7 | 44 | 15 |
| P36 | 411911 | 4497979 | 5 | 45 | 15 |
| P37 | 411911 | 4497989 | 8 | 45 | 15 |
| P38 | 411895 | 4498011 | 5 | 47 | 15 |
| P39 | 411877 | 4498033 | 6 | 47 | 19,2 |
| P40 | 411877 | 4498029 | 7 | 48 | 3,8 |
| P41 | 411872 | 4498024 | 5 | 49 | 15 |
| P42 | 411854 | 4498056 | 5 | 50 | 15 |
| P43 | 411852 | 4498062 | 5 | 50 | 15 |
| P44-P1 | 411830 | 4498070 | 6 | 52 | 14,8 |

Based on GPS handheld Garmin eTrex Vista mapping was depicted originally forty four (44) points of the perimeter of the park Observatory (Table 3). Table 3 with bright color displays brand GPS-handheld points Garmin e-Trex Vista mapping, which are common with those of surveying with GPS android EGSA'87 (Table 4) and the KTIMATOLOGIO AE (Table 5).

The surveying results of GPS android EGSA'87 are given in details at Table 4 and also for the same points are given the coordinates from the official webpage of KTIMATOLOGIO AE in the Table 5.

*Table 4. Surveying points by applying Android «EGSA'87».*

| POINT | X | Y | Accuracy (m) |
|---|---|---|---|
| A1 | 411831,672 | 4498069,596 | 1,07 |
| A2 | 411746,692 | 4497979,239 | 1,13 |
| A3 | 411807,981 | 4497899,158 | 1,03 |
| A4 | 411926,187 | 4497932,895 | 1,51 |
| A5 | 411919,105 | 4497932,879 | 1,39 |
| A6 | 411931,789 | 4497952,415 | 1,69 |
| A7 | 411898,324 | 4497997,910 | 1,33 |
| A8 | 411878,154 | 4498031,797 | 1,55 |
| A9 | 411833,456 | 4498072,987 | 1,31 |

*Table 5. "True values" KTIMATOLOGIO AE.*

| POINTS | X | Y |
|---|---|---|
| K1 | 0411830,51 | 4498073,02 |
| K2 | 0411740,55 | 4497980,41 |
| K3 | 0411810,67 | 4497891,78 |
| K4 | 0411926,42 | 4497930,14 |
| K5 | 0411625,10 | 4497936,75 |
| K6 | 0411928,41 | 4497947,34 |
| K7 | 0411902,61 | 4497998,93 |
| K8 | 0411876,15 | 4498025,39 |
| K9 | 0411832,50 | 4498073,68 |

Within the Observatory Park in order to geo-referenced we counted another three points given in Table 6.

*Table 6. Mapping of specific internal points with GPS Garmin eTrex Vista mapping.*

| POINTS | X | Y | Accurasy (m) | ELEVATION (m) |
|---|---|---|---|---|
| Observatory Park | 0411782 | 4497977 | 6 | 38 |
| Israeli Monument | 0411820 | 498037 | 8 | 43 |
| Proposed Park's Position 0 | 0411825 | 4497931 | 6 | 37 |

We worked the field measurements by the method of the theory of errors, involving comparison between field measurements and the "true value" which led us to the following tables (7, 8, 9), the corresponding diagrams (3, 4, 5) and of course the appropriate conclusions in each case.

Initially we compared each GPS handheld with the "true value" of the KTIMATOLOGIO AE and then all together to find the differences and decide what is the most suitable for low-cost surveys in the urban fabric.

***Table 7.*** *Mean square error of measurement between KTHMATOLOGIO AE & GPS Android EGSA'87.*

| points | KTIMATOLOGIO Coordinates | | GPS Android EGSA '87 Coordinates | | Difference U | | Average Square Error $\mu\tau = \pm ((\upsilon\upsilon)/(n-1))^{0,5}$ | |
|---|---|---|---|---|---|---|---|---|
| | $E_{\Theta}$ | $N_{\Theta}$ | $E_{GPS}$ | $N_{GPS}$ | $E_{\Theta}$- $E_{GPS}$ | $N_{\Theta}$ - $N_{GPS}$ | E | N |
| K1-A1 | 411830 | 4498073 | 411831,67 | 4498069,6 | -1,672 | 3,404 | | |
| K2-A2 | 411740 | 4497980 | 411746,69 | 4497979,24 | -6,692 | 0,761 | | |
| K3-A3 | 411810 | 4497891 | 411807,98 | 4497899,16 | 2,019 | -8,158 | | |
| K4-A4 | 411926 | 4497930 | 411926,19 | 4497932,9 | -0,187 | -2,895 | | |
| K5-A5 | 411925 | 4497936 | 411919,11 | 4497932,88 | 5,895 | 3,121 | $\mu_{\tau E}$ = 0,738 | $\mu_{\tau N}$ = 1,408 |
| K6-A6 | 411928 | 4497947 | 411931,79 | 4497952,42 | -3,789 | -5,415 | | |
| K7-A7 | 411902 | 4497998 | 411898,32 | 4497997,91 | 3,676 | 0,09 | | |
| K8-A8 | 411876 | 4498025 | 411878,15 | 4498031,8 | -2,154 | -6,797 | | |
| K9-A9 | 411832 | 4498073 | 411833,46 | 4498072,99 | -1,456 | 0,013 | | |
| | | | | | -4,36 | -15,876 | | |

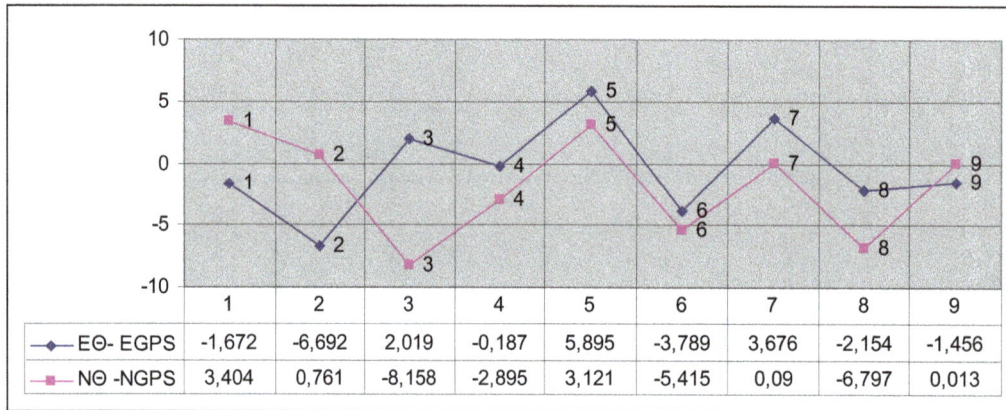

| | 1 | 2 | 3 | 4 | 5 | 6 | 7 | 8 | 9 |
|---|---|---|---|---|---|---|---|---|---|
| E⊖- EGPS | -1,672 | -6,692 | 2,019 | -0,187 | 5,895 | -3,789 | 3,676 | -2,154 | -1,456 |
| N⊖ -NGPS | 3,404 | 0,761 | -8,158 | -2,895 | 3,121 | -5,415 | 0,09 | -6,797 | 0,013 |

***Figure 4.*** *Evaluation of accuracy of E and N coordination for GPS Android EGSA'87.*

***Table 8.*** *Mean square error of measurement between KTHMATOLOGIO AE & GPS Garmin eTrex Vista mapping.*

| Points | KTIMATOLOGIO Coordinates | | GPS GARMIN-eTrex Vista Mapping Coordinates | | Difference U | | Average Square Error $\mu\tau = \pm ((\upsilon\upsilon)/(n-1))^{0,5}$ | |
|---|---|---|---|---|---|---|---|---|
| | $E_{\Theta}$ | $N_{\Theta}$ | $E_{GPS}$ | $N_{GPS}$ | $E_{\Theta}$-$E_{GPS}$ | $N_{\Theta}$ -$N_{GPS}$ | E | N |
| K1-P1 | 411830 | 4498073 | 411829 | 4498070 | 1 | 3 | | |
| K2-P12 | 411740 | 4497980 | 411747 | 4497983 | -7 | -3 | | |
| K3-P22 | 411810 | 4497891 | 411810 | 4497905 | 0 | -14 | | |
| K4-P31 | 411926 | 4497930 | 411923 | 4497934 | 3 | -4 | | |
| K5-P32 | 411925 | 4497936 | 411922 | 4497943 | 3 | -7 | $\mu_{\tau E}$ = 0,353 | $\mu_{\tau N}$ = -1,118 |
| K6-P33 | 411928 | 4497947 | 411919 | 4497940 | 9 | 7 | | |
| K7-P37 | 411902 | 4497998 | 411911 | 4497989 | -9 | 9 | | |
| K8-P40 | 411876 | 4498025 | 411877 | 4498029 | -1 | -4 | | |
| K9-P44 | 411832 | 4498073 | 411830 | 4498070 | 2 | 3 | | |
| | | | | | 1 | -10 | | |

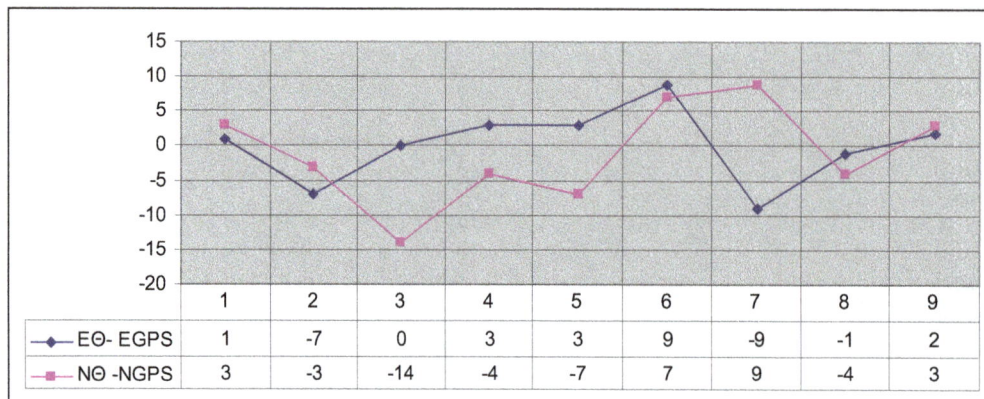

| | 1 | 2 | 3 | 4 | 5 | 6 | 7 | 8 | 9 |
|---|---|---|---|---|---|---|---|---|---|
| E⊖- EGPS | 1 | -7 | 0 | 3 | 3 | 9 | -9 | -1 | 2 |
| N⊖ -NGPS | 3 | -3 | -14 | -4 | -7 | 7 | 9 | -4 | 3 |

***Figure 5.*** *Evaluation of accuracy between E & N coordinates GPS Garmin eTrex Vista.*

The gaps between the field measurements with GPS Android EGSA'87 per location and the corresponding "true values" of KTIMATOLOGIO AE, do not exceed the acceptable error factory 5 meters relative to East, but as to the north we have two points diverge significantly. The first gap was at point: K3 ~ A3; due to canopy cover and the second (K8 ~ A8) due to the phenomenon of multi-reflection, because it is between the buildings of AHEPA's Hospital and the Faculty of Engineering.

The gaps between the field measurements of GPS Garmin eTrex Vista mapping and the corresponding "true values" from KTIMATOLOGIO AE, mostly do not exceed acceptable factory error of 5 meters to the East as in most respects except one (K7 ~ P37) because it is located between the buildings of AHEPAN's Hospital and the Polytechnic School and because of the phenomenon of multi-reflection. In the North we have two points diverge significantly. The first gap was at point K3 ~ A22 due to the canopy cover and the second gap was at point K7 ~ P37 due to the phenomenon of multi-reflection; because it is located between the buildings of AHEPA's Hospital and the Polytechnic School, and also due to the symmetry of satellites.

**Table 9.** *Comparison between field measurement of GPS Garmin eTrex Vista mapping & GPS Android EGSA'87 & KTHMATOLOGIO AE in East.*

| E_GPS-EGSA'87 | E_GPS-GARMIN eTrex | X KTHMATOLGIO AE |
|---|---|---|
| 411831,672 | 411829 | 411830,51 |
| 411746,692 | 411747 | 411740,55 |
| 411807,981 | 411810 | 411810,67 |
| 411926,187 | 411923 | 411926,42 |
| 411919,105 | 411922 | 411625,1 |
| 411931,789 | 411919 | 411928,41 |
| 411898,324 | 411911 | 411902,61 |
| 411878,154 | 411877 | 411876,15 |
| 411833,456 | 411830 | 411832,5 |

**Comparison between different handheld GPS types at East coordinates**

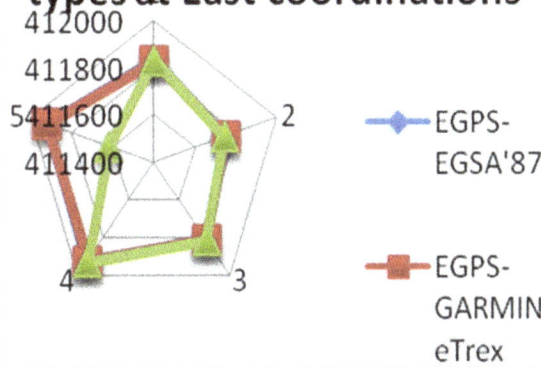

**Figure 6.** *Deviations between field measurements of two different types of handheld GPS and KTIMATOLOGIO AE (East).*

In Table 9 which is expressed in Figure 6, we observe price differentials relative to E or X, of the two GPS and with that of the KTIMATOLOGIO. We realize that differences are minimal except point 5.

**Table 10.** *Comparison between field measurment of GPS Garmin eTrex Vista mapping & GPS Android EGSA'87 & KTHMATOLOGIO AE in North.*

| N_GPS-GARMIN | N_GPS-Android | N_Θ |
|---|---|---|
| 4498070 | 4498069,596 | 4498073 |
| 4497983 | 4497979,239 | 4497980 |
| 4497905 | 4497899,158 | 4497891 |
| 4497934 | 4497932,895 | 4497930 |
| 4497943 | 4497932,879 | 4497936 |
| 4497940 | 4497952,415 | 4497947 |
| 4497989 | 4497997,91 | 4497998 |
| 4498029 | 4498031,797 | 4498025 |
| 4498070 | 4498072,987 | 4498073 |

**Comparison between different handheld GPS at Norht coordinations**

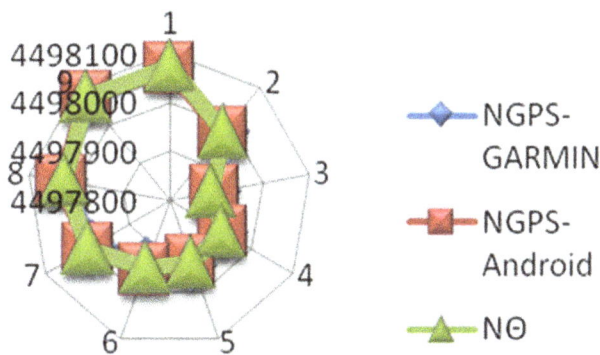

**Figure 7.** *Deviations between field measurements of two different types of handheld GPS and KTIMATOLOGIO AE (North).*

Observing Table 10, which is expressed by the Figure 7, and where field measurements are captured between the two GPS handheld with that of the KTIMATOLOGIO AE as a northwards (N or Y), we find that deviations are not displayed.

Based on the above diagrams and tables we find that there are no significant deviations from the true value of the KTIMATOLOGIO AE. However, differences were shown in the mean squared error of both GPS handheld with the KTIMATOLOGIO AE, which does not exceed 0.75 with respect to E and 1.5 regarding N. Therefore we can say that both handheld GPS are acceptable to be used for mapping urban green when variations of measurements' are within the factory errors organs.

### 5.2. Observatory Park Redevelopment Study

The sustainable proposal was based on a combination of methods related to urban green in order to achieve a recovery of the Park Observatory with minimal interventions and low cost. The general concept was to design an architectural plan based on Greek aromatic plants and painted rocks in order to recreate the landscape and designe a galactic realistic depiction of our planetary system. The construction of that kind of park will give the opportunity to the students of Physics but also to the visitors of the AHEPA's Hospital to have an educational time with not many lectures. The next architectural plan gives all our sustainable proposals concerning the Observatory Park in AUTH Campus.

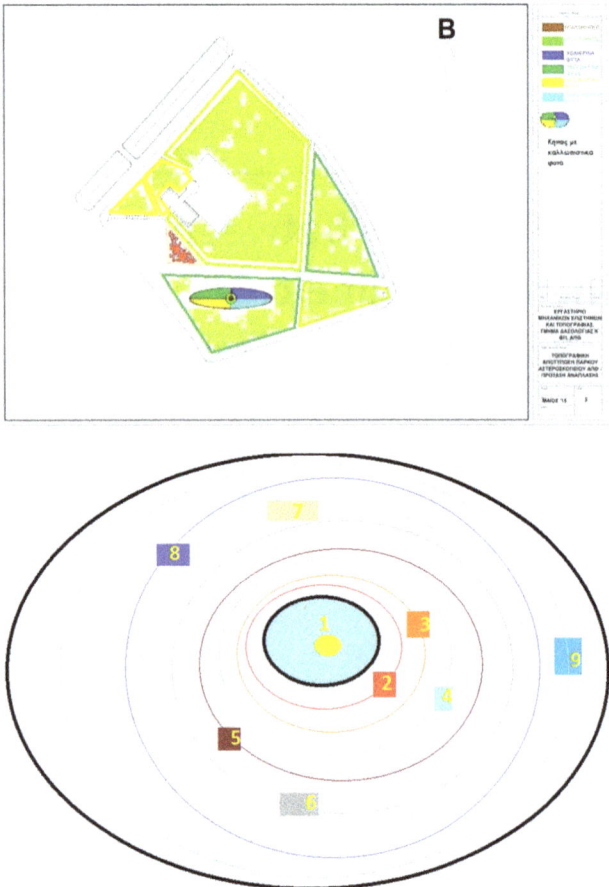

*Figure 8. The Order of rocks by the orbit of the planets of our solar system around the sun.*

*Figure 9. Colors and sizes which are corresponding to the planets, as they will be placed into rock garden.*

The sustainable redevelopment proposal concerns: 1) the construction of a rock garden which is with the solar system and where we put rocks painted in matching colors of the planets and in positions similar to their actual orbit around the sun. In the following images (5 & 6), the formation is given analytically.

The proposed "Four Season Garden" (Fig. 10), located in the Northeast side of the Park, is proposed to import plants to bloom in all four seasons and edible shrubs like lavender, thyme, mint and herbs like lemongrass, evening primrose, jasmine, honeysuckle, etc, so that the rotation of colors, aromas and flowering creates a calm atmosphere for the students, employees teachers of Aristotle University, and also patients and their companions in the adjacent hospital AHEPA. The plan following indicates the details of the proposed garden.

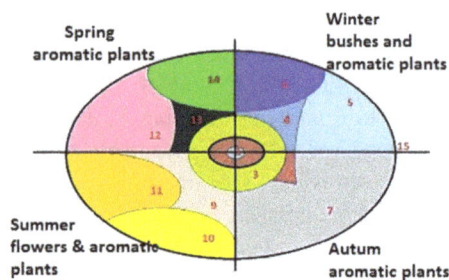

| No | TYPE | BLOOMING PERIOD |
|----|------|-----------------|
| 1 | FOUNTAIN | |
| 2 | ROCK-GARDEN | |
| 3 | GRASS-MATTERIAL | |
| 4 | WINTER FLOWERS | WINTER |
| 5 | CAMELIA | WINTER |
| 6 | GIASMEN | WINTER |
| 7 | CYCLAMEN | AUTUM |
| 8 | CHRYSANTHEMUM | AUTUM |
| 9 | ANTIRRHINUM MAJUS | SUMMER |
| 10 | HELIANTHUS | SUMMER |
| 11 | PETOUNIA | SUMMER |
| 12 | DIANTHUS | SPRING |
| 13 | VIOLA | SPRING |
| 14 | HYAKINTHUS | SPRING |
| 15 | LEILAND BUSHES | ALL SEASON |

*Figure 10. The "Four Season Garden".*

# 6. Discussion and Conclusions

This research was initiated in two levels, the use of low cost handheld GPS for mapping apark within the urban fabric, and field measurements were evaluated based on the "true value", as given by the KTIMATOLOGIO AE and upon this surveying we proposed a Sustainable Architecture Landscape Study based on a Mild Regenerating and renewing the existing part and creating educational rock- and four seasons garden.

The results of comparisons between fields measurements with two different handheld GPS and true values, were found satisfactory and acceptable, since it was within the organ factory errors. So the use of low cost handheld GPS is suggested for the architectural landscape design in the urban fabric.

The Observatory Park redevelopment proposals is based on the use of shrubs and herbs that can be purchased from the Departments of Agriculture, Forestry and Natural Environment, and therefore the construction cost will be lower and the species proposed are available and have many requirements in water and planting operations (pruning, fertilization, etc.).

At a time when the cost - benefit analysis will be shown that is more relevant than ever we believe that such architectural landscape study is easy to be implemented and the results are going to be shown within a decade. In any case a renewal sustainable proposal of utilization a university park is always an educational statement.

# Acknowledgements

We need to thank for the completion of this paper the Technical Service office of AUTH for providing the original topographic plan, the KTIMATOLOGIO AE for free access to the survey data and as well as all the scientists whose writings were consulted to draw up this scientific and research work.

This research has been co-financed by the European Union (European Social Fund – ESF) and Greek national funds through the Operational Program "Education and Lifelong Learning" of the National Strategic Reference Framework (NSRF) - Research Funding Program: Thales - Investing in knowledge society through the European Social Fund.

# References

[1]    Doucas K., 2002. Forest and Agricultural Lands, Giachoudi Publications, pp. 403, Thessaloniki.

[2]    Kassios K., 2003. "Planning and management of urban and suburban green areas in Athens." Chapter in the collective volume entitled "A future for Athens" curated S. Ceca. Athens: Papazisi.

[3]    Karameris, A., 2008. Forest Recreation. University notes AUTH.

[4]    Law 2508/1997, 1997. Sustainable urban development of the country's towns and villages and other provisions, http://nomoi.info/FEK-A-124-1997-pp-1.html

[5]    Dafis, Sp., 2001. Urban Forestry. Publications Art of Text, Thessaloniki.

[6]    Papadimitriou A., A. Stergiadou and B. Drosos, 2007. Comparative Performance Site plan between three-dimensional model Analog and Digital, Minute Book 14 Forestry Congress, SAD Publications, Thessaloniki.

[7]    Stergiadou Anastasia, 2001. Utilization, development and opening of mountainous areas (areas of Grevena, Ioannina and Kilkis), PhD at the Faculty of Geotechnical Sciences, School of Forestry and Natural Environment, Department of Forest Techniques and Water Works, Aristotle, http: //phdtheses.ekt. gr / eadd / handle / 10442/22812.

[8]    Stergiadou A., Stergiadis Ch., Giannoulas V. & Doukas K., 2003. The contribution of different GPS type's in sustainable development (Case: mountain Chortiatis - Kissos), Proceedings of 8th International Conference on Environment Science and Technology, Limnos 8 - 10 September 2003, pp. 774-781, Athens.

[9]    Fuller R. and Gaston K., 2009. _The scaling of green space coverage in European cities.Biol. Lett., Vol. 5, no3: 352-355.

[10]   Fotiou A., Pikridiou Ch., 2006. GPS and surveying purposes. Ziti, Thessaloniki.

# 9

# R-290 vapor compression heat pump for recovering and upgrading waste heat of air-conditioner by using spiral coil tank

**Nattaporn Chaiyat***, **Natthawud Dussadee**

School of Renewable Energy, Maejo University, Chiang Mai, Thailand

**Email address:**

benz178tii@hotmail.com (Chaiyat N.), natthawu@yahoo.com (Dussadee N.)

**Abstract:** In this study, a concept of using a vapor compression heat pump for recovering and upgrading waste heat of an air-conditioner has been presented. R-290 has been selected due to its high efficiency and the environmental impact. R-290 heat pump at heating capacity 3 kW has been constructed to recover waste heat from the discharge refrigerant leaving compressor of R-134a air-conditioner at cooling capacity 1 TR. From the study results, it could be seen that the modified unit gives better $EER_{AC}$ when the cooling water does not over 43 °C. A set of simplified model has been developed to predict the system performance and the simulated results agree quite well with the measured data. Moreover, profile of hot water consumption in the department of children's hospital room, Maharaj Nakorn Chiang Mai Hospital is chosen to study. It was found that the hospital requires hot water is 0.815 $m^3$/d at 50 °C temperature, one unit of R-290 heat pump is enough to generate hot water with the economic results of saving cost and payback period around 765.46 USD/y and 1.97 y, respectively.

**Keywords:** Air-Conditioner, Vapor Compression Heat Pump, Performance Curve, Heat Recovery, Spiral Coil Tank

## 1. Introduction

Vapor compression heat pump (VCHP) is one technology for upgrading a low temperature heat to a higher temperature level. In a conventional VCHP, the low temperature heat is supplied at the evaporator and delivered the higher temperature at the condenser. Theoretical and experimental studies of the VCHP have been reported by various literatures. Chaiyat and Chaichana [1] simulated the natural working fluid used in vapor compression heat pump for generating hot water at 70 °C. The heat source was hot spring at the temperature around 40-50 °C. Five working fluids, R-290 (Propane) R-600 (Butane) R-600a (Isobutane) R-1270 (Propylene) and R-717 (Ammonia) were selected. The considered parameters were the unit heat transfer, mass flow rate, specific volume, maximum pressure, maximum temperature and COP of the cycle. It was found that the suitable natural working fluid was R-290 for the geothermal heat pump. Chaiyat and Kiatsiriroat [2] presented the experimental study of R-123 vapor compression heat pump to recover waste heat from water-cooled condensers. Waste heat was upgraded and generated hot water up to 70 °C. The

EER of the water-cooled air-conditioner increased around 20% higher than that of the air-cooled unit.

Heat pump has also been used to recover heat from extracted vapor of the agricultural product during drying and heat could be upgraded to a higher temperature and supplied to generate hot air. The heat pump drying was found to be an effective equipment for drying high quality produce with low energy consumption as reported by Singharajwarapan and Chaiyat [3], Pendyala et al. [4], Chou et al. [5], Clements et al. [6], Young et al. [7] and Sadchang [8].

There are some reports on solar-boosted heat pump for hot water production. Heat pump extracts heat from low temperature solar heat by using flat-plate solar collector at its evaporator and the high temperature heat is upgraded and supplied through condenser for generating hot water. A 80 kW heat pump unit was installed at a building in Maharaj Nakorn Chiang Mai Hospital as reported by Burapha and Kiatsiriroat [9]. The heat pump unit consumed one-third of electrical power compared with the electrical water heater. For economic result, the payback was around 3.6 y. The vapor compression heat pump could act as a heat recovery unit. The evaporator could recover a low temperature waste heat or other low temperature heat source such as solar energy or geothermal energy. The low quality heat could be developed to a higher temperature at the condenser and used in other processes. In this study, the heat pump is used to recover waste heat of the refrigerant leaving compressor by the spiral coil water tank at the series connection with the air-cooled air-conditioner to generate hot water.

The objective of this research is to study the possibility for using the vapor compression heat pump which is used to recover waste heat from the air-conditioner and upgraded heat to generate hot water compared with the electrical water heater in the department of children's hospital room, Maharaj Nakorn Chiang Mai.

## 2. System Description

Generally, heat transfers from the high temperature heat source to the low temperature heat sink. For process to reverse heat, heat pump is needed and a power consumption from the external source is used to transfer heat from the low temperature heat source to the high temperature heat sink as shown in Fig. 1.

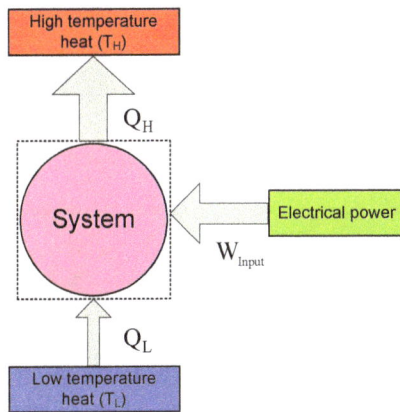

*Figure 1. The heat pump concept.*

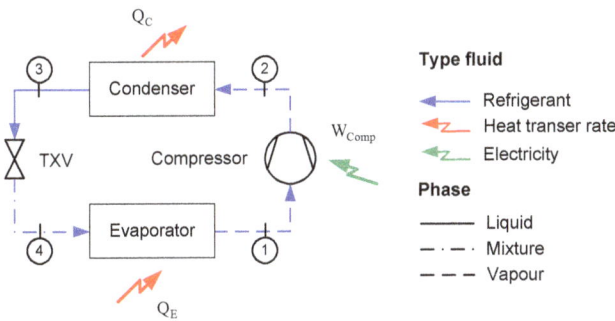

*Figure 2. Vapor compression heat pump cycle.*

The main components of vapor compression are compressor, condenser, evaporator and expansion valve as shown in Fig. 2. The working fluid has a low boiling temperature at a low pressure ($P_L$). At state 1, the fluid in vapor phase is boiled at the evaporator and compressed in the compressor to state 2 to a high pressure ($P_H$). The vapor at the high temperature and pressure is condensed in the condenser to be liquid at state 3. The liquid is then throttled to low

pressure at the expansion valve at state 4. Temperature of fluid will drop down, thus, the fluid could be absorbed low temperature heat at the evaporator again and the new cycle restarts.

The basic equations for the behavior of each component in the VCHP cycle as presented in Fig. 2 are as follows:

Evaporator

$$Q_E = \dot{m}_{ref}(h_1 - h_4) \tag{1}$$

$$\dot{m}_{ref} = \dot{m}_1 = \dot{m}_2 = \dot{m}_3 = \dot{m}_4 \tag{2}$$

$$Q_E = \dot{m}_{da}(h_{a,i} - h_{a,o}) - M_w h_w \tag{3}$$

Compressor

$$W_{Comp} = \dot{m}_{ref}(h_2 - h_1) \tag{4}$$

$$s_1 = s_2 \quad \text{(Isentropic process)} \tag{5}$$

$$\eta_{Comp} = (h_2' - h_1)/(h_2 - h_1) \tag{6}$$

Condenser

$$Q_C = \dot{m}_{ref}(h_2 - h_3) \tag{7}$$

$$Q_C = \dot{m}_{HW}Cp_{bulk,HW}(T_{HW,o} - T_{HW,i}) \tag{8}$$

Expansion valve

$$h_3 = h_4 \quad \text{(Throttling process)} \tag{9}$$

Energy efficiency ratio (EER) for heating

$$EER_{Heating} = Q_C / W_{Sys} \tag{10}$$

Energy efficiency ratio (EER) for cooling

$$EER_{Cooling} = Q_E / W_{Sys} \tag{11}$$

## 3. Material and Method

In this study, the vapor compression heat pump is used for recovering rejected heat from the air-conditioner which is modified to add a spiral coil water tank in the air-conditioner cycle. The diagram to improve the air-conditioner by using the heat pump and the spiral coil tank are given in Fig. 3.

The cool water tank with having the spiral coil absorbs heat from a superheat refrigerant leaving the compressor at high temperature around 60-80 °C. After that, refrigerant at lower temperature is sent to the air-cooled condenser which means a low condensing temperature ($T_C$) and a low power consumption of compressor. In this study, R-290 or propane is the working fluid of heat pump system which refers from the study results of Chaiyat and Chaichana [1] as the high efficiency and the environmental impact. Table 1 shows the properties of R-290 compared with the conventional refrigerants in heat pump system. It could be found that R-290 is a high efficiency in term of heat of vaporization and a friendly environment in terms of atmospheric life time, ozone depletion potential and global warming potential. The component descriptions of the air-conditioner and the heat pump are given in Table 2 and Table 3, respectively.

**Table 1.** *Properties of various refrigerants [10-12].*

| Refrigerant | R-22 | R-290 | R-134a |
|---|---|---|---|
| Critical temperature (°C) | 96.14 | 96.68 | 101.06 |
| Critical pressure (MPa) | 4.99 | 4.25 | 4.06 |
| Critical density (kg/m³) | 523.84 | 218.50 | 511.90 |
| Boiling point (°C) | -40.81 | -42.09 | -26.07 |
| Heat of vaporization (kJ/kg) | 164.24 | 302.30 | 160.88 |
| Flammability | NO | YES | NO |
| Toxicity | NO | NO | NO |
| Atmospheric life time (y) | 13.3 | < 1 | 14 |
| ODP[1] (CO$_2$-ralated) | 0.034 | ~ 0 | 0.0015 |
| GWP[2] (100 y) | 1780 | 0 | 1320 |

Note: [1] Ozone depletion potential [2] Global warming potential

The air-conditioner in testing room is tested its thermal performance firstly with the air-cooled condenser and then the result is compared with the unit with spiral coil tank and heat pump. For the latter case, the heat pump recovered heat from the cooling water in the spiral coil tank and upgraded to generate hot water in the hot water tank. The experiments in each case are tested as shown the details of the testing procedures in Fig. 3.

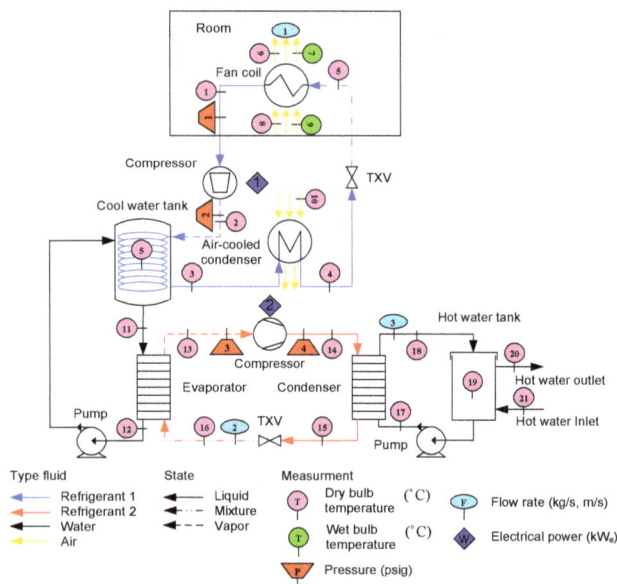

**Figure 3.** *R-290 Vapor compression heat pump for recovering and upgrading waste heat from the air-conditioner.*

**Table 2.** *Descriptions of the air-conditioner components.*

| Devices | Properties |
|---|---|
| Fan coil | Fin and tube heat exchanger |
| | Capacity 3.516 kW (1 TR) |
| | Tube size (OD) 5.0 mm |
| | Fins per inch 18 FPI |
| | No. of rows & column 2R, 15C |
| Compressor | Hermetic (rotary) compressor R-134a |

| Devices | Properties |
|---|---|
| | Capacity 1.434 kW |
| | Compression ratio 6.0 Max |
| Condenser | Fin and tube heat exchanger |
| | Capacity 5.275 kW |
| | Tube size (OD) 7.0 mm |
| | Fins per inch 18 FPI |
| | No. of rows & column 2R, 36C |
| Expansion valve | Capacity 3.516 kW |
| | Pressure ratio 3.00 |
| | Orifice type thermo static |

**Table 3.** *Descriptions of the heat pump components.*

| Devices | Properties |
|---|---|
| Compressor | Reciprocating compressor R-22 |
| | Power input 1.3 kW |
| Evaporator | Plate heat exchanger |
| | Capacity 4.00 kW |
| | Area 0.65 m² |
| Condenser | Plate heat exchanger |
| | Capacity 5.00 kW |
| | Area 0.88 m² |
| Expansion valve | Thermo static orifice 02 |
| | Capacity 5.00 kW |
| | Pressure ratio 3.00 |
| Spiral coil tank | Capacity 150 L |

# 4. Results and Discussion

## 4.1. Performance Curve of the Air-Conditioner

For this study, a performance curve of vapor compression cycle is developed which is a mathematical correlation between of an energy efficiency ratio (EER) and a different temperature of air entering at condenser and evaporator ($T_{a,C,i} - T_{a,E,i}$).

Fig. 4 shows a cooling energy efficiency ratio for cooling ($EER_{AC}$) of the tested air-conditioner before improvement with the temperature difference between the air entering at condenser and evaporator ($T_{a,C,i} - T_{a,E,i}$). It could be seen that the $EER_{AC}$ decreases when the temperature difference increases which follows the Carnot efficiency concept.

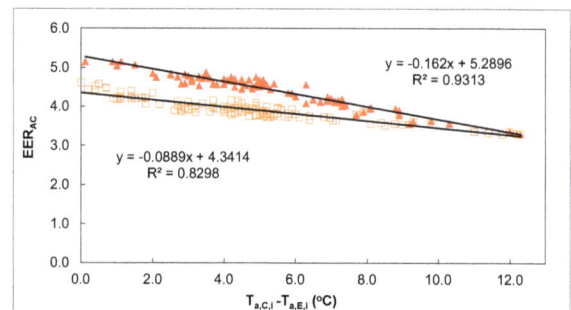

**Figure 4.** *Performance curve of the normal air-conditioner and the modified air-conditioner by using the spiral coil tank and the heat pump unit.*

For the air-conditioner with spiral coil tank and heat pump, the rejected heat is recovered by the spiral coil tank at temperature around 40 °C and upgraded to hot water at temperature around 55-60 °C. It could be found that the $EER_{AC}$ of the modified air-conditioner is higher than the normal air-conditioner which means that the modified unit consumes the lower electrical power compared with that of the normal unit at the same cooling capacity as shown in Fig. 4. However, for using the spiral coil tank, when the rejected heat is recovered and accumulated as hot water, at the value of $T_{a,C,i} - T_{a,E,i}$ around 12 °C or water temperature in the spiral coil tank is around 43 °C, the $EER_{AC}$ tends to be lower than that of the normal air-cooled condition. This result is corresponding with the study results of Chaiyat and Kiatsiriroat [2] which the maximum water temperature for water-cooled condenser is not over than 45 °C. It could be recommended that when the cooling water is over this value, the heat pump unit should extract heat out of the spiral coil tank.

The equations of $EER_{AC}$ and temperature difference of heat source and heat sink for the normal and modified units are as follows:

$$EER_{AC,Normal} = -0.162(T_{a,C,i} - T_{a,E,i}) + 5.2896 \qquad (12)$$

$$EER_{AC,Modified} = -0.0889(T_{a,C,i} - T_{a,E,i}) + 4.3414 \qquad (13)$$

### 4.2. Performance Curve of R-290 Vapor Compression Heat Pump

For vapor compression heat pump, the $EER_{VCHP}$ is described to be function of the temperature difference between the water temperature entering condenser ($T_{HW,i}$) and the water temperature entering evaporator ($T_{CW,i}$) as shown in Fig. 5. It could be seen that the $EER_{VCHP}$ decreases when the temperature difference between heat source and heat sink increases which is similarly with the performance curve of air-conditioner. The empirical correlation could be found in the form of:

$$EER_{VCHP} = -0.0587(T_{HW,i} - T_{CW,i}) + 2.9341. \qquad (14)$$

From the testing results of R-290 heat pump unit, it could be seen that R-290 heat pump unit could be generated hot water temperature around 57 °C continuously at supplied heat source temperature around 40 °C as also shown the temperature profiles of heat pump unit in Fig. 5.

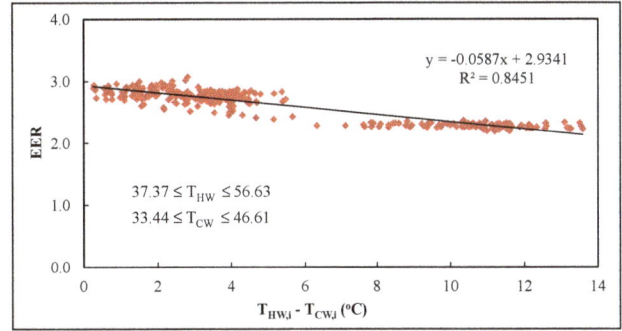

**Figure 5.** *Performance curves of the R-290 VCHP from the experimental results.*

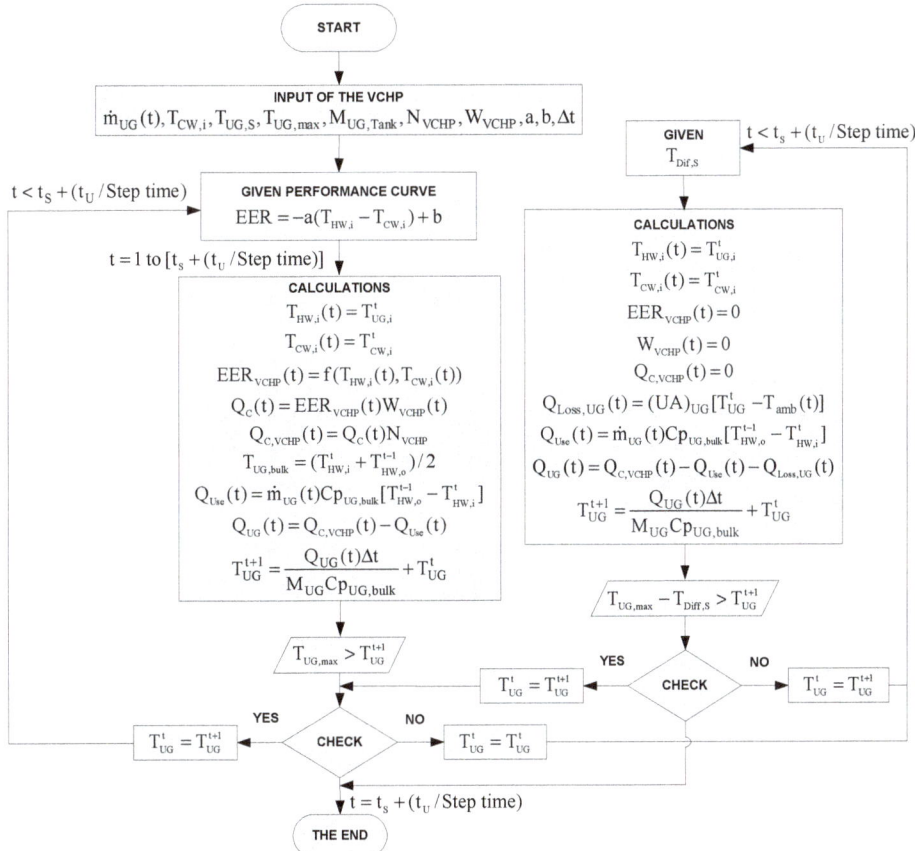

**Figure 6.** *Flow chart for simulation of the vapor compression heat pump.*

## 4.3. Verification

Fig. 6 shows steps for calculation of the heat pump cycle with the simplified model. Performance correlations of the EERVCHP with the operating temperatures are given. The inputs data are the operating conditions and the useful hot water temperature profiles. The upgraded hot water temperature in hot water tank and leaving condenser are the outputs of calculation.

Fig. 7 shows the simulated results of the $EER_{AC}$ and the heating capacity at condenser during ($Q_C$) when hot water in tank is used at the steady state condition of useful hot water flow rate around 0.03 L/s. $EER_{AC}$ and $Q_C$ are nearly constants at hot water temperature in the storage tank constant at around 55 °C. The simulated results agree well with the measured data. It could be seen that the simplified models could be used to simulate the performances of heat pump system that is used to recover the waste heat from the discharge refrigerant of air-conditioner. Thus the model will be used to predict the possibility in using this concept for generating hot water in the department of children's hospital room, Maharaj Nakorn Chiang Mai. The details are given in the next part.

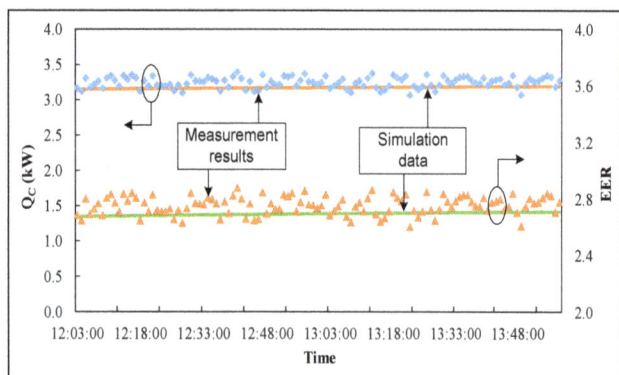

**Figure 7.** *Comparison results of the measured data and the simulation results of hot water temperature from R-290 VCHP.*

## 4.4. Generating Hot Water by Useful Temperature Profile

Profile of hot water consumption in the department of children's hospital room, Maharaj Nakorn Chiang Mai Hospital is selected for the simulation. The required hot water consumption is around 815 L/d and the working temperature is about 50 °C. The conditions for the simulation are as follows:

- Profile of hot water consumption ($\dot{m}_{UG,t}$) is shown in Table 4.
- Initial temperature in hot water storage ($T_{UGS}$) is at 30 °C.
- Maximum temperature of heat pump ($T_{UGmax}$) is 55 °C.
- Fill-in water temperature ($T_{HW,i}$) is at 30 °C.
- Hot water tank ($V_{UG,Tank}$) is 300 L.
- Heat loss of hot water storage tank ($UA_{UG}$) is 5 w/m$^2$.
- The ambient temperature ($T_{amb}$) is 30 °C.

From the simulation, it could be seen that one unit of R-290 heat pump could recover waste heat from 1 unit of air-condenser to generate hot water in this hospital. Fig. 8

shows comparison of the temperature history during a day between R-290 heat pump and electrical heater when the storage tank is 300 L. Around 12-13 o'clock, there is a high rate of water consumption thus the temperature slightly drops down. Fig. 8 also shows the advantage point of R-290 VCHP which the average hot water temperature is higher than that of using the electrical heater. Moreover, R-290 VCHP consumes the electrical power at around 1/3 time compared with the normal electrical heater. Including that R-290 is the organic type of refrigerant which is friendly with the environment.

**Table 4.** *Profile of using hot water in the hospital.*

| Time | Quality (L) | Flow rate (L/s) |
|------|-------------|-----------------|
| 8.00 | 105 | 0.029 |
| 9.00 | 25 | 0.007 |
| 10.00 | 65 | 0.018 |
| 11.00 | 50 | 0.014 |
| 12.00 | 60 | 0.017 |
| 13.00 | 165 | 0.046 |
| 14.00 | 70 | 0.019 |
| 15.00 | 110 | 0.031 |
| 16.00 | 140 | 0.039 |
| 17.00 | 25 | 0.007 |
| Total | 815 | 0.198 |

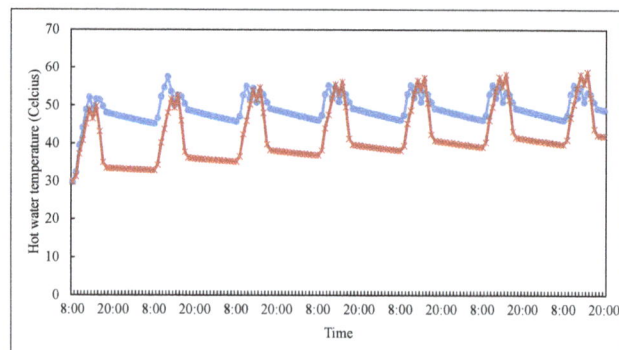

**Figure 8.** *The results of simulation from the hospital data in 84 h for storage tank 300 L.*

## 4.5. Economic Analysis

For the economic result for generating hot water in the hospital, a simple payback of the hot water system from the waste heat of air-conditioner is compared with the unit having the electrical water heater for hot water generation. The details of general conditions for economic analysis are:

A. The operating day is 365 d/y.
B. The operating time is 8 h/d.
  During 9.00-22.00 o'clock is 1 h/d.
  During 22.00-9.00 o'clock is 7 h/d.
C. Electricity charge (Time of Use Rate: TOU) are as:
  During 9.00-22.00 o'clock is 0.1144 USD/kWh.
  During 22.00-9.00 o'clock is 0.0677 USD/kWh.
  Ft (Fuel adjustment charge at the given time) is 0.0215 USD/kWh.
  Peak demand charge is 9.7104 USD/kWh·m.
D. Maintenance cost is 3 % of initial investment.
From the testing result, it is found that one R-290 heat pump

unit uses the electrical power consumed around 1.18 $kW_e$ while that of the electrical heater is 3 $kW_e$. Thus the payback could be calculated by:

$$\text{Payback period} = \text{Total investment} / \text{Annual saving.} \quad (15)$$

Table 5 shows the economic results for using R-290 VCHP to produce hot water temperature around 55 °C compared with the conventional heater. It could be seen that the saving cost of the electrical power consumption from using R-290 VCHP compared with the electrical heater is around 765.46 USD/y [13] and payback period of the modified system is around 1.94 y.

**Table 5.** *The economic results for using R-290 VCHP and the electrical heater.*

| Descriptions | R-290 VCHP | Electrical heater |
|---|---|---|
| The electrical power consumption (kW) | 1.18 | 3.00 |
| Operating time (h/d) | 8.00 | 8.00 |
| The rate of electrical power (kWh/y) | 3,445.60 | 8,760.00 |
| Cost of the electrical power (USD/y) | 496.28 | 1,261.74 |
| Investment cost (USD) | 1,551.85 | 155.19 |
| Payback period (y) | 1.94 | - |

## 5. Conclusions

From the study results, the conclusions are as follows:
1  R-290 is selected as the working fluid for the heat pump in term of its high efficiency and the environmental impact.
2  The air-conditioner with spiral coil tank and heat pump gives better $EER_{AC}$ when the cooling water is not over 43 °C.
3  The simulated results agree quite well with the measured data.
4  For the hospital at required hot water 0.815 $m^3/d$ and 50 °C, the saving cost and the payback period of R-290 heat pump are 765.46 USD/y and 1.97 y, respectively.

## Acknowledgements

The authors would like to thank the School of Renewable Energy, Maejo University for supporting testing facilities. Highly acknowledge to the TSUS Innovation Intelligence Ltd for the budget support.

## Nomenclature

| | |
|---|---|
| A | Area, $(m^2)$ |
| Cp | Specific heat capacity, (kJ/kg·K) |
| h | Enthalpy, (kJ/kg) |
| ṁ | Mass flow rate, (kg/s) |
| N | Number, (Unit) |
| P | Pressure, (bar) |
| Q | Heat rate, (kW) |
| R | Refrigerant, (-) |
| v | Specific volume, $(m^3/kg)$ |
| V | Volume, $(m^3)$ |
| s | Entropy, (kJ/kg·K) |
| t | Time, (s) |
| T | Temperature, (°C) |
| U | Overall heat transfer coefficient, $(W/m^2 \cdot K)$ |
| W | Work, (kW) |
| **Greek Symbol** | |
| η | Efficiency, (%) |
| **Subscript** | |
| a | Air |
| AC | Air-conditioner |
| amb | Ambient |
| C | Condenser |
| Comp | Compressor |
| CW | Cooling water |
| da | Dry air |
| dif | Difference |
| e | Electrical |
| E | Evaporator |
| FC | Fan coil |
| H | High |
| HS | Heat source |
| HW | Hot water |
| i | Inlet |
| L | Low |
| max | Maximum |
| min | Minimum |
| o | Outlet |
| ref | Refrigerant |
| S | Start |
| SP | Spiral coil |
| SW | Supplied water |
| Sys | System |
| th | Thermal |
| U | Used |
| UG | Upgraded |
| VCHP | Vapor compression heat pump |
| W | Water |

## References

[1]  N. Chaiyat, C. Chaichana, "Working fluid selection for geothermal heat pump", Engineering Journal Chiang Mai University, 2006, 13: 27–32.

[2]  N. Chaiyat, T. Kiatsiriroat, "Recovering and upgrading waste heat of air-conditioner by combining R-123 vapor compression heat pump," In: Proceeding of the 9th Conference on Energy Heat and Mass Transfer in Thermal Equipment, Prachuap Khiri Khan, Thailand, March 11–12, 2010.

[3]  F. S. Singharajwarapan, N. Chaiyat, "Vapor compression heat pump using low temperature geothermal water: A case study of northern Thailand", In: Proceeding of the 10th Asian Geothermal Symposium, Tagaytay, Philippines, September 22–24, 2013.

[4]  V. R. Pendyala, S. Devotta, V. S. Patwardhan, "Heat pump assisted dryer Part 2: Experimental study", International Journal of Energy Research, 1990, 14: 493–507.

[5]   S. K. Chou, M. N. A. Hawlader, J. C. Ho, N. E. Wijeysundera, and S. Rajasekar, "Heat pump in the drying food products", The Science and Technology Information Network of the Philippines, 1994, 8: 1–4.

[6]   S. Clements S, S. Jia, P. Jolly, "Experimental verification of a heat pump assisted continuous dryer simulation model", International Journal of Energy Research, 1993, 17: 19–28.

[7]   G. S. Young, S. Birchall, R. L. Mason, "Heat pump drying of food products prediction of performance and energy efficiency", In: Proceeding of the 4th ASEAN Conference on Energy Technology, Bangkok, Thailand, October17–20, 1995.

[8]   A. Sadchang, "Design and construction of heat pump dryer prototype for small industry," Master Engineering Dissertation, Chiang Mai University, Chiang Mai, Thailand, 2006.

[9]   M. Burapha, T. Kiatsiriroat, "Simplified model of solar water heating with heat pump assisted", The journal of industrial technology, 2008, 2: 15–23.

[10]  American Society of Heating, Refrigerating and Air-Conditioning Engineers, Inc, "ASHRAE Handbook fundamentals," Refrigerant, American, 2009.

[11]  Properties of R-32, <http://www.refrigerants.com/> [Accessed 1 June 2014].

[12]  Properties of R-410a, <http://www.nationalref.com/> [Accessed 1 June 2014].

[13]  Exchange rate, <http://www.bot.or.th/> [Accessed 1 June 2014].

# 10

# Effect of solar-induced water temperature on the growth performance of African sharp tooth catfish (*Clarias gariepinus*)

**Wirawut Temprasit[1], Alounxay Pasithi[1], Suthida Wanno[1], Supannee Suwanpakdee[1], Sudaporn Tongsiiri[1], Natthawud Dussadee[2], Niwooti Whangchai[1,*]**

[1]Faculty of Fisheries Technology and Aquatic Resources, Maejo University, Sansai, Chiang Mai 50290, Thailand
[2]School of Renewable Energy, Maejo University, Sansai, Chiang Mai 50290, Thailand

**Email address:**
niwooti@hotmail.co.th (N. Whangchai)

**Abstract:** The effect of solar-induced temperature on the growth performance of African sharp tooth catfish (*Clarias gariepinus*) was studied based on a completely randomized design (CRD). Fishes with an average initial weight of 4.07±0.58 g were cultured for 90 days in 3 treatments with 3 replications, outdoor plastic lining ponds (treatment 1), outdoor cement ponds (treatment 2) and indoor cement ponds (treatment 3). The study investigation revealed that water temperature was significantly different among treatments ($p<0.05$) and the highest value was observed in treatment 3 (30.91±1.60 °C), followed by treatment 1 (29.19±1.54 °C) and treatment 2 (27.58±1.58 °C), respectively. Results of the experiment further showed that the differences in temperatures affected the growth and survival rate of the fishes. After 90 days of culture, fishes in treatment 1 had significantly higher weight (298.75±4.32 g/fish), growth rate (3.32±0.05 g/day) and survival rate (95.0±2.0) than treatment 2 (198.40±5.25 g/fish, 2.20±0.06 g/day and 89.0±2.0) and treatment 3 (198.40±5.25 g/fish, 2.20±0.06 g/day and 87.6±2.1) ($p<0.05$). The results indicate that the application of plastic greenhouse to increase the temperature is an alternative that could be applied for aquaculture, especially during winter when temperature is unsuitably lower.

**Keywords:** Catfish, Plastic Greenhouse, Temperature, Solar Energy

## 1. Introduction

Careers aquaculture has gained attention in Thailand. Freshwater fish have continued to expand in 2010, Thailand has an area of 1,007,709 acres of freshwater fish to produce 496,599 tones 532,487 farms with a total farm value of approximately 23,544,932 million Thai-baht, which is tilapia and catfish around 16,392 million [1]. At present, the catfish have much successful business due to the development of improved varieties. Because, these fishes that have good growth and resistance to disease. So each year the yield is getting higher.

However, in the Northern part of Thailand, an area where the temperature is relatively low. In winter, the air temperature drops below 15°C and the temperature difference between day and night about 15-20 °C. Farmers must manage multiple methods, such as the cost of food.

Farmers have to manage the environment in the pond, water quantity and quality. The effects of season, temperature will affect the animals directly. Low water temperatures will affect the culture and productivity of aquatic. Maxwell [2] studied the water temperature in the pond. By using greenhouses could help to raise the temperature of water in the pond. And can raise the temperature of water in ponds up to 3.1-5.7 °C raising fish in the winter, with growth rates well above the pond. A study by Tiwari et al. [9] reported that catfish in ponds that control the water temperature in the range 30-32 °C using Solar-Heat compared which an uncontrolled water temperature. The water temperature is in the range of 25-29 °C was cultured in the winter and rainy seasons. Found the fish in controlled water temperature pond have a growth rate better than uncontrolled water temperature pond clearly. The growth higher rate of about 1.6-1.7 times, however have a high cost.

The purpose of this research to get knowledge and detailed

information of using solar energy as a source of low-cost energy to keep the water temperature increases using plastic greenhouses to raise the temperature. This is another way to increase productivity of fish and potential to develop and usage of solar power to benefit and enhance the growth rate of fish in the winter.

# 2. Materials and Methods

## 2.1. Preparation of Ponds

The experiment was divided into three units including outdoor plastic lining ponds (treatment 1) measuring 25 sq., outdoor cement ponds (treatment 2) measuring 25 sq. and indoor cement ponds (treatment 3) measuring 25 sq. Added water to depth 1.5 m, then spread a poly cage net, nets 2 cm. in size 1x1x1.5 meters (length x width X height) which was conducted at the School of Renewable Energy, Maejo University.

## 2.2. Experimental Design

The experimental design was randomized (Complete Randomized Design; CRD).Catfish of farms in Chiang Mai average weight 4.07 ± 0.58 g/fish. Take catfish to adapt in cages measuring 15 sq. for 24 hr. Counting and weighing catfish before start experiment 100 fish/cage in the pond (three treatments and three replicates). The amount of food eaten was recorded during the 90-day trial period (Figure 1).

*Figure 1. Experiment ponds, plastic lining ponds (A), outdoor cement ponds (B), indoor cement ponds (C).*

## 2.3. Food and Feeding Food

Used in the experiment was floating food. Protein no has less than 30 percent to 2 times daily at 09.00-10.00 am; and 04:00- 05:00 pm; to satiation throughout the trial period of 90 days. The water discharge 10 percent of the water every 10 days.

## 2.4. Temperature

Temperature used Thermometer Model TA318 depth of 30 cm of water in each treatment. The storage temperature of the pond outside distance was away from each pond 50 cm. Storage time 2 times a day at 09.00 am and 05.00 pm. Data was recorded every day until 90 days reached.

## 2.5. Data Collection

Data collection was done every 15 days for each dietary treatment. Fish length and weight were determined to monitor fish growth. Daily feed consumption, growth performance, feed conversion ratio, survival rate and yield productivity were also determined.

## 2.6. Statistical Analysis

Data are analysis of variance ANOVA for CRD and compare the difference of averages by Duncan, s New Multiple Range test at a confidence level of 95 % confidence level. All statistical analyses were performed using SPSS version 11.5.

# 3. Result and Discussion

## 3.1. Temperature Difference

Changes of water temperature inside ponds and external ponds during 09.00 am and 05.00 pm., outdoor plastic lining ponds (treatment 1), outdoor cement ponds (treatment 2) and indoor cement ponds (treatment 3) for a period of 90 days, However, the highest recorded mean the water temperature at 09.00 am was difference is significant different ($p < 0.05$). The water temperature at indoor cement ponds (treatment 3) 29.89 ± 0.97 °C, outdoor plastic lining ponds (treatment 1) 27.00 ± 1.53 °C and outdoor cement ponds (treatment 2) 26.46 ± 1.24 °C, respectively. However the water temperature in experiment at 05.00 pm was difference is significant different (p<0.05). Show that, indoor cement ponds (treatment 3) 31.93 ± 1.40 °C, outdoor plastic lining ponds (treatment 1) 29.20 ± 1.46 °C and outdoor cement ponds (treatment 2) 28.93 ± 0.98 °C , respectively (Table 1, Figure 2 and 3).

Effect of temperature in outside ponds at 09.00 am was difference is significant different (p <0.05), the temperature outside ponds the highest indoor cement ponds (treatment 3) average 34.23 ± 3.20 °C and the temperature outside ponds outdoor plastic lining ponds (treatment 1) and outdoor cement ponds (treatment 2) the same temperature average 30.71 ± 3.20 °C. And temperature outside ponds at 05:00 pm. was difference is significant different (p <0.05) the temperature.

external ponds highest indoor cement ponds (treatment 3) average 41.40 ± 4.48 °C and the temperature outside ponds outdoor plastic lining ponds (treatment 1) and outdoor cement ponds (treatment 2). The same temperature average was 35.71 ± 2.88 °C (Table 1, Figure 2 and 3). Young [7] presented that the water temperature in the greenhouse increased 4.67-5.83 °C compared to the water temperature outside the greenhouse. The temperature of water as a result of the covering of the greenhouse which is consistent with literature information [8] that were tested in the greenhouse pond or pond-based plastics that can cause water temperatures to raise 2.8-4.4 °C.

*Table 1. Water and air temperature (X±SD) at 09.00 am. and 05:00 pm.*

| pond | Temperature(°C) | | | |
| --- | --- | --- | --- | --- |
| | In pond 09.00 am. | In pond 05.00 pm. | Out pond 09.00 am. | Out pond 05:00 pm. |
| Outdoor plastic lining ponds | 27.00±1.53[b] | 29.20±1.46[b] | 30.71±3.20[a] | 35.71±2.88[a] |
| Outdoor cement ponds | 26.46±1.24[a] | 28.93±0.98[a] | 30.71±3.20[a] | 35.71±2.88[a] |
| Indoor cement ponds | 29.89±0.97[c] | 31.93±1.40[c] | 34.23±3.20[b] | 41.40±4.48[b] |

Values are mean of three replicates. Treatment means within a row followed by a different letter are significantly different (P≤0.05)

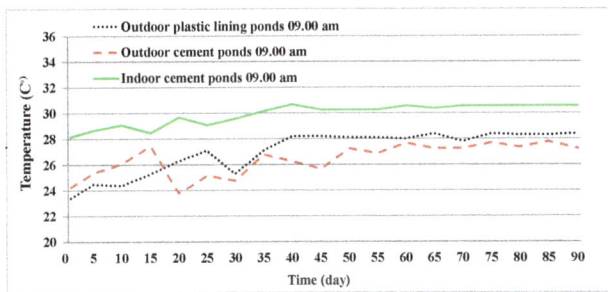

*Figure 2. The temperature of different ponds (Indoor cement ponds, Outdoor cement ponds and plastic lining ponds) at 09.00 am from 0-90 days.*

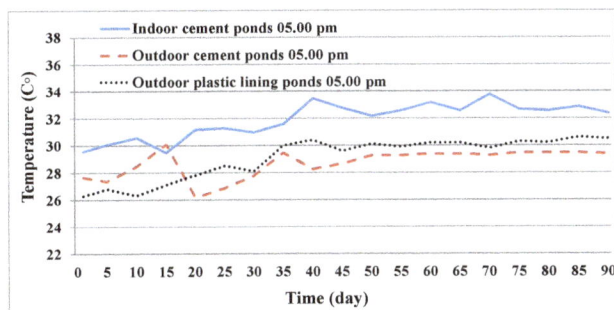

*Figure 3. The temperature of different ponds (cement pond in the green house, cement pond and plastic pond) at 05.00 pm from 0-90 days.*

## 3.2. Growth Performance of African Sharp Tooth Catfish (Clarias Gariepinus)

The average weight of African sharp tooth catfish (*Clarias gariepinus*) for a period of 90 days has increased is significant different (p <0.05). At 15 days has average weight 15.14 ± 2.15 to 8.49 ± 1.53 g/fish indoor cement ponds (treatment 3) was increased the average weight 15.14 ± 2.15 g, outdoor cement ponds (treatment 2) 10.91 ± 1.64 g and outdoor plastic

lining ponds (treatment 1) 8.49 ± 1.53 g, respectively. However, at 90 day has increased is significant different (p <0.05) was in the range of 298.75 ± 4.32 to198.40 ± 5.25 g/fish, indoor cement ponds (treatment 3) the highest average weight 298.75 ± 4.32 g, outdoor plastic lining ponds (treatment 1) 200.79 ± 7.26 and outdoor cement ponds (treatment 2) 198.40 ± 5 g, respectively (Figure 4). The affects high temperature water the process of fish especially eating and digestion. The protease function effectively to the growth of fish [5] found that the increase water temperature has effect on the activity of digestive enzymes accelerate the digestion of nutrients, thus the growth of the fish. Elissa [4] reported that for cobia that are growing rapidly. Yorozu et al. [6] presented the nourish *Macrobrahim roesenbergii* in the winter using cover plastic ponds to maintain temperature that high growth rate of *Macrobrahim roesenbergii* than not cover plastic ponds.

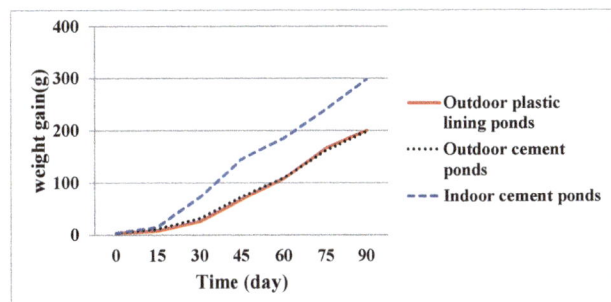

*Figure 4. The weight gain (g) of catfish culture with different pond during 90 day culture period.*

## 3.3. Feed Conversion Ratio

Feed conversion ratio (FCR) in African sharp tooth catfish (*Clarias gariepinus*) are not significantly different (P>0.05).

Outdoor plastic lining ponds (treatment 1), outdoor cement ponds (treatment 2) and indoor cement ponds (treatment 3) was

$1.80 \pm 0.04$ $1.83 \pm 0.06$ $1.78 \pm 0.06$, respectively (Table 2).

*Table 2. Growth parameters (Mean±SD) and feed intake of catfish for plastic lining ponds (T1), Outdoor cement ponds (T2) and Indoor cement ponds on 90 days.*

| pond | Temperature ($^0$C) | SGR (%/day) | Feed intake (%/fish/day) | Survival rate (%) | FCR |
|---|---|---|---|---|---|
| Outdoor plastic lining ponds | 29.19±1.54[b] | 3.77±0.16[a] | 3.52±0.10[a] | 87.00±1.17[a] | 1.80±0.04[a] |
| Outdoor cement ponds | 27.58±1.58[a] | 3.76±0.16[a] | 3.58±0.09[a] | 89.00±1.17[a] | 1.83±0.06[a] |
| Indoor cement ponds | 30.91±1.60[c] | 4.16±0.16[b] | 5.61±0.07[b] | 95.00±1.17[b] | 1.78±0.06[a] |

Values are mean of three replicates. Treatment means within a row followed by a different letter are significantly different (P≤0.05)

### 3.4. Survival Rate

Fish survival rate of African sharp tooth catfish (*Clarias gariepinus*) is significant different (P<0.05) after 90 days the highest is indoor cement ponds (treatment 3) 95.00 ± 1.17 %, outdoor cement ponds (treatment 2) 89.00 ± 1.17 %, and outdoor plastic lining ponds (treatment 1) 87.00 ± 1.17 %, respectively.

### 3.5. Water Temperature

Significant different (p <0.05) in 3 treatment is values ranging from 30.91 ± 1.60 to 27.58 ± 1.58 °C, the average daily temperature indoor cement ponds (treatment 3) is maximum is 30.91 ± 1.60 °C, outdoor plastic lining ponds (treatment 1) average daily temperature was 29.19 ± 1.54 °C and the average daily minimum temperature is outdoor cement ponds (treatment 2) 27.58 ± 1.58 °C, respectively. [3] Studied the effects associated with the estimation of parameters related to greenhouse pond. Productivity of fish ponds in greenhouses showed a better performance compared to the open ponds. [9] Studied the water temperature in the culture ponds by using solar energy can control water temperature in the ponds and the fish are growing well compared to wells without temperature control.

## 4. Conclusion

This study raveled that the effect of solar-induced water temperature on the growth performance of African sharp tooth catfish (*Clarias gariepinus*). The results indicated that the application of plastic greenhouse to increase the temperature is an alternative that could be applied for aquaculture, especially during winter when temperature and indicate growth rate of African sharp tooth catfish (*Clarias gariepinus*); it show that the solar energy can be used to control the temperature of the water in the culture ponds.

## Acknowledgements

This study was funded by School of Renewable Energy and Faculty of Fisheries Technology and Aquatic Resources, Maejo University in generosity place of study.

## References

[1] G. Eason, B. Noble, and I. N. Sneddon, "On certain integrals of Lipschitz-Hankel type involving products of Bessel functions", Philosophical transactions of the Royal Society of London. Ser. A, 1955, 247: 529–551.

[2] J. C. Maxwell, "A Treatise on Electricity and Magnetism", 3rd ed., vol. 2. Oxford: Clarendon, 1892, pp.68–73.

[3] I. S. Jacobs and C. P. Bean, "Fine particles, thin films and exchange anisotropy," in Magnetism, vol. III, G. T. Rado and H. Suhl, Eds. New York: Academic, 1963, pp. 271–350.

[4] K. Elissa, "Title of paper if known," unpublished.

[5] R. Nicole, "Title of paper with only first word capitalized," J. Name Stand. Abbrev., in press.

[6] Y. Yorozu, M. Hirano, K. Oka, and Y. Tagawa, "Electron spectroscopy studies on magneto-optical media and plastic substrate interface," IEEE Transl. J. Magn. Japan, vol. 2, pp. 740–741, August 1987 [Digests 9th Annual Conf. Magnetics Japan, p. 301, 1982].

[7] M. Young, The Technical Writer's Handbook. Mill Valley, CA: University Science, 198.

[8] Department of Fisheries (DOF). 2011. Fisheries statistics of Thailand 2010. Information technology center: Department of Fisheries Ministry of Agriculture and Cooperatives 9/2011. [in Thai]

[9] G.N., Tiwari, B. Sarkar, L. Ghosh, "Observation of Common carp (*Cyprinus carpio*) fry-fingerlings rearing in a greenhouse during winter period", Agricultural Engineering International. 2006. 43:37–48.

[10] L., Ghosh, G.N. Tiwari., T, Das, B. Sarkar, "Modeling the Thermal Performance of Solar Heated Fish Pond: An Experimental Validation", Asian Journal of Scientific Research", 2008, 1: 338–350.

[11] L., Sun, H., Chen, "Effects of water temperature and fish size on growth and bioenergetics of cobia (Rachycentron canadum)", Aquaculture, 2014, 426–427: 172–180.

[12] M. A. Shcherbina, O. P. Kazlauskene, "Water temperature and digestibility of nutrient substances by carp", Hydrobiologia. 1971, 9: 40–44.

[13] N. Whangchai, T. Ungsethaphan., C. Chitmanat, K. Mengumphan, S. Uraiwan, "Performance of Giant Freshwater Prawn (Macrobrachium rosenbergii de Man) Reared in Earthen Ponds Beneath Plastic Film Shelters", Chiang Mai Journal of Sciences, 2007, 34:89–96.

[14] S. Zhu, J. Detour, S. Wang, "Modeling the thermal characteristics of greenhouse pond systems", Aquaculture Engineering, 1998, 18: 201–217.

[15] S. L. Klemetson, G. L. Rogers, "Aquaculture pond temperature modeling", Aquaculture Engineering, 1985, 14:191–208.

[16] W. Khonwat, K.Tanongkiat, "Study the possibility of using solar heating systems extra heat pumps to control temperature in fish ponds", The 2nd Agricultural Engineering and Environmental Science, Phayao University, Phayao, Thailand. [in Thai], 2010.

# 11

# Dye-Sensitized Solar Cell Using Copper and Nitrogen Co-doped Titania as Photoanode

**Purnima Dashora, Chetna Ameta, Rakshit Ameta, Suresh C. Ameta**[*]

Department of Chemistry, PAHER University, Udaipur, (Raj.) India

**Email address:**

purnimadashora@gmail.com (P. Dashora), ameta_sc@yahoo.com (S. C. Ameta)

**Abstract:** Energy crisis is a burning problem in the present scenario, as natural energy resources will be exhausted very soon, due to their rapid utilization. Solar cells have attracted the attention of researchers, as using these devices sunlight can be converted into electricity, which is freely available to us. DSSCs is one of the important and new type of solar cell, which deliver higher photoelectric conversion efficiency and low production cost by combining wide-band gap semiconductor electrode, dye as sensitizer, a counter electrode and redox electrolyte like iodide and triiodide ions between them. In the present work, a comparison is made for the efficiency of pure $TiO_2$ and Cu/N co-doped $TiO_2$ fabricated DSSCs. Pure $TiO_2$ and Cu/N co-doped $TiO_2$ were prepared through sol-gel process. These electrodes were also characterized by X-ray diffraction (XRD), scanning electron microscopy (SEM), fourier transform infrared (FTIR), transmission electron microscopy (TEM) and diffuse reflectance spectra (DRS) techniques to know about their morphology, band gap, particle size etc. The cell was prepared by coating of $Cu/N–TiO_2$ film on the conductive side of FTO glasses using Rhodamine B dye as sensitizer. Liquid electrolyte $I^-/I_3^-$ redox couple and carbon (graphite) as counter electrode and light intensity 60 $mWcm^{-2}$ were used. The observations revealed that Cu/N doped electrode showed maximum conversion efficiency with an open circuit voltage ($V_{oc}$) = 395.0 mV, short circuit current ($i_{sc}$) = 0.0339 mA, $V_{pp}$ = 66.2 mV and $i_{pp}$ = 0.0209 mA with fill factor = 0.10 and the power conversion efficiency ($\eta$) = 0.0023%, which is higher than that of pure $TiO_2$. The results showed that the doping of $TiO_2$ by copper and nitrogen improved the efficiency of this solar cell 38 times more in compare to pure $TiO_2$.

**Keywords:** Titania, Photoanode, Dye-Sensitized Solar Cell, Co-doped, Photovoltaic Performance

## 1. Introduction

In coming few decades, renewable energy especially solar energy, has attracted much attention because it offers a clean, environment-friendly, abundant, and infinite energy resource to us. A solar cell directly converts solar energy into electrical power using a semiconductor. Among all the organic cells, dye sensitized solar cells (DSSCs) provide more effective method because of their high conversion efficiency, low cost, easy manufacturing process, and non-toxicity. DSSC has five main components — transparent conducting glass, semiconductor materials, dyes as a sensitizer, redox-couple electrolyte and counter electrode. The principle of DSSCs was firstly reported by Gratzel and O' Regan using $TiO_2$ films. The electrical conversion efficiency in simulated solar light is about 7.1-7.9% and 12% in diffuse daylight [1]. Chiba et al obtained 11.1% conversion efficiency using titania electrodes with different haze. It indicates that conversion efficiency of the cell increases with

increase in the haze of $TiO_2$ electrodes [2]. Different approaches have been attempted to achieve higher power conversion efficiency such as co-sensitization, core-shell microspheres, composite nanoparticles [3, 4, 5].

Various doping process have been used to enhance the activity of the semiconducting materials. A strong red shift in the visible light region has been observed by doped semiconductors as compared to undoped semiconductor [6]. Gu et al. reported that there is improvement in open-circuit voltage due to an upward shift of the Fermi level, but the oxygen defects generated retard the negative shift of the Fermi level [7]. Modification of $TiO_2$ electrode by doping with metal and non-metal such as chromium, nickel, nitrogen, fluorine, and iodine doped $TiO_2$ was made by Xie et al. [8], Chou et al. [9] and Niu et al. [10]. A scattering layer and a nano-crystalline $TiO_2$ electrode layer was fabricated using of $Er^{3+}$ and $Yb^{3+}$ co-doped $TiO_2$, which showed better light scattering as well as lower transmittance property. The light scattering layer increased the efficiency of the DSSC to by 15.6% [11]. Sn/F dual doping with $TiO_2$ was proposed by Duan et al. and the power conversion efficiency was about 8.89% which is higher than the undoped $TiO_2$ [12].

In the present work, Cu/N-$TiO_2$ was synthesized by sol-gel method and the electric output of the cell was examined using different parameters such as dye concentration, electrolyte concentration, light intensity, and surface area of semiconductor.

# 2. Experimental

## 2.1. Preparation of Pure and Doped TiO₂

Both pure and co-doped Cu/N–$TiO_2$ were synthesized by using sol-gel method. It is schematically presented in Fig 1.

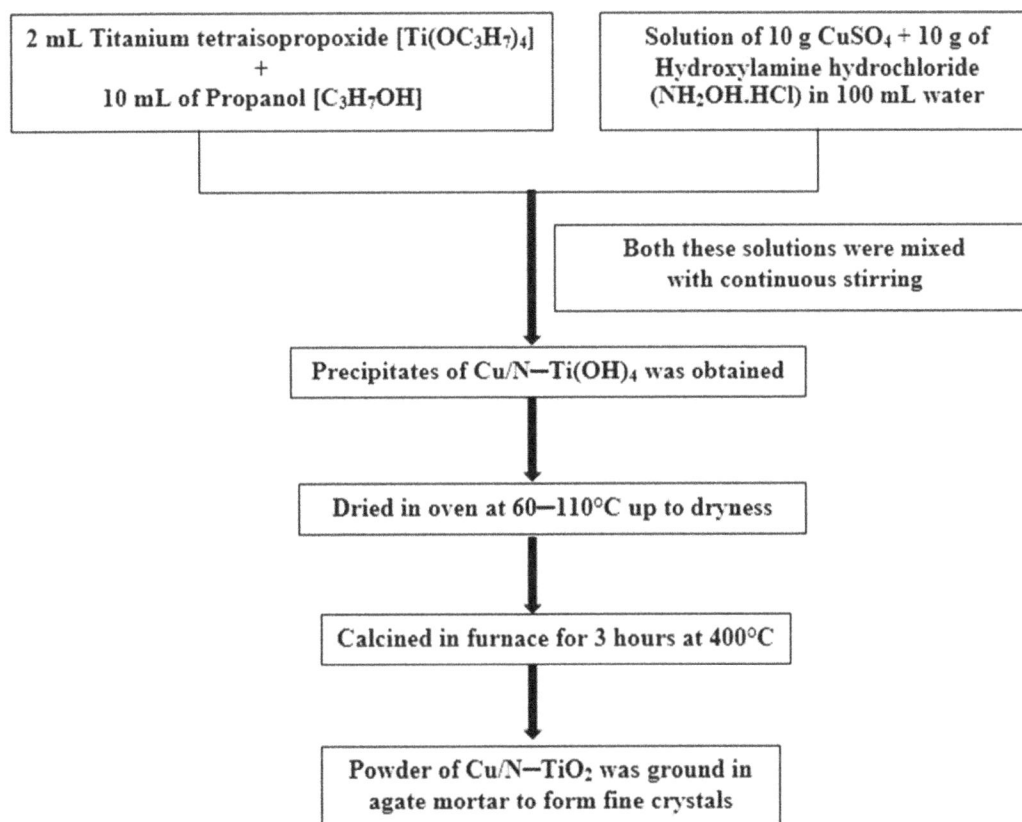

**Fig. 1.** Sol-gel method for the preparation of doped $TiO_2$.

The same procedure was followed to prepare for undoped $TiO_2$ with the only difference that in this case, no dopant was added.

## 2.2. Fabrication of Cell

The working electrode was prepared with Cu/N–$TiO_2$ paste in acetic acid with few drops of dishwashing liquid as a surfactant. This paste was coated on the FTO glass (2.2 mm thickness, 7-9 ohm $cm^{-2}$, L 25 mm x W 25 mm, Shilpa Enterprises, Nagpur, India) by doctor blade method and left for a few minutes to let it dry. Then the glass was heated on a hot plate at 100°C for 45 min. Rhodamine B (1.0 x $10^{-3}$ M) dye solution was prepared in ethanol and used as a sensitizer for the working electrode. The working electrode was dipped in dye solution for 15 min. and then rinsed with ethanol to remove extra dye. A counter electrode, coated with carbon (graphite) was clipped onto the top of the working electrode. 0.47 M iodine and 0.51 M potassium iodide was dissolved in 10 mL of ethylene glycol. This $I^-/I_3^-$ redox couple was used as liquid electrolyte.

# 3. Characterization of Semiconductors

## 3.1. FTIR Spectra of Doped Cu/N–TiO₂

The Fourier transform infrared (FTIR) spectrum of the synthesized samples were recorded in potassium bromide (KBr) pellets on a Perkin Elmer Spectrum RX1 spectrometer in the range from 4000 $cm^{-1}$ to 400 $cm^{-1}$ at a scanning rate of $cm^{-1}$/min.

The FTIR spectrum of Cu/N–$TiO_2$ in given Fig. 2. The broad band at 3365-3318 $cm^{-1}$ was due to O–H stretching. The peak between at 1630-1622 $cm^{-1}$ was attributed to O–H bending vibration of adsorbed water molecule and hydroxyl groups on the surface of $TiO_2$ [13]. The peaks at 494-433 $cm^{-1}$ and 731-702 $cm^{-1}$ are due to bending and stretching mode of Ti–O–Ti [14]. In the low frequency region, the bands around 585-525 $cm^{-1}$ were attributed to the Cu–O and Cu–Ti stretching vibration and 604-625 $cm^{-1}$ to Cu–N–O bending [15, 16]. A typical absorption band occurred around 1092-1016 $cm^{-1}$ belonging to Ti–N stretch vibrations and new peaks in the range of 1153-1115 $cm^{-1}$ and 1118-1198 $cm^{-1}$ appeared due to N–$TiO_2$ [17, 18]. The bands of O–Ti–O and Ti–O appeared at 625-616 $cm^{-1}$ and 660-626 $cm^{-1}$ [19].

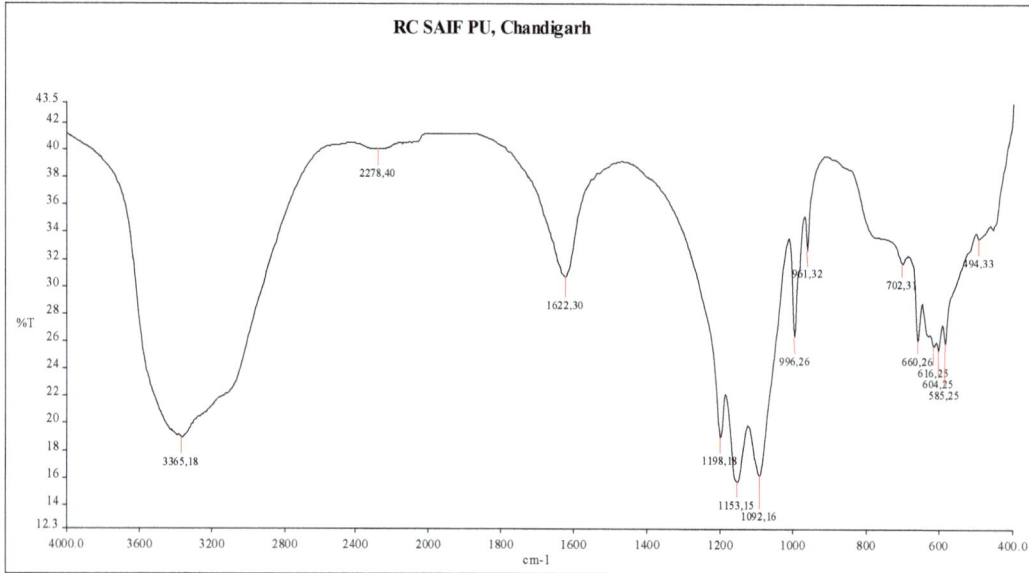

**Fig. 2.** *FTIR spectrum of doped Cu/N–$TiO_2$.*

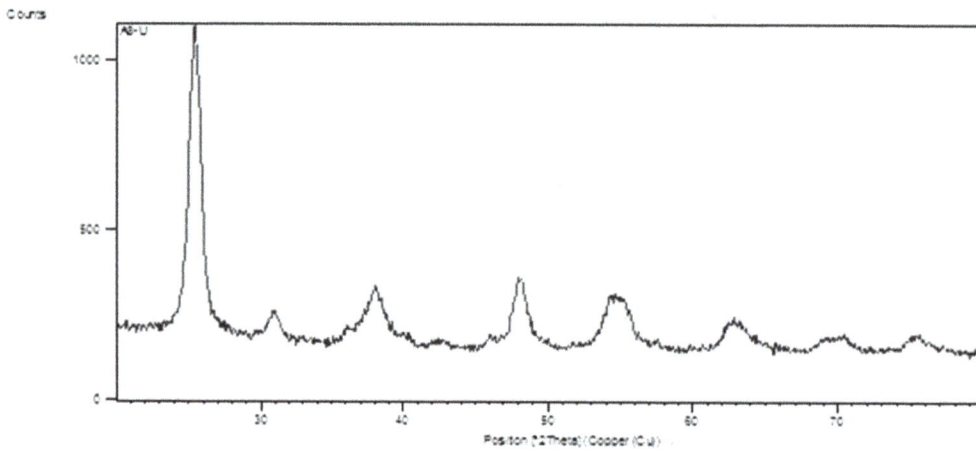

**Fig. 3.** *XRD of undoped $TiO_2$.*

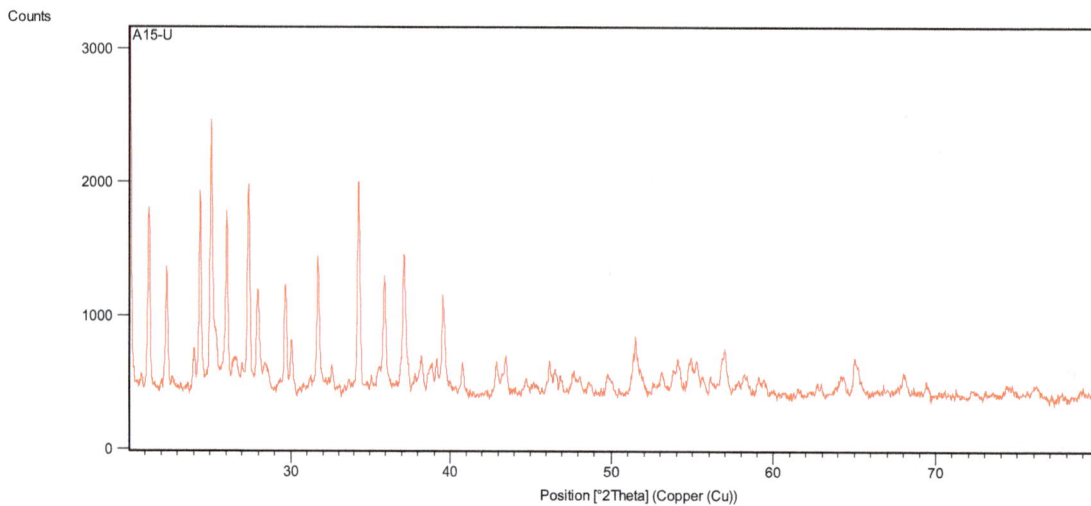

**Fig. 4.** *XRD of Cu/N–$TiO_2$.*

## 3.2. X-ray Diffraction

The X-ray diffraction of undoped $TiO_2$ and $Cu/N–TiO_2$ samples are given in Fig. 3 and 4. The XRD patterns were recorded on Panlytical X' pert Pro model X-ray diffraction using Cu $K\alpha$ radiation as the X-ray source. The diffractograms were recorded in the $2\theta$ range of 20-80°. The average crystalline size (D) of the undoped and $Cu/N–TiO_2$ material can be calculated from the Debye–Scherrer formula:

$$D = \frac{K\,\lambda}{\beta\,Cos\theta} \qquad (1)$$

Where D is the crystalline size (nm), $\lambda$ is the wavelength of X-ray source ($\lambda = 0.1540$ nm for CuK$\alpha$), $\beta$ is the full width at half maximum intensity (FWHM–in radian), and $\theta$ is the Bragg diffraction angle (°). It is also confirmed by the spectra that both; undoped and doped $TiO_2$ are crystalline in nature. The average crystalline size of the pure $TiO_2$ was 10.15 nm and $Cu/N–TiO_2$ was 86.62 nm. It was observed that after doping, the particle size increases.

## 3.3. Scanning Electron Microscopy (SEM)

**Fig. 5.** SEM of Cu/N–TiO₂

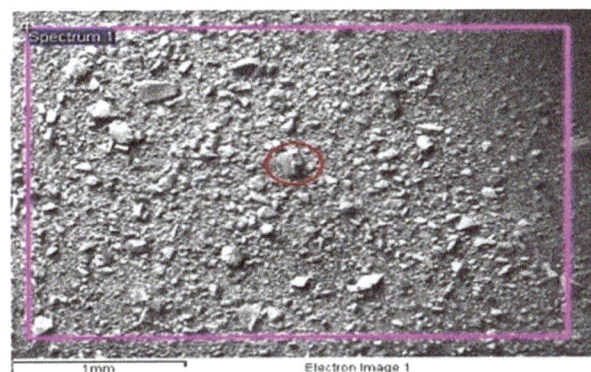

**Fig. 6.** SEM of Undoped TiO₂.

Fig. 5 and 6 show SEM images of Cu–N doped $TiO_2$ and undoped $TiO_2$. SEM has been used to observe the morphological changes caused by loading of metal and non-

metal on the surface of titania. It was observed that combination of the dopants Cu and N on $TiO_2$ increases particle size.

## 3.4. Transmission Electron Microscopy (TEM)

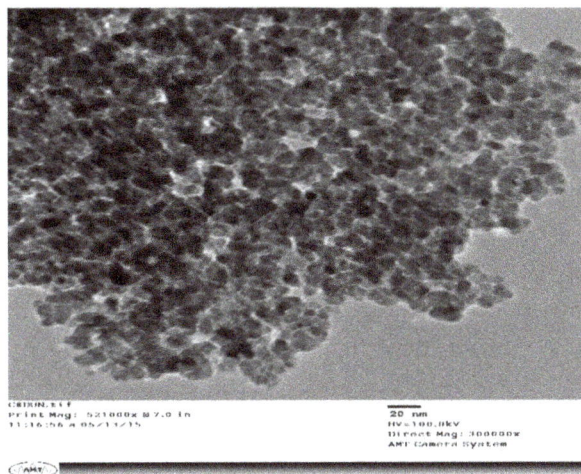

**Fig. 7.** TEM of undoped TiO₂.

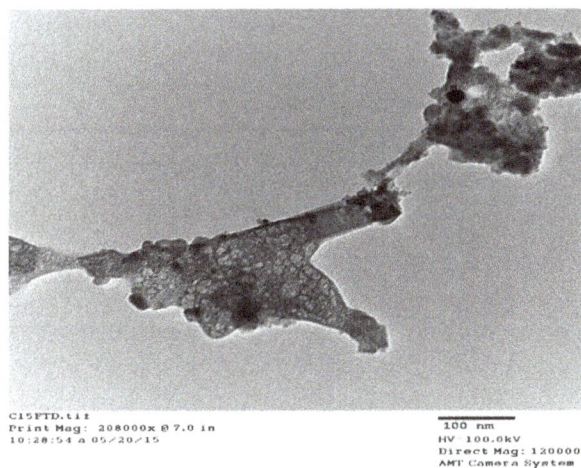

**Fig. 8.** TEM of Cu/N–TiO₂.

Fig. 7 and 8 show TEM images of pure and co-doped $TiO_2$. Doping of $TiO_2$ is clearly observed as evident from some lumps in co-doped $TiO_2$, which is absent in undoped $TiO_2$

## 3.5. Diffuse Reflectance Spectrum (DRS)

The diffuse reflectance spectrum was scanned between 200-800 nm using UV Vis-3000 + spectrophotometer. It shows an intense absorption in the visible region at 739 nm. The band gap of co-doped $TiO_2$ was calculated to be 1.67 eV, which is much less than undoped $TiO_2$ (3.2 eV).

# 4. Performance of Dssc

## 4.1. Variation of Potential with Time

The effect of potential with time on the electrical output of the cell was observed and results are reported in Table 1. The cell was placed in dark and the potential was measured after it becomes stable. A change in potential with the time was observed, when the cell was exposed in light (60 mWcm$^{-2}$).

The potential was measured with digital multimeter (Mastech-M830bZ). The cell was charged for 70 min. and then light was cut off. It was observed that the potential was increased with increasing time.

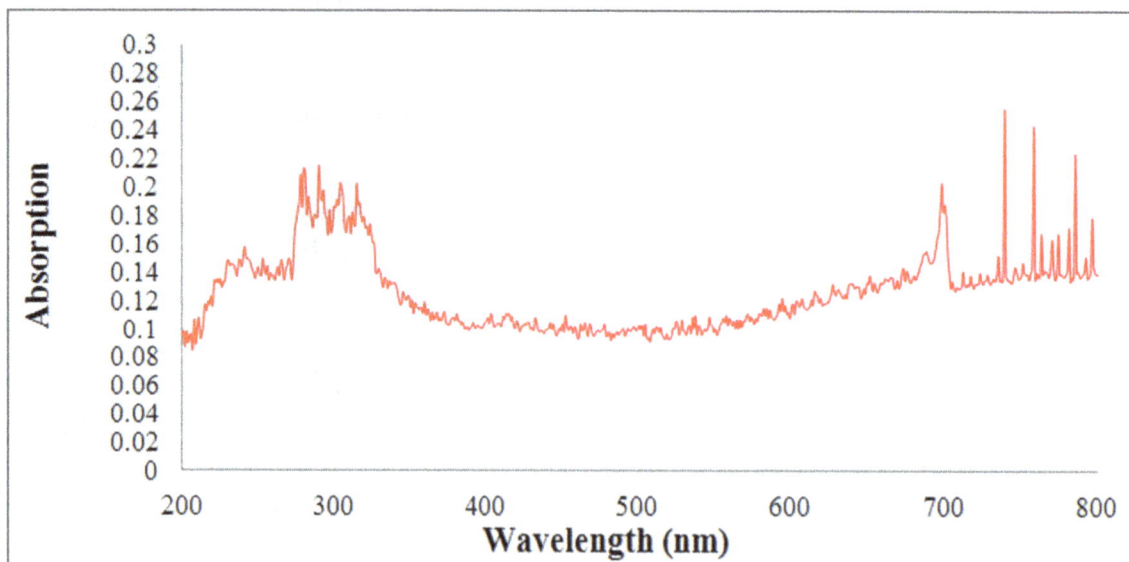

**Fig. 9.** *DRS of Cu/N–TiO₂.*

**Table 1.** *Variation of potential with time.*

| Time (min.) | Potential (mV) | Time (min.) | Potential (mV) |
|---|---|---|---|
| 0.0 | -395.0 | 75 | -18.7 |
| 5 | -253.2 | 80 | -17.8 |
| 10 | -230.7 | 85 | -17.5 |
| 15 | -203.6 | 90 | -17.2 |
| 20 | -187.8 | 95 | -16.9 |
| 25 | -118.5 | 100 | -16.5 |
| 30 | -91.9 | 105 | -15.9 |
| 35 | -66.0 | 110 | -14.8 |
| 40 | -48.9 | 115 | -14.1 |
| 45 | -39.6 | 120 | -13.7 |
| 50 | -34.0 | 125 | -13.7 |
| 55 | -29.1 | 130 | -12.2 |
| 60 | -22.0 | 135 | -12.2 |
| 65 | -16.1 | 140 | -12.2 |
| 70 (Light off) | -16.1 | | |

[Rhodamine B] = 1.0 x 10$^{-3}$M; Electrolyte [I₂] = 0.47 M, [KI] = 0.51 M; Exposed surface area = 1.0 x 1.0 cm$^2$; Light intensity = 60 mWcm$^{-2}$

**Table 2.** *Variation of current with time.*

| Time (min.) | Current (μA) | Time (min.) | Current (μA) |
|---|---|---|---|
| 0.0 | 12.8 | 75 | 14.3 |
| 5 | 20.6 | 80 | 13.7 |
| 10 | 41.2 | 85 | 12.9 |
| 15 | 76.6 (i$_{max.}$) | 90 | 12.3 |
| 20 | 56.6 | 95 | 12.1 |
| 25 | 51.7 | 100 | 10.4 |
| 30 | 46.5 | 105 | 9.2 |
| 35 | 40.1 | 110 | 8.7 |
| 40 | 34.9 | 115 | 8.5 |
| 45 | 29.0 | 120 | 8.3 |
| 50 | 25.8 | 125 | 8.1 |
| 55 | 21.5 | 130 | 8.0 |
| 60 | 17.6 | 135 | 8.0 |
| 65 | 15.9 | 140 | 8.0 |
| 70 (Light off) | 15.0 (i$_{eq.}$) | | |

[Rhodamine B] = 1.0 x 10$^{-3}$ M; Electrolyte [I₂] = 0.47 M, [KI] = 0.51 M; Exposed surface area = 1.0 x 1.0 cm$^2$; Light intensity = 60 mWcm$^{-2}$

### 4.2. Variation of Current with Time

The effect of current with time on the electrical output of the cell was observed and results are summarized in Table 2. The current was also measured by digital multimeter (Mastech-M830bZ). The current of cell increases rapidly and in few minutes, it reaches its maximum, which is represented as i$_{max}$. When irradiation time was increased further, current starts decreasing steadily and reaches almost a stable (constant value). This value is represented as i$_{eq}$, when the source of light was removed, the current starts decreasing further.

### 4.3. Effect of Dye Concentration

The potential and current of the cell was observed using different concentrations (1.6 x 10$^{-3}$ M – 0.3 x 10$^{-3}$ M) of dye. The results are given in Table 3. It was observed that as the concentration of dye was increased, the number of sensitizer molecules also increases but above 1.0 x 10$^{-3}$M, both the current as well as potential were found to decrease.

**Table 3.** *Effect of dye concentration.*

| Dye ($10^3$M) | Potential (mA) | Current ($\mu$A) |
|---|---|---|
| 0.3 | 292.4 | 15.9 |
| 0.6 | 319.8 | 27.8 |
| 1.0 | 395.0 | 33.9 |
| 1.3 | 379.7 | 31.5 |
| 1.6 | 318.2 | 23.6 |

Electrolyte [$I_2$] = 0.47 M, [KI] = 0.51 M; Exposed surface area = 1.0 x1.0 cm$^2$; Light intensity = 60 mWcm$^{-2}$

## 4.4. Effect of Electrolyte Concentration

The effect of the concentration of liquid electrolyte ($I_2$ and KI) on the performance of the DSSC was observed, and the results are presented in Table 4. The concentration of one component was kept constant and other was varied to know the effect of components of redox couple, $I_2$ and KI. As the concentration of iodine was increased; both, the potential and current were increased but above 0.47 M, the current and potential were found to decrease. When the potassium iodide concentration was increased, both the current and potential were increased. The optimum conditions for $I_2$ and KI were obtained as 0.47 M and 0.51 M.

**Table 4.** *Effect of $I_2$ concentration.*

| $I_2$ (M) | Potential (mA) | Current ($\mu$A) |
|---|---|---|
| 0.39 | 220.1 | 11.7 |
| 0.43 | 312.0 | 23.3 |
| 0.47 | 395.0 | 33.9 |
| 0.51 | 219.6 | 28.1 |
| 0.55 | 153.7 | 13.0 |

Exposed surface area = 1.0 x1.0 cm$^2$; Light intensity = 60 mWcm$^{-2}$; [Rhodamine B] = 1.0 x $10^{-3}$M

**Table 5.** *Effect of KI concentration.*

| KI (M) | Potential (mA) | Current ($\mu$A) |
|---|---|---|
| 0.44 | 129.9 | 9.7 |
| 0.46 | 197.5 | 15.9 |
| 0.48 | 216.4 | 22.9 |
| 0.50 | 273.9 | 26.3 |
| 0.51 | 395.0 | 33.9 |

Exposed surface area = 1.0 x1.0 cm$^2$; Light intensity = 60 mWcm$^{-2}$; [Rhodamine B] = 1.0 $\times$ $10^{-3}$M

## 4.5. Effect of Surface Area of Electrode

**Table 6.** *Effect of exposed surface area.*

| Area (cm$^2$) | Potential (mA) | Current ($\mu$A) |
|---|---|---|
| 0.6 x 0.6 | 203.1 | 24.2 |
| 1.0 x 1.0 | 395.0 | 33.9 |
| 1.3 x 1.3 | 257.0 | 19.4 |
| 1.6 x 1.6 | 121.2 | 14.8 |

[Rhodamine B] = 1.0 x $10^{-3}$ M; Light intensity = 60 mWcm$^{-2}$; Electrolyte [$I_2$] = 0.47 M, [KI] = 0.51 M.

The performance of the cell may also be affected by the area of the semiconductor. Its effect was observed by changing the surface area of the electrode. The results are reported in Table 6. The potential and current were increased with increasing surface area, but above 1.0 $\times$ 1.0 cm$^2$ area, both the current as well as potential were decreased.

## 4.6. Effect of Light Intensity

Light intensity may also affect the electrical output of the cell and therefore, the effect of light intensity was also observed. The results are shown in Table 7. Light intensity was varied from 30 to 70 mWcm$^{-2}$. An increasing trend was observed because an increase in light intensity will increase the number of photon per unit area. The current and potential were found to decrease above 60 mWcm$^{-2}$.

**Table 7.** *Effect of light intensity.*

| Light Intensity (mWcm$^{-2}$) | Potential (mA) | Current ($\mu$A) |
|---|---|---|
| 30 | 136.1 | 12.8 |
| 40 | 177.8 | 15.8 |
| 50 | 235.3 | 22.5 |
| 60 | 395.0 | 33.9 |
| 70 | 346.2 | 25.0 |

[Rhodamine B] = 1.0 x $10^{-3}$ M; Electrolyte [$I_2$] = 0.47 M, [KI] = 0.51 M; Exposed surface area = 1.0 x 1.0 cm$^2$

## 4.7. i–V Characteristics of the Cell

The open circuit voltage ($V_{oc}$), keeping the circuit open and short circuit current ($i_{sc}$), keeping the circuit closed, were measured by a digital multimeter. The values of photocurrent and photopotential were observed with the help of a carbon pot (log 470 K) connected in the circuit by applying an external load. The results of variation of potential and current are represented in Table 8. and graphically in Fig. 8

**Table 8.** *i–V characteristics.*

| Potential (mV) | Current ($\mu$A) | Fill factor |
|---|---|---|
| 395.0 | 0.0 | |
| 300.0 | 0.5 | |
| 253.0 | 1.7 | |
| 199.1 | 2.5 | |
| 145.7 | 4.7 | |
| 133.2 | 8.6 | |
| 120.8 | 10.3 | |
| 112.8 | 12.1 | |
| 86.1 | 13.5 | |
| 77.7 | 14.7 | |
| 76.1 | 15.4 | |
| 74.9 | 16.1 | |
| 73.1 | 17.5 | |
| 68.9 | 19.7 | |
| 66.2 | 20.9 | 0.10 |
| 62.0 | 21.7 | |
| 58.6 | 23.4 | |
| 49.1 | 27.1 | |
| 34.7 | 30.8 | |
| 18.2 | 32.4 | |
| 0.0 | 33.9 | |

[Rhodamine B] = 1.0 x $10^{-3}$ M; Electrolyte [$I_2$] = 0.47 M, [KI] = 0.51 M; Exposed surface area = 1.0 x 1.0 cm$^2$; Light intensity = 60 mWcm$^{-2}$

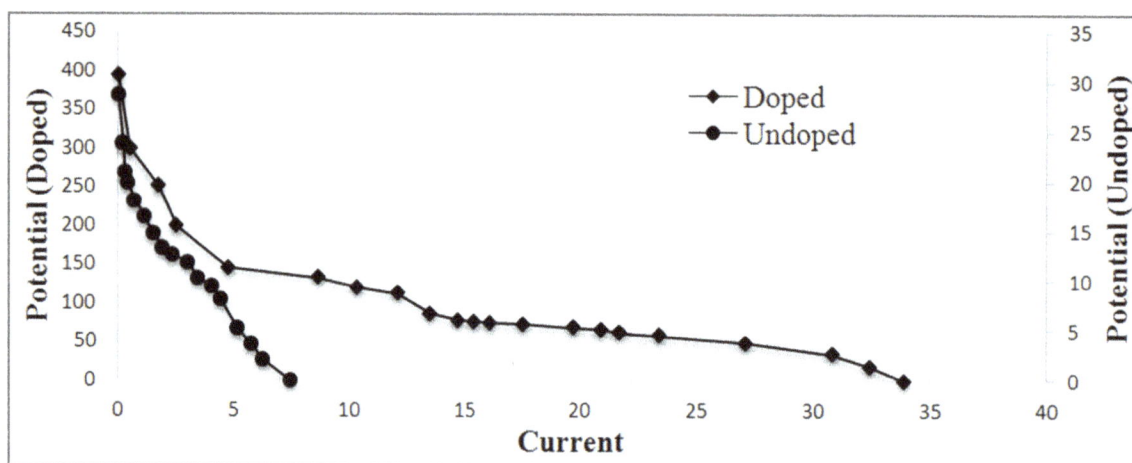

*Fig. 10. Photocurrent-voltage curve of the cell.*

Value of $V_{oc}$, $i_{sc}$, $V_{pp}$ (voltage at power point) and $i_{pp}$ (current at power point) were determined with the help of this plot. The maximum voltage at open circuit ($V_{oc}$) was 395.0 mV and the maximum current at short circuit ($i_{sc}$) was 0.0339 mA. The ($V_{pp}$) = 66.2 mV and ($i_{pp}$) = 0.0209 mA were obtained at power point. Fill factor was calculated by using equation (2) and it was found to be 0.10.

$$\text{Fill Factor} = \frac{V_{pp} \times i_{pp}}{V_{oc} \times i_{sc}} \qquad (2)$$

## 5. Cell Efficiency

The power conversion efficiency ($\eta$) of cell is the ratio of electric output at the power point and power of incident radiation ($P_{in}$). The power conversion efficiency ($\eta$) was determined by following equation:

$$\eta = \frac{FF \times V_{oc} \times i_{sc}}{P_{in}} \qquad (3)$$

This cell exhibited 0.0023% overall power conversion efficiency. The comparative study of electrical parameters, fill factor and conversion efficiency of co-doped and undoped DSSC are given in Table 9.

*Table 9. Comparative results of pure and Cu/N co-doped TiO₂ in DSSC.*

| Sample | $i_{pp}$ (mA) | $V_{pp}$ (mV) | $i_{sc}$ (mA) | $V_{oc}$ (mV) | FF | $\eta \times 10^2$ (%) |
|---|---|---|---|---|---|---|
| Pure TiO$_2$ | 0.0040 | 9.5 | 0.0074 | 28.7 | 0.17 | 0.0060 |
| Cu/N–TiO$_2$ | 0.0209 | 66.2 | 0.0339 | 395.0 | 0.10 | 0.23 |

## 6. Conclusion

It is reported that doping of a semiconductor enhances its activity as it lowers its band gap. In the present investigation, two different DSSCs were fabricated using pure $TiO_2$ and Cu/N–$TiO_2$. These electrodes were prepared in the laboratory by sol-gel process. The FTIR analysis of these semiconductors confirmed O–H, Ti–O–Ti, Cu–O, Cu–Ti, Ti–N stretching and Cu–N–O bending. XRD and SEM analysis showed that the particle size of the semiconductor was increased on doping $TiO_2$ by copper and nitrogen. DRS data also confirmed lowering of band gap of $TiO_2$ when doped with copper and nitrogen. The observations confirmed that the photocatalytic performance of DSSC fabricated with Cu/N–$TiO_2$ was found better than pure $TiO_2$. It is also worth mentioning that the performance of Cu/N–$TiO_2$ cell was enhanced 38 times as compared to pure $TiO_2$.

## Acknowledgement

The authors are thankful to PAHER University, Udaipur for providing necessary laboratory facilities. We are also thankful to Director, SAIF, Punjab University, Chandigarh; University of Kota, Kota; M. S. University of Baroda, Vadodara for providing XRD, TEM, FTIR, DRS and SEM analysis.

## References

[1]  O'Regan, B., Grätzel, M. A low-cost, high-efficiency solar cell based on dye-sensitized colloidal TiO2 films, Nature, Vol. 353, pp. 737-740, 1991.

[2]  Chiba, Y., Islam, A., Watanabe, Y., Komiya, R., Koide, N., Han, L., Dye-sensitized solar cells with conversion efficiency of 11.1%. Japn. J. Appl. Phys. Part 2: Letters, Vol. 45, pp. L638-L640, 2006.

[3]  Elangovan, R., Venkatachalam, P. Co-sensitization Promoted Light Harvesting for Dye-Sensitized Solar Cells, J. Inorg. Organomet. Polymer Mater., Vol. 25, pp. 823-831, 2015.

[4]  Pang, A. Sun, X. Ruan, H. Li, Y. Dai, S. Wei, M. Highly efficient dye-sensitized solar cells composed of TiO2@SnO2 core-shell microspheres, Nano Energy, Vol. 5, pp. 82-90, 2014.

[5]  Eom, T.S., Kim, K.H., Bark, C.W., Choi, H.W., Influence of Fe₂O₃ doping on TiO2 electrode for enhancement photovoltaic efficiency of dye-sensitized solar cells, Mol. Cryst. Liq. Cryst., Vol. 600, pp. 39-46, 2014.

[6]  J. Y. Park, K.H. Lee , B. S. Kim, C. S. Kim , S. E. Lee, K. Okuyama, H. D. Jang, and T. O. Kim, Enhancement of dye-sensitized solar cells using Zr/N-doped TiO2 composites as photoelectrodes, RSC Adv., vol. 4, pp. 9946-9952, 2014.

[7]  F. Gu, W. Huang , S. Wang, X. Cheng ,Y.Hu, and P. S. Lee, Open-circuit voltage improvement in tantalum-doped TiO2 nanocrystals, Phys. Chem. Chem. Phys., vol. 16, no. 47, pp. 25679-2568, 2014.

[8]  Yanan Xie, Niu Huang, Sujian You, Yumin Liu, Bobby Sebo, Liangliang Liang, Xiaoli Fang, Wei Liu, Shishang Guo, Xing-Zhong Zhao, Improved performance of dye-sensitized solar cells by trace amount Cr-doped TiO2 photoelectrodes, J. Power Sources, vol. 224, pp. 168-173, 2013.

[9]  Chuen-Shii Chou, Yan-Hao Huang, Ping Wuc, Yi-Ting Kuo, Chemical-photo-electricity diagrams by Ohm's law – A case study of Ni-doped TiO2 solutions in dye-sensitized solar cells, Appl. Energy, vol. 118, pp. 12–21, 2014.

[10]  Niu, M. Cui, R. Wu, H. Cheng, D. Cao, D. Enhancement mechanism of the conversion efficiency of dye-sensitized solar cells based on nitrogen, fluorine, and iodine doped TiO2 photoanodes, J. Phys. Chem. C, Vol. 119, pp. 13425-13432, 2015.

[11]  C-H. Han, H-S. Lee,K-W. Lee, S-D. Han, and I. Singh, Synthesis of Amorphous $Er^{3+}$-$Yb^{3+}$ Co-doped TiO2 and Its Application as a Scattering Layer for Dye-sensitized Solar Cells, Bull. Korean Chem. Soc., Vol. 30, pp. 219-223, 2009.

[12]  Duan, Y., Zheng, J., Xu, M., Song, X., Fu, N., Fang, Y., Zhou, X., Lin, Y., Pan, F., Metal and F dual-doping to synchronously improve electron transport rate and lifetime for $TiO_2$ photoanode to enhance dye-sensitized solar cells performances, J. Mater. Chem. A, vol. 3, pp. 5692-5700, 2015.

[13]  R. Ocwelwang and L. Tichagwa, Synthesis and Characterization of Ag and Nitrogen Doped $TiO_2$ Nanoparticles Supported on A Chitosan-PVAE Nanofibre Support, Int. J. Adv. Res. Chem. Sci., vol. 1, pp. 28-37, 2014.

[14]  R. Sharmila Devi, Dr. R. Venckatesh, and Dr. Rajeshwari Sivaraj, Synthesis of Titanium Dioxide Nanoparticles by Sol-Gel Technique, Int. J. Innov. Res. Sci. Eng. Technol., vol. 3, pp. 15206-15211, (2014).

[15]  B. V. Rao, A. D. P. Rao and V. Raghavendra Reddy, Influence of $Mo^{6+}$ On Ftir And Mössbauer Spectroscopic Properties of Copper Ferrite, Int. J. Innov. Res. Sci. Eng. Technol, vol. 2, pp. 7768-7779, 2013.

[16]  H. Tachikawa, T. Iyama and T. Hamabayashi, Metal–ligand interactions of the Cu–NO complex at the ground and low-lying excited states: an ab initioMO study, Electron. J. Theor. CH., vol. 2, pp. 263–267, 1997.

[17]  A. Rahmati, Nitrification of Reactively Magnetron Sputter Deposited Ti-Cu Nano-Composite Thin Films, Soft Nanoscience Letters, vol. 3, pp. 14-21, 2013.

[18]  H. Diker, C. Varlikli, K. Mizrak, and A. Dana, Characterizations and photocatalytic activity comparisons of N-doped nc-$TiO_2$ depending on synthetic conditions and structural differences of amine sources, Energy, vol. 36, pp. 1243-1254, 2011.

[19]  K. Lv, H. Zuo, J. Sun, K. Deng, S. Liu, X. Li, and D. Wang, (Bi, C and N) codoped $TiO_2$ nanoparticles, J. Hazard. Mater., vol. 161, pp. 396–401, 2009.

# 12

# Experimental analysis on thermal efficiency of evacuated tube solar collector by using nanofluids

**Hashim A. Hussain[1], Qusay Jawad[2, *], Khalid F. Sultan[3]**

[1]Electromechanical. Eng. Dept, University of Technology, Baghdad, Iraq
[2]Elect. Eng. Dept, University of Technology, Baghdad, Iraq
[3]Electromechanical. Eng. Dept, University of Technology, Baghdad, Iraq

**Email address:**

doctorhashim2004@yahoo.com (H. A. Hussain), qqaajj92@gmail.com(Q. Jawad), ksultan61@yahoo.com (K. F. Sultan)

**Abstract:** This research is to study performance of a evacuated tube solar collector when silver (Ag(30nm)) + distilled water and oxide titanium ($ZrO_2$(50nm)) + distilled water nanofluids was taken as the working fluid. With higher thermal conductivity of the working fluid the solar collector performance could be enhanced compared with that of distilled water. The two types of nanoparticles are used to investigate at different concentration (i.e. 0, 1, 3 and 5 % vol), mass flow rate (30,60 and 90 lit/hr m$^2$) and the based working fluid was distilled water. The effect of different nanoparticle concentrations of Ag and $ZrO_2$ mixed with distilled water as base fluid was examined on solar collector efficiency for different mass flow rates (30, and 90 lit/hr m$^2$). The area under the curve as an index was used for comparing the effects of mass flow rates and nanoparticle concentrations on the collector total efficiency. The experimental results indicated that the concentration at 1%vol showed insignificant results compared with distilled water. As well as The nanofluids (Ag + DW), at concentrations (1, and 5%vol) and mass flow rates (30, and 90 lit/hr m$^2$), the thermal solar characteristics values of $F_R(\tau\alpha)$, $- F_R U_L$ were 0.488, 1.168 W/m$^2$.k, 0.593 and 1.252 W/m$^2$.k, while the nanofluid ($ZrO_2$ + DW) 0.437,1.025 W/m$^2$.k ,0.480 and 1.140 W/m$^2$.k respectively. Whereas in the case of distilled water at mass flow rates 30 lit/hr m$^2$ and 90 lit/hr m$^2$ were 0.413,0.973 W/m$^2$.k,0.442 and, 1.011 W/m$^2$.k respectively. Moreover use of nanofluids (Ag(30nm) + + distilled water) and ($ZrO_2$(50nm) + distilled water) as a working fluid could improve thermal performance of flat plate collector compared with distilled water, especially at high inlet temperature. The solar collector efficiency for nanofluid (Ag(30nm) + distilled water) was greater than nanofluid ($ZrO_2$(50nm) + distilled water) due to small particle size for the silver compared with zirconium oxide as well as high thermal conductivity for silver. The type of nanofluid is a key factor for heat transfer enhancement, and improve performance of evacuated tube solar collector.

**Keywords:** Evacuated Tube Solar Collector, Thermal Performance, Metal and Oxide Metal, Nanofluid

## 1. Introduction

The most important benefit of renewable energy systems is the decrease of environmental pollution. The crisis of the energy cost and its demand increases exponentially with fossil energy nearing exhaustion for present and future time as well as the environmental and air pollution are being more severe, so the strong demand to use or produce a new or renewable, clean and low cost energy is raised to confront this crisis Ali [1]. Renewable energy sources such as sun energy can be substituted for exceeding human energy needs, Taki et al. [2] . Solar energy as one of the most significant forms of renewable energy sources has drawn a lot of attention as there is a belief it can play a very important role

in meeting a major part of our futures' need to energy Hedayatizadeh et al. [3] . However solar energy as an eternal and widespread energy source has low density and is frequently changing as well as the gap between the time of radiation and consumption is the main disadvantage. Hence, collecting and storage of solar energy during radiation time is required for the consuming period. Water is a good material for receiving and storage of solar energy and the solar water heater (SWH) is one of the fastest growing technologies in the renewable energies sector Kumar and Rosen, [4]. Water heating by solar energy is the most important application of direct solar energy use in the world today Wongsuwan and Kumar [5], while Flat Plate Solar Water Heater (FPSWH) is a well – known  technology. The thermal efficiency of the solar water heaters has improved by using some techniques Rezania, Taherian, & Ganji, [6]. Up to now, many studies have been done in order to improve the thermal efficiency of SWHs Koffi et al.[7]; Jaisankar et al.[8]; Jaisankar et al. [9]; Alshamaileh [10]; Kumar and Rosen [11]. The many ways of increasing heat transfer through heat exchangers can be divided into two categories: Passive and active methods. Contrasts to active techniques, passive methods do not need an external force. Using nanofluids as heat transfer medium is a passive method for increasing heat transfer. In spite of many scientific works studying the effect of nanofluids application on thermal efficiency of heat exchangers, there exists very limited information about the study of nanofluids effect on flat-plate solar collectors. Das et al.,[12]; expressed that the nanofluids could be utilized to enhance heat transfer from solar collectors to storage tanks and to increase the energy density. Natarajan and Sathish [13] also believed the novel approach of increasing the efficiency of solar water heater through the introduction of nanofluids instead of conventional heat transfer fluids. Tiwari et al.,[14]; investigated the effect of using $Al_2O_3$ nanofluid as an absorbing medium in a flat-plate solar collector theoretically. They also studied the effect of mass flow rate and particle volume fraction on the efficiency of the collector. Their results showed that using the optimum particle volume fraction 1.5% of $Al_2O_3$ nanofluid increases the thermal efficiency of solar collector in comparison with water as working fluid by 31.64%. Otanicar and Golden [15] reported the experimental results on solar collector based on nanofluids composed of a variety of nano particles (carbon nano tubes, graphite, and silver). The efficiency improvements were up to 5% in solar thermal collectors by utilizing nanofluids as the absorption mechanism. The experimental and numerical results demonstrated an initial rapid increase in efficiency with volume fraction, followed by a leveling off in efficiency as volume fraction continues to increase. Yousefi et al. [16,17] studied the effect of $Al_2O_3$ and MWCNT water nanofluid on the efficiency of a FPSC (flat plate solar collector) experimentally. The results showed that using $Al_2O_3$ and MWCNT water nanofluids in

comparison with water as working fluid increased the efficiency up to 28.3% and 35%, respectively. Taylor et al. [18] investigated on applicability of nanofluids in high flux solar collectors. Experiments on a laboratory-scale nanofluid dish receiver suggest that up to 10% increase in efficiency is possible-relative to a conventional fluid – if operating conditions are chosen carefully for 0.125% volume fraction of graphite. Anyway, up to now just a few studies have been done on nanofluids application in SWH, especially FPSC. Since FPSCS are the most commonly used systems in the renewable energies sector, any attempt for improving the rate of energy harvest seems very effective. Considering the previous studies, nanofluid is a new candidate for this aim.

In the present study, the main purpose of this work is to study the effect of silver (Ag) and zirconium oxide ($ZrO_2$) – distilled water nanofluids, mass flow rate, concentration, and nanoparticle size on solar collector performance more over efficiency of the collector.

## 2. Preparation of Silver and Zirconium Oxide Nanofluids

The studied nanofluid is formed by silver (Ag (30 nm)) and zirconium oxide ($ZrO_2$ (50 nm)) nanoparticles and Two – step method was applied by by dispersing pre – weighed quantities of dry nanoparticles in base fluid. In a typical procedure, the pH of each nanofluids a mixture was measured .The mixtures were then subjected to ultrasonic mixing [100 kHz, 300 W at 25 – 30 $^0$C, Toshiba, England] for two hour to break up any particle aggregates. The acidic pH is much less than the isoelectric point [iep] of these particles, thus ensuring positive surface charges on the particles. The surface enhanced repulsion between the particles, which resulted in uniform dispersions for the duration of the experiments. The prepared nanofluids could stay stable for 4 hours at least. The figure (1) shows nanofluids which containing (Ag (30 nm)) and zirconium oxide ($ZrO_2$ (50 nm)). Nanofluids with different volume fractions ($\Phi$= 1, 3, and 5vol %) are used.

*Fig 1. Show nanofluids for $ZrO_2$ + distilled water ,Ag+ distilled water and DW*

## 3. Experimental Setup with Twenty Riser Tubes

In Fig (2) the schematic view of set up is shown. Three

temperature measurements are required for solar collector testing i.e. ambient air temperature and the nanofluid temperature at the collector inlet and outlet. The surrounding air temperature measured by temperature sensor. The specification of evacuated tube solar collector indicated in details are summarized in Table 1.  Fig (3) reveal evacuated    tube solar collector used in the experiments.

***Table 1.*** *Specification of solar collector with 20 riser.*

| GTC – Solar Specification | | | |
|---|---|---|---|
| **Out tank material** | **White color steel** | **Capacity** | **120 L** |
| Vacuum tube | 47x1500 mm series glass tube | Insulation | High density pressure |
| Frame material | 40 degree gavernzied steel | Series No | 0811JS |
| Inner tank material | SUS 3042b food grade 0.41mm | Manufacture date | NOV.18th.2008 |

It is a glazed (one cover) solar collector that is exposed to south with tilt angle 48° are inlet and outlet of heat transfer fluid to the FPSC, respectively. Two mercury bar thermometers at the inlet and outlet of solar collector measured the temperature of heat transfer nanofluid , respectively with accuracy of 0.1 °C. The bulbs of thermometers were placed inside the tubes completely. Simultaneously, temperatures of the three mentioned points were also measured by PT100 sensors for gaining higher accuracy. Pump [Bosch 2046 – AE], German carried distilled water and nanofluid  through the collector, two control valves after pump and solar storage. Mass flow rate was measured directly by flow meter type  Dwyer series MMA mini – Master flow meter. To fulfill the quasi – steady state conditions, it was tried to have a slow change in inlet fluid temperature, hence a heat exchanger was applied. Solar radiation (I) was measured by a TES 1333 solar power meter (Houston Texas) as shown in Fig. (4) with accuracy typically within ±10 W/m$^2$ and resolution 0.1 W/m$^2$. The Prova AVM – 03 anemometer as shown in Fig. (5) also provided the accurate measurements of wind velocity with ± 3.0% accuracy.

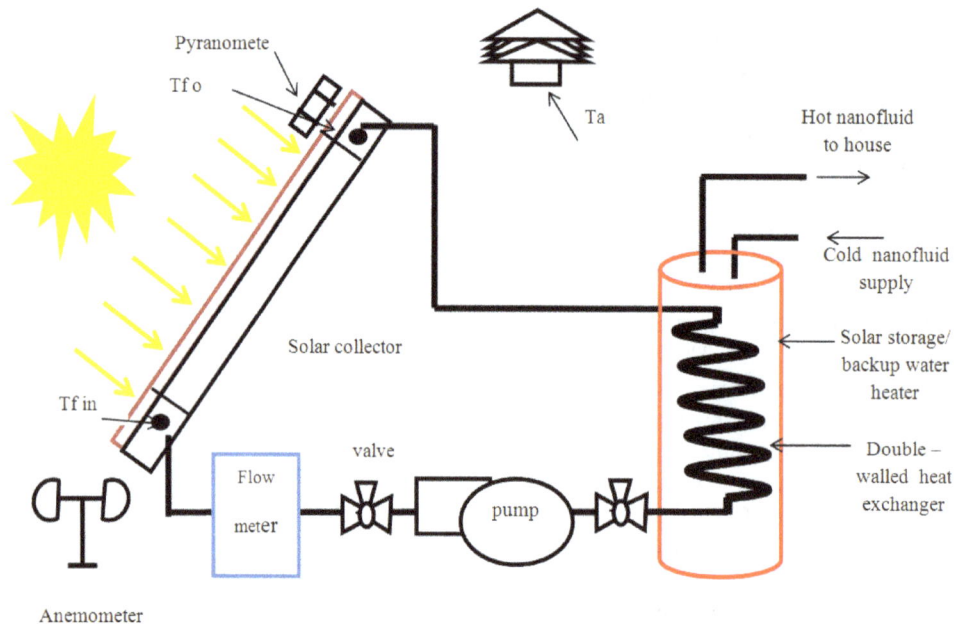

***Fig 2.*** *The experimental set up schematic*

***Fig 3.*** *The experimental evacuated tube solar collector*

***Fig 4.*** *Solar power meter*

*Fig 5. Anemometer*

# 4. Measurement of Nanofluid Thermal Properties

All physical properties of the nanofluids (Ag, $ZrO_2$ + DW) and distilled water needed to calculate the useful heat energy, thermal energy ,collector efficiency and the convective heat transfer are measured. The dynamic viscosity ($\mu$) is measured using brook field digital viscometer model DV – E. The thermal conductivity, specific heat and density are measured by Hot Disk Thermal Constants Analyzer (6.1), specific heat apparatus (ESD – 201) as well as the measurement of density was carried out by weighing a sample and volume. The thermal properties of nanofluids dynamic viscosity ($\mu$) , thermal conductivity, specific heat and density are measured with different volume concentrations at 0,1%, , 3%, and 5 %vol.

# 5. Estimation of Nanofluid Thermo – Physical Properties

The empirical relation used in this study to comparison with the practical measurements for nanofluid properties. The thermo physical properties of nanofluid were calculated at the average bulk temperature of the nanofluid by the following equations.

The volume fraction ($\Phi$) of the nanoparticles is defined by[ 19].

$$\varphi = \frac{v_p}{v_p + v_f} = m \frac{\pi}{6} d_p^{-3} \qquad (1)$$

Density [20].

$$\rho_{nf} = \Phi \rho_{nf} + (1-\Phi)\rho_{Dw} \qquad (2)$$

Viscosity [20].

$$\mu_{nf} = (1-\Phi)\mu_{Dw} + \Phi\mu_{Dw} \qquad (3)$$

Specific heat [20].

$$Cp_{nf}\, \rho_{nf} = \Phi(\rho_s\, Cp_s) + (1-\Phi)(\rho_{Dw} Cp_{Dw}) \qquad (4)$$

Recently Chandrasekar et al.[21] presented an effective thermal conductivity model (Eq.5)

$$\frac{k_{nf}}{k_{Dw}} = \left[\frac{Cp_{nf}}{Cp_{Dw}}\right]^{-0.023} \left[\frac{\rho_{nf}}{\rho_{Dw}}\right]^{1.358} \left[\frac{\mu_{Dw}}{\mu_{nf}}\right]^{0.126} \qquad (5)$$

*Table 2. Thermo – physical properties of the nanofluids employed [22 ].*

| Nano sized particles | $\rho$ (Kg/m³) | Cp (J/kg k) | k (W/m k) | Mean diameter (nm) |
|---|---|---|---|---|
| silver (Ag) | 10500 | 235 | 429 | 30 |
| Zirconium oxide (ZrO₂) | 5890 | 278 | 22.7 | 50 |

# 6. Data Analysis and Validation

The temperature distribution, useful heat energy and the collector efficiency were calculated by using the equations (6) to (13) which were derived from [7], [8], [1] and [9]. To obtain the relationship between the temperature and the mass flow rate (m), the equation for useful heat energy (Qu), as:

$$Q_u = A_c F_R \left[I\alpha\tau - U_L\left(T_f - T_a\right)\right] \qquad (6)$$

The heat energy is converted into thermal energy of water in the pipes, as:

$$Q = \dot{m}\, Cp\left(T_{fo} - T_{fi}\right) \qquad (7)$$

Then

$$\dot{m}\, Cp\left(T_{fo} - T_{fi}\right) = A_c F_R\left[I\alpha\tau - U_L\left(T_f - T_a\right)\right] \qquad (8)$$

Therefore,

$$\left(T_{fo} - T_{fi}\right) = \left(\frac{A_c F_R}{\dot{m}\, Cp}\right)\left[I\alpha\tau - U_L\left(T_f - T_a\right)\right] \qquad (9)$$

$F_R$ may be obtained from

$$F_R = \frac{\dot{m}\, Cp}{A_c U_L}\left[1 - \exp\left(\frac{U_L F' A_c}{\dot{m}\, Cp}\right)\right] \qquad (10)$$

Then the collector efficiency is obtained by using the relation,

$$\eta = \frac{Q_u}{A_c I} \qquad (11)$$

Substitution of Eqs. (7) and (9) in Eq. (10) yields,

$$\eta = F_R \left[ \alpha\tau - \frac{U_L \left( T_f - T_a \right)}{I} \right] \qquad (12)$$

Since $F_R$, $\alpha \tau$ & $U_L$ are constant,

$$\eta \, \alpha \left[ \frac{\left( T_f - T_a \right)}{I} \right] \qquad (13)$$

Therefore , the plots of instantaneous efficiency ($\eta_i$) versus $\frac{\left(T_i - T_a\right)}{I}$ would be straight lines with intercept $F_R(\tau\alpha)$ and slope $- F_R U_L$. In spite of these difficulties, long – time performance estimate of many solar heating systems, collectors can be characterized by the intercept and slope ( i.e., by $F_R(\tau\alpha)$, and $- F_R U_L$) [23]. Using curve fitting tool box of Mat lab, a line was fitted to experimental data of thermal efficiency versus the reduced temperature parameters, $\frac{\left(T_i - T_a\right)}{I}$, for each case. Goodness of fitting was determined by $R^2$. Finally the area under curves as index of collector total efficiency was used for comparing the cases.

## 7. Results and Discussion

At first of all the collector was tested for distilled water as working fluid. The experimental results as shown in Figs.(7 – 12) and Table 3. It can be seen the performance curves of the solar collectors under the ASHRAE Standard with silver (Ag) and zirconium oxide ($ZrO_2$) nanofluids at concentrations (0,1, 3 and 5%vol) and mass flow rates (30,60 and 90 lit/hr m$^2$). It was found that the collector efficiency of the nanofluids (Ag+ DW) and,( $ZrO_2$+DW) at 5% vol were higher than that for distilled water due high thermal conductivity compared with distilled water. Again, the nanofluids (Ag + DW) and,( $ZrO_2$ + DW) at 1%vol still gave similar result with distilled water. The nanofluids (Ag + DW), at concentrations (1, and 5%vol) and mass flow rates (30, and 90 lit/hr m$^2$), the thermal solar characteristics values of $F_R(\tau\alpha)$, $- F_R U_L$ were 0.488,1.168 W/m$^2$.k, 0.593 and 1.252 W/m$^2$.k, while the nanofluid ($ZrO_2$+DW) were 0.437,1.025 W/m$^2$.k ,0.480 and 1.140 W/m$^2$.k respectively. Whereas in the case of distilled water at mass flow rates 30 lit/hr m$^2$ and 90 lit/hr m$^2$ were 0.413,0.973 W/m$^2$.k, 0.442 and ,1.011 W/m$^2$.k respectively. This meant that using of nanofluids (Ag + DW) and,( $ZrO_2$ + DW) as a working fluid were able to increase solar collector performance. Then the evacuated tube solar collector could operate at higher temperature compared with distilled water. Since slopes of models are negative, one can see that increasing (Ti – Ta), causes the efficiency to zero ( in $X_{max}$ ).

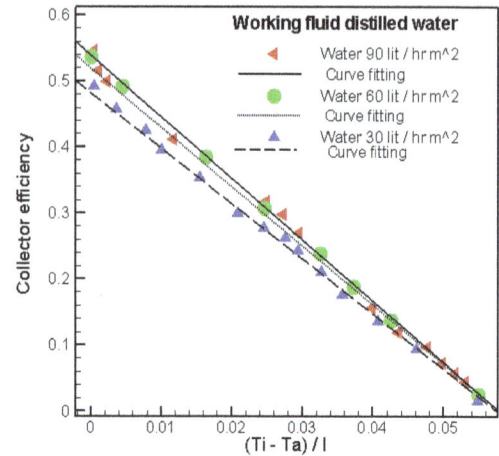

*Fig 6. Collector efficiency for three mass flow rate of water as working fluid in*

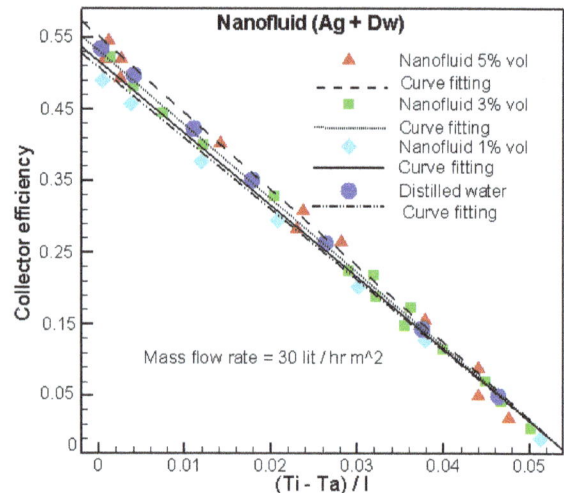

*Fig 7. Collector efficiency at different Φ for nanofluid (Ag + DW)*

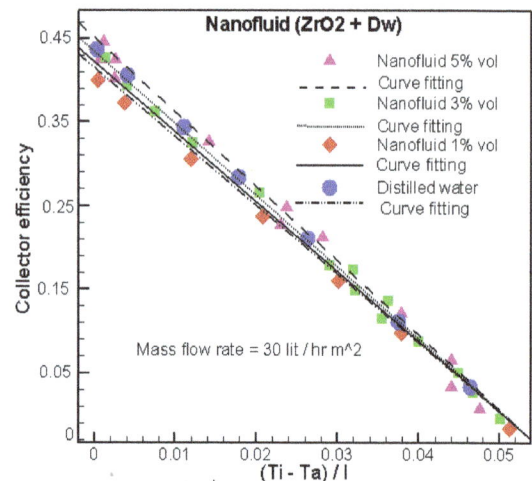

*Fig 8. Collector efficiency at different Φ for nanofluid (ZrO2 + DW)*

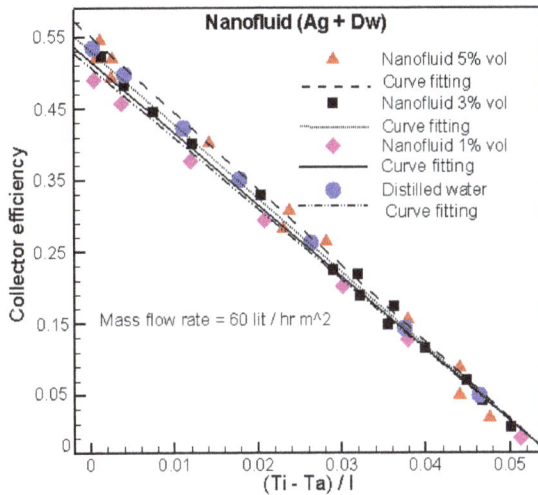

Fig 9. Collector efficiency at different Φ for nanofluid(Ag +DW)

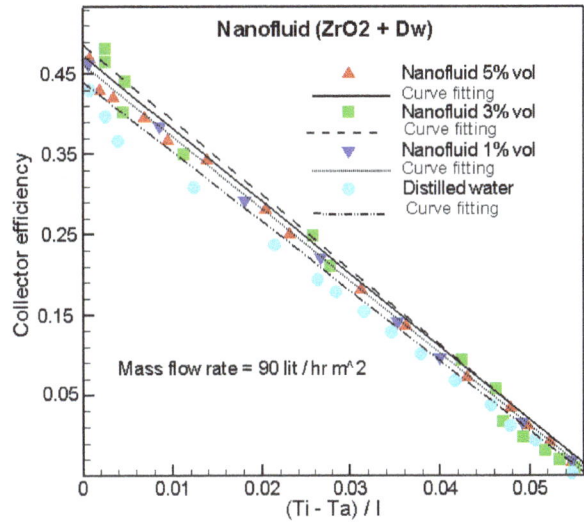

Fig 12. Collector efficiency at different Φ for nanofluid(ZrO2 +DW)

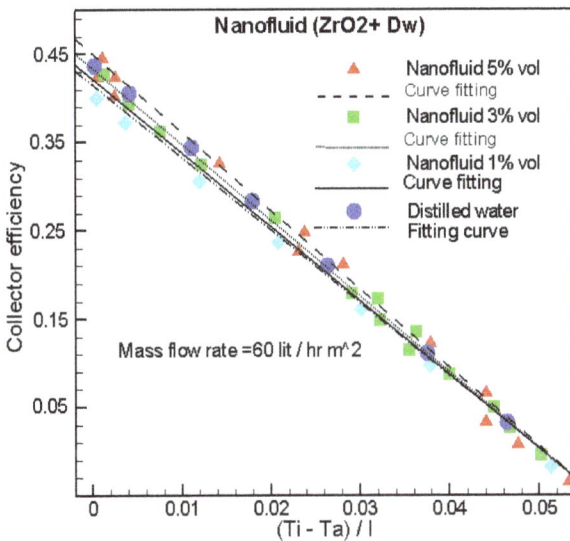

Fig 10. Collector efficiency at different Φ for nanofluid(ZrO2 +DW)

Diffusion and relative movement of nanoparticles near tube wall lead to rapid heat transfer from wall to nanofluid [24]. The slopes became steeper for nanofluids comparing to water which shows the effect of using nanofluids in enhancement of the collector heat removal factor ($F_R$). Another parameter for comparing the collector efficiency is 'A' (Area under the curve×100) that has been brought in Table 2. It represents the entire range of the collector efficiency (from X = 0 to $X_{max}$). Amounts of 'A' for three mass flow rate of distilled water are 1.29, 1. 31 and 1.33, respectively which proves the 1.55 % and 3.1% increase of second and third mass flow rate relevant to first mass flow rate. Also 'A' for three mass flow rates of nanofluid ($ZrO_2$ + DW) at 5%vol in comparison with mass flow rate of distilled water has increased by 6.20, 8.39 and 10.52%, respectively. While nanofluid (Ag + DW) at 5%vol were increased by 17.82, 19.09 and 21.05% respectively. Increasing mass flow rate or using nanofluids instead of base fluid are methods for increasing collector efficiency factor through increasing of heat transfer coefficient inside the tube ( hfi ) [23].

Figs.(13 – 18) indicated the temperature difference between inlet and outlet solar collector and the mass flow rate for the two types of nanofluids (Ag + DW) and ( $ZrO_2$ + DW) with volume fraction (1, 3 and 5%vol) and distilled water, respectively. Since higher the concentration of nanoparticles, higher thermal conductivity of the working fluid was obtained then the fluid could get more heat rate from the solar collector. It could be seen that when the nanofluid concentration was increased, the temperature difference between inlet and outlet would be lower compared with that of distilled water. However, at 1% vol of silver and zirconium oxide nanoparticles, the insignificant results were obtained. It could be noted that for (3%vol , 5%vol) of silver and zirconium nanoparticles, especially for low mas flow rate (30 lit/hr m$^2$) and high inlet temperature, the temperature difference was more deviated from that of distilled water. This meant that the two types of nanofluids could get more heat rate thus the heat loss from the collector was less

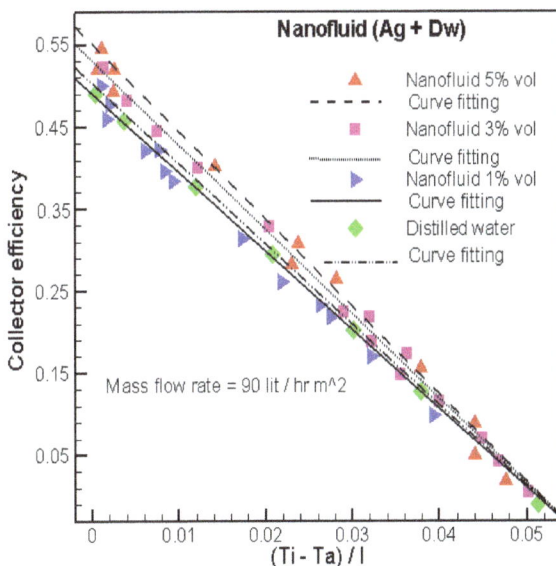

Fig 11. Collector efficiency at different Φ for nanofluid(Ag +DW)

compared with that of distilled water. It was un doubtful that when the mass flow rate and the inlet temperature increased the temperature difference decreased.

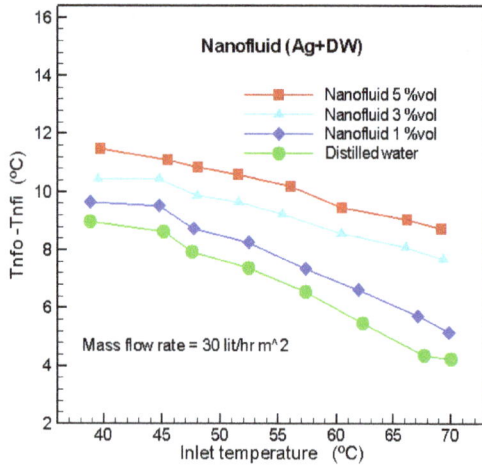

**Fig 13.** *Variation of temperature between inlet and outlet Collector solar for (Ag + DW) at different Φ*

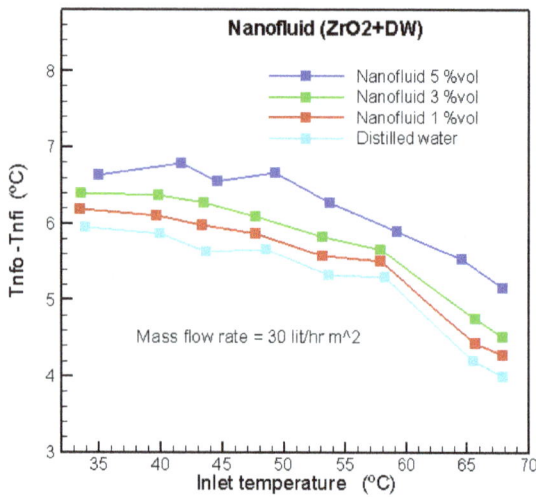

**Fig 14.** *Variation of temperature between inlet and outlet Collector solar for (ZrO2 + DW) at different Φ*

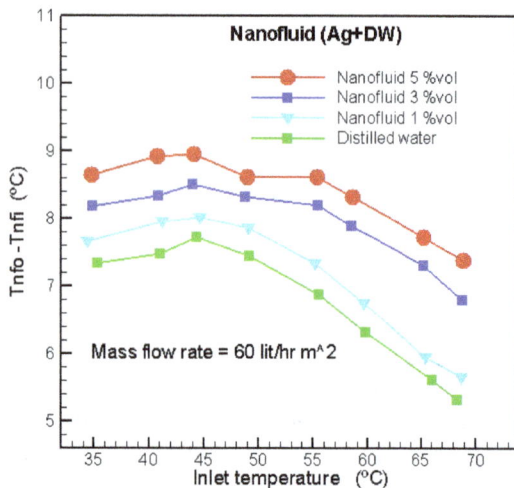

**Fig 15.** *Variation of temperature between inlet and outlet Collector solar for (Ag + DW) at different Φ*

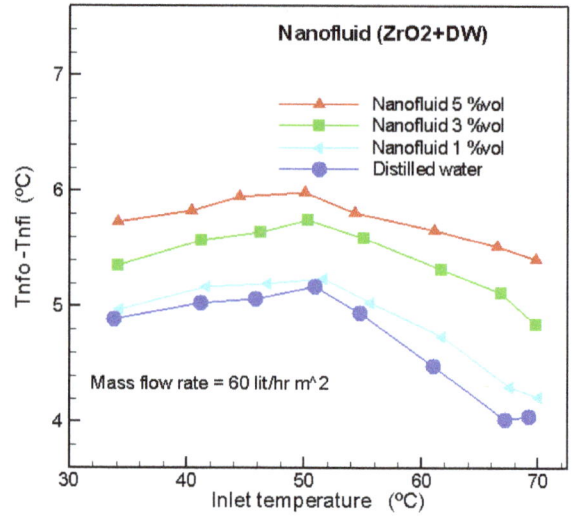

**Fig 16.** *Variation of temperature between inlet and outlet Collector solar for (ZrO2 + DW) at different Φ*

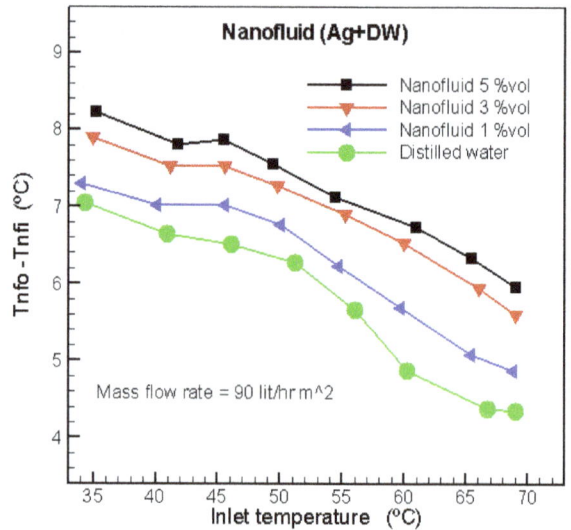

**Fig 17.** *Variation of temperature between inlet and outlet Collector solar for (Ag + DW) at different Φ*

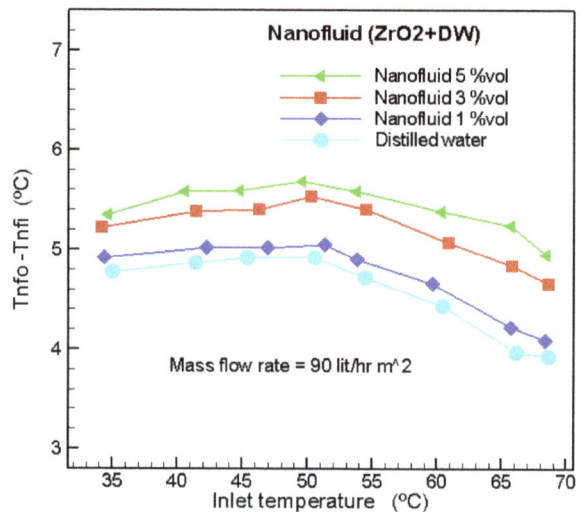

**Fig 18.** *Variation of temperature between inlet and outlet Collector solar for (ZrO2 + DW) at different Φ*

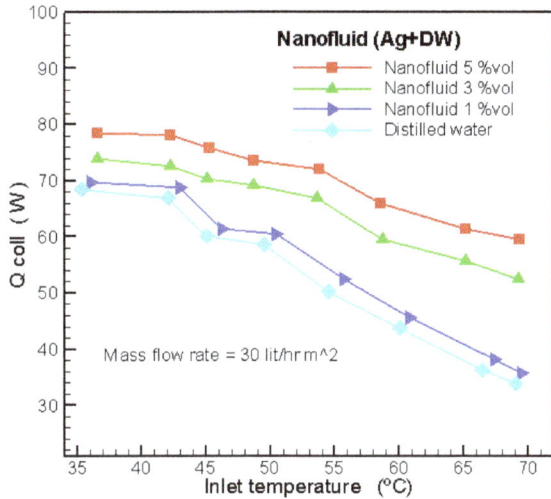

*Fig 19. Variation of useful heat gain from Collector solar At various Φ for nanofluid (Ag + DW)*

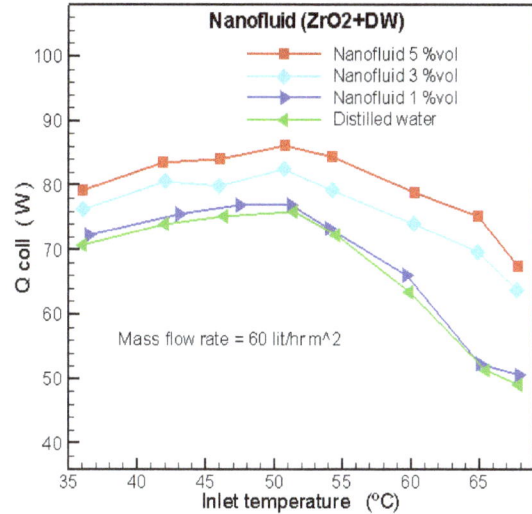

*Fig 22. Variation of useful heat gain from Collector sola At various Φ for nanofluid (ZrO2 + DW)*

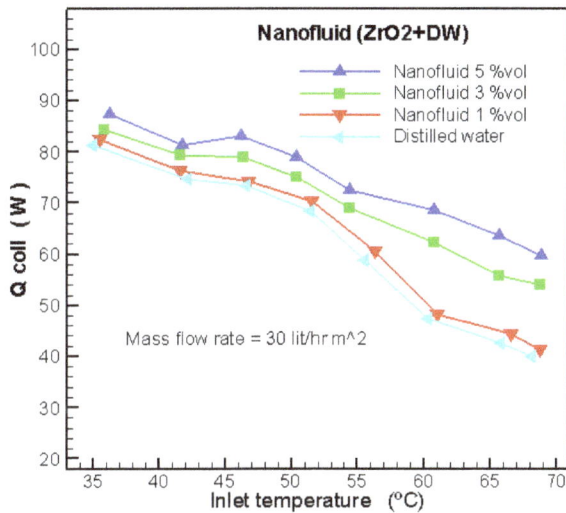

*Fig 20. Variation of useful heat gain from Collector sola At various Φ for nanofluid (ZrO2 + DW)*

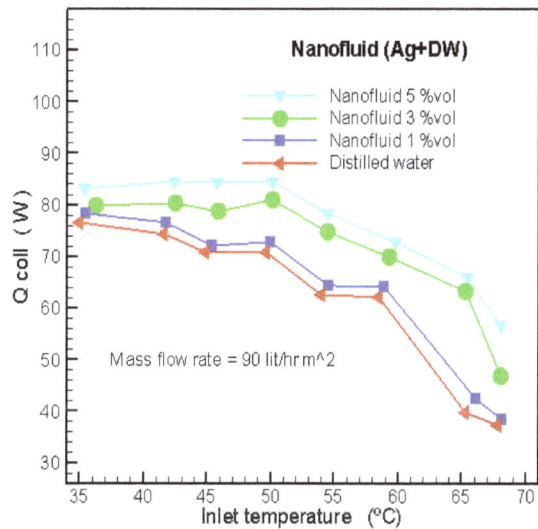

*Fig 23. Variation of useful heat gain from Collector solar At various Φ for nanofluid (Ag + DW)*

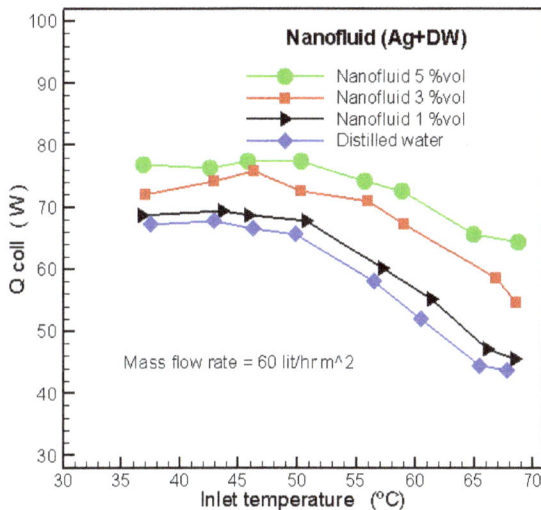

*Fig 21. Variation of useful heat gain from Collector solar At various Φ for nanofluid (Ag + DW)*

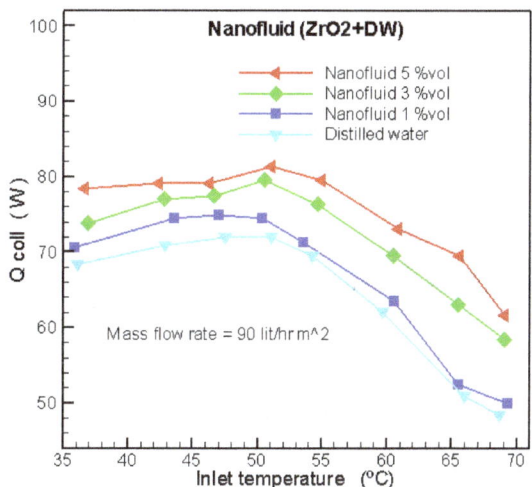

*Fig 24. Variation of useful heat gain from Collector solar At various Φ for nanofluid (ZrO2 + DW)*

*Table 3. The experimental results*

|  | Volume fraction % vol | $\dot{m}$ Lit/hr.m² | Model | Area under curve X100 (A) | $R^2$ |
|---|---|---|---|---|---|
| Distilled Water (DW) | 0 | 30 | $\eta = -0.973\ X + 0.413$ | 1.29 | 0.992 |
|  | 0 | 60 | $\eta = -0.989\ X + 0.435$ | 1.31 | 0.982 |
|  | 0 | 90 | $\eta = -1.011\ X + 0.442$ | 1.33 | 0.996 |
| Nanofluid (ZrO₂+DW) | 1%vol | 30 | $\eta = -1.025\ X + 0.437$ | 1.35 | 0.988 |
|  | 3%vol | 30 | $\eta = -1.044\ X + 0.442$ | 1.36 | 0.977 |
|  | 5%vol | 30 | $\eta = -1.064\ X + 0.449$ | 1.37 | 0.983 |
|  | 1%vol | 60 | $\eta = -1.070\ X + 0.456$ | 1.38 | 0.994 |
|  | 3%vol | 60 | $\eta = -1.078\ X + 0.463$ | 1.41 | 0.985 |
|  | 5%vol | 60 | $\eta = -1.085\ X + 0.469$ | 1.42 | 0.993 |
|  | 1%vol | 90 | $\eta = -1.09\ 0X + 0.472$ | 1.44 | 0.99 |
|  | 3%vol | 90 | $\eta = -1.114\ X + 0.475$ | 1.46 | 0.996 |
|  | 5%vol | 90 | $\eta = -1.140\ X + 0.480$ | 1.47 | 0.992 |
| Nanofluid (Ag+DW) | 1%vol | 30 | $\eta = -1.168\ X + 0.488$ | 1.49 | 0.994 |
|  | 3%vol | 30 | $\eta = -1.175\ X + 0.492$ | 1.5 | 0.991 |
|  | 5%vol | 30 | $\eta = -1.198\ X + 0.512$ | 1.52 | 0.995 |
|  | 1%vol | 60 | $\eta = -1.212\ X + 0.534$ | 1.53 | 0.985 |
|  | 3%vol | 60 | $\eta = -1.222\ X + 0.555$ | 1.55 | 0.992 |
|  | 5%vol | 60 | $\eta = -1.231\ X + 0.563$ | 1.56 | 0.991 |
|  | 1%vol | 90 | $\eta = -1.236\ X + 0.578$ | 1.57 | 0.979 |
|  | 3%vol | 90 | $\eta = -1.239\ X + 0.585$ | 1.59 | 0.983 |
|  | 5%vol | 90 | $\eta = -1.252\ X + 0.593$ | 1.61 | 0.983 |

The useful heat gains from the solar collectors at various inlet temperature, mass flow rate (30, 60 and 90 lit/hr m²) and volume fraction (1, 3 and 5%vol) are shown in Figs (19 – 24 ). The changes were similar to those shown in Figs. (13 – 18). Moreover the nanofluids (Ag + DW, ZrO₂ + DW) at 5%vol showed better performance compared with distilled water while the nanofluid at 1%vol gave similar results with distilled water. The solar collector efficiency for nanofluid (Ag(30nm) was greater than nanofluid (ZrO₂(50nm)) due to small particle size for the silver compared with oxide zirconium as well as high thermal conductivity for silver. The type of nanofluid is a key factor for heat transfer enhancement, and improve performance of evacuated tube solar collector.

# 8. Conclusion

The thermal enhancement of solar collector performance was investigated with nanofluids (Ag(30nm)+DW) and (ZrO₂(50nm)+DW) as working fluid. The two types of nanoparticles are used in investigate with four particles concentration ratios (i.e. 0, 1, 3 and 5 % vol) and the based working fluid was distilled water. The summary results are as follows:

1.  The nanofluids (Ag + DW), at concentrations (1, and 5%vol) and mass flow rates (30, and 90 lit/hr m²), the thermal solar characteristics values of $F_R(\tau\alpha)$, $- F_R U_L$ were 0.488, 1.168 W/m².k , 0.593 and 1.252 W/m².k, while the nanofluid (ZrO₂ + DW) 0.437,1.025 W/m².k ,0.480 and 1.140 W/m².k respectively. Whereas in the case of distilled water at mass flow rates 30 lit/hr m² and 90 lit/hr m² were o.413,0.973 W/m².k,0.442 and ,1.011 W/m².k respectively .

2.  Use of nanofluids (Ag(30nm) and ZrO₂(50nm)) as a working fluid could improve thermal performance of evacuated tube solar collector compared with distilled water, especially at high inlet temperature.

3.  The solar collector efficiency for nanofluid (Ag(30nm) was greater than nanofluid (ZrO₂(50nm)) due to small particle size for the silver compared with oxide zirconium as well as high thermal conductivity for silver.

# References

[1]   Ali, M. H. "Analysis Study of Solar Tower Power Plant & Its Configuration Effects on Its Performance in Iraq (Baghdad City)". Modern Applied Science, 7(4), 55 – 69, 2013.

[2]   Taki, M., Ajabshirchi, Y., Behfar, H., & Taki, M. "Experimental Investigation and Construction of PV Solar Tracker Control System Using Image Processing". Modern Applied Science, 5(6), 237 – 244,2011.

[3]  Hedayatizadeh, M., Ajabshirchi, Y., Sarhaddi, F., Safavinejad, A., Farahat, S., & Chaji, H. "Thermal and Electrical Assessment of an Integrated Solar Photovoltaic Thermal (PV/T) Water Collector Equipped with a Compound Parabolic Concentrator (CPC)". International Journal of Green Energy, 10(5), 494 – 522 ,2013.

[4]  Kumar, R., & Rosen, M. A. "Integrated collector – storage solar water heater with extended storage unit". Applied Thermal Engineering, 31, 348 – 354 , 2011.

[5]  Wongsuwan, W., & Kumar, S. "Forced circulation solar water heater performance prediction by TRNSYS and ANN". International Journal of Sustainable Energy, 24(2), 69 – 86 ,2005.

[6]  Rezania, A., Taherian, H., & Ganji, D. D. "Experimental Investigation of a Natural Circulation Solar Domestic Water Heater Performance Under Standard Consumption Rate". International Journal of Green Energy, 9(4), 322 – 334, 2012.

[7]  Koffi, P. M. E., Andoh, H. Y., Gbaha, P., Toure, S., & Ado, G. "Theoretical and experimental study of solar water heater with internal exchanger using thermo siphon system". Energy Conversion and Management, 49, 2279 – 2290 ,2008.

[8]  Jaisankar, S., Radhakrishnan, T. K., & Sheeba, K. N. " Experimental studies on heat transfer and friction factor characteristics of thermosyphon solar water heater system fittedith spacer at the trailing edge of twisted tapes". Applied Thermal Engineering, 29, 1224 – 1231 ,2009a.

[9]  Jaisankar, S., Radhakrishnan, T. K., & Sheeba, K. N. " Experimental studies on heat transfer and friction factor characteristics of forced circulation solar water heater system fitted with helical twisted tapes". Solar Energy, 83, 1943 – 1952 ,2009b.

[10]  Alshamaileh, E. "Testing of a new solar coating for solar water heating applications. Solar Energy", 84,1637 – 1643, 2010.

[11]  Kumar, R., & Rosen, M. A.. Thermal performance of Integrated collector-storage solar water heater with corrugated absorber surface. *Applied Thermal Engineering, 30*, 1764-1768,2010.

[12]  Das, S. K., Choi, S. U. S., Yu, W., & Pradeep, T. "Nanofluid Science and Technology". John Wiley & Sons, Inc., Publication, 2007.

[13]  Natarajan, E., & Sathish, R. "Role of nanofluids in solar water

heater". The International Journal of Advanced Manufacturing Technology,23,1876 – 8 ,2009.

[14]  Tiwari, A., K. Pradyumna Ghosh, P., Sarkar, J. "Solar Water Heating Using Nanofluids –A Comprehensive Overview And Environmental Impact Analysis". International Journal of Emerging Technology and Advanced Engineering, 3(Special Issue 3), 221 – 224. . 2013.

[15]  Otanicar, T., & Golden, J. "Comparative environmental and economic analysis of conventional and nanofluid solar hot water technologies". Environ. Sci. Technol., 43, 6082e7,2009.

[16]  Yousefi, T., Veysy, F., Shojaeizadeh, E., & Zinadini, S. An experimental investigation on the effect of Al2O3-H2O nanofluid on the efficiency of flat-plate solar collectors. *Renewable Energy, 39*, 293 – 298 ,2012a.

[17]  Yousefi, T., Veysy, F., Shojaeizadeh, E., & Zinadini, S. "An experimental investigation on the effect of MWCNT – H2O nanofluid on the efficiency of flat – plate solar collectors". Experimental Thermal and Fluid Science, *39*, 207 – 212, 2012b.

[18]  Taylor, R. A., Phelan, P. E., Otanicar, T. P., Walker, C. A., Nguyen, M., Trimble, S., & Prasher, R. "Applicability of Nanofluids in High Flux Solar Collectors". Renewable and Sustainable Energy, 3(2), 023104,2011.

[19]  Khanafer K, Vafai K, A Critical syntesis of thermophysical Characteristics of Nanofluids, International Journal of Heat and Mass transfer(Under Press), 2011.

[20]  Kumar, R., & Rosen, M. A.. Thermal performance of Integrated collector – storage  solar water heater with corrugated absorber surface. Applied Thermal Engineering, 30, 1764 – 1768,2010.

[21]  Chandrasekar M., Suresh S., Chandra Bose A., Experimental investigations and theoretical determination of thermal conductivity and viscosity of $Al_2O_3$/ water nanofluids. Exp. Thermal and Fluid Sci.34 , 210, 2010.

[22]  J.P.Holman, Heat transfer,8th ed. 2008.

[23]  Duffie, J. A., & Beckman, W. A. "Solar engineering of thermal processes". Wiley publication. Hedayatizadeh, M., Ajabshirchi, Y., Sarhaddi, F., Safavinejad, A., Farahat, S.,1991.

[24]  Kahani, M., Zeinali Heris, S., & Mousavi, S. M." Comparative study between metal oxide nanopowders on thermal characteristics of nanofluid flow through helical coils". Powder Thechnology, 246, 82 – 92 ,2013.

# 13

# High performance for real portable charger through low-power PV system

**Yousif I. Al-Mashhadany[1], Hussain A. Attia[2]**

[1]Electrical Engineering Dept., College of Engineering, University of Al-Anbar, Al-Anbar, Iraq
[2]Electronics and Communications Eng. Dept., American University of Ras Al Khaimah, Dubai, UAE

**Email address:**

yousif.almashhadany@uoanbar.edu.iq (Y. I. Al-Mashhadany), hattia@aurak.ae (H. A. Attia)

**Abstract:** This paper proposes a novel design for a solar-powered charger for low-power devices. The level of the charging current is controllable and any residue power is saveable to a rechargeable 9V battery. Two power sources (AC and solar) are used, and two charging speeds are possible. Quick charging is 20% of the battery output current (almost 180mA/hr) so the current is limited to 34 mA. Two types of cellular batteries (5.7V and 3.7V) can be charged. Normal charging is 10% of the cellular battery output current (almost 1,000mA/hr), so the charging current is limited to 100mA. The design uses only a few components so the system is cost effective besides being highly portable. It was simulated on MultiSim Ver. 11 before being implemented practically to validate it. The results from the simulation and the experiment show the design's sufficient feasibility for practical implementation.

**Keywords:** PV Energy System, Portable Charger, Current Limiting

## 1. Introduction

The emergence of alternative energy sources and new forms of exploration will rise with the technological evolution and development of societies. In the Eighteenth Century with rudimentary technology there was renewable energy. At the First Industrial Revolution, was the discovery of coal associated with the steam machine? In the Nineteenth Century is the World War I with the discovery of the principles of thermodynamics, development of transport, discovery of oil and natural gas in the Mid-twentieth Century. With World War II, nuclear power appears later computer science, robotics, which together gives rise to the Third Industrial Revolution in the last decades of the twentieth century [1-3].

In the recent past, many countries have strengthened their efforts to increase the fraction of electricity produced from renewable energy sources in order to reduce the greenhouse gas emissions from fossil fuel power plants. While most modern electrical appliances receive their power directly from the utility grid, a growing number of everyday devices require electrical power from batteries in order to achieve greater mobility and convenience. Rechargeable batteries store electricity from the grid for later use and can be conveniently recharged when their energy has been drained. Appliances that use rechargeable batteries include everything from low-power cell phones to high-power industrial fork lifts. The sales volume of such products has increased dramatically in the past decade [4-6].

The system used to draw energy from the grid, store it in a battery, and release it to power a device is called a battery charger system. While designers of battery charger systems often maximize the energy efficiency of their devices to ensure long operation times between charging, they often ignore how much energy is consumed in the process of converting ac electricity from the utility grid into dc electricity stored in the battery. Significant energy savings are possible by reducing the conversion losses associated with charging batteries in battery-powered products [3].

Mobility and mobile computing systems are systems that can easily be physically moved or may be performed while being moved. The small size, limited memory and processing power, low power consumption and limited connectivity are the main characteristics of mobile devices.

The reliance on small devices and the ones which makes easier the daily tasks increases every day. With advances in solar technology, batteries and electronics in general, have been increasingly possible to develop new devices that require less energy to operate. Of course all mobile devices (without network connection) could have, because the power required for solar modules is suitable for battery use. Furthermore, mini charge regulators that reach the maximum power devices without causing overload of the batteries.

Portable devices (mobile phones, tablets, notebooks and netbooks) have become increasingly popular, especially with the proliferation of access to wireless technology. One of its main characteristics is to rely on battery power for its operations. Techniques that allow energy savings, therefore, have been researched to meet the need for immediacy in which the world currently requires [3].

Batteries are nowadays the main energy provider to portable devices. They are used for their high power density and ease of use. Their disadvantages, however, limit their application. Their energy density can drop to as low as 200Wh/kg and their technology seem to improve slower than do other technologies [5-8].

Depleting fossil fuel and increased demand for energy have spurred the search for other sources of energy such as solar, wind, ocean thermal, tidal, biomass, geothermal, nuclear energy, etc. The abundance and widespread availability of solar energy, however, make it the most attractive among other energies that can be feasibly extracted. It can be converted into electricity through low-power PV energy systems, for portable applications (charging of mobile phones) and used in rural areas (solar lamps). The high cost of PV panels and their low efficiency, however, reduce solar energy's competitiveness in the energy market as a major source of power generation. It still, however, is better than conventional energy sources where portability is required [9-12].

This paper considers a novel design for, and the physical implementation of, a solar-charger-based PV energy system for charging of cellular and rechargeable batteries. The charger current can be controlled and any residue power saved in a rechargeable battery (9V). Sources for the design are a solar panel (3W, 18V) and an AC power supply. Two charging speeds are possible (slow and fast). The paper next presents the design of the novel system and its simulation, the experiment results, and the practical implementation [13-16].

## 2. Battery Charging

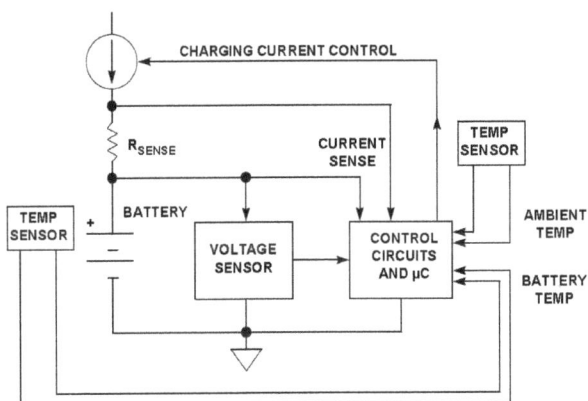

**Fig. 1.** *Generalized Battery Charging Circuit[6]*

A generalized battery charging circuit is shown in Fig. 1. The battery is charged with a constant current until fully charged. The voltage developed across the RSENSE resistor is used to maintain the constant current. The voltage is continuously monitored, and the entire operation is under the control of a microcontroller which may even have an on-chip A/D converter. Temperature sensors are used to monitor battery temperature and sometimes ambient temperature.

This type of circuit represents a high level of sophistication and is primarily used in fast-charging applications, where the charge time is less than 3 hours. Voltage and sometimes temperature monitoring is required to accurately determine the state of the battery and the end-of-charge. Slow charging (charge time greater than 12 hours) requires much less sophistication and can be accomplished using a simple current source.

## 3. Novel Design of the Solar-Powered Portable Charger with Current Limiter

Fig. 2 is a block diagram of the proposed charger. The solar and dc power sources join through two decoupling diodes. The meeting point provides the dc supply voltage to the main part of the design, which has two charging circuits of different specifications. One charging circuit delivers suitable voltage and (limited) charging current to a rechargeable battery, whereas the other is for charging of two types of mobile devices (3.7V and 5.7V).

In general the battery chargers operate in three modes:
1. Active charge mode, during which the battery is being charged from a discharged state. Most battery chargers draw the most power from the outlet during this mode.
2. Maintenance charge mode, during which the battery charge state is being maintained at a fully-charged state. A battery charger typically draws less power in this mode than in active charge mode.
3. No battery mode, during which no battery is connected to the charger at all. Many chargers continue to draw a current in this mode, even though they are doing no useful work [5,6].

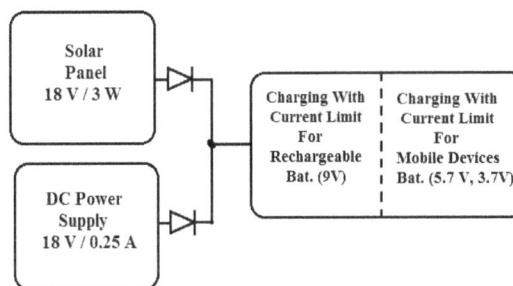

**Fig. 2.** *The proposed portable charger*

- DC Power Supply Circuit

Fig. 3 shows an 18V/250mA dc power source supplying two successive charging circuits. The power supply circuit is a full-wave rectifier with a step-down transformer (T1: 220 / 15V, 250mA).

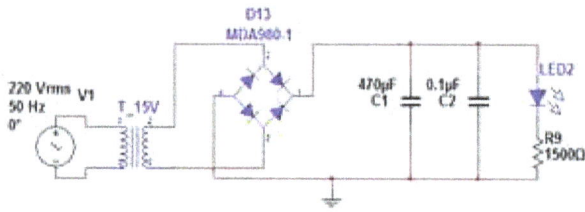

**Fig. 3.** *The 18V / 0.25A, DC Power Supply*

• Charging with the Current Limiter Circuit

The circuit delivers the higher power supply between the two to the next part of the circuit. Its second function is to provide a suitable charging voltage to a 9V rechargeable battery and supply a high level of charging current (20% of the battery output current, i.e., almost 180mA/hr, so the proposed design limits the current to 34mA, for which the shunt resistor controlling the charging limit should be R3=20Ω. The maximum voltage Vbe must be 0.7V. Of the transistor, R3 = Rbe = (0.7V / 34mA) = 20.5Ω.

The second part of the circuit provides charging voltages to 5.7V and 3.7V cellular batteries during suitable selection of the Zener diode connections D2 (ZDP7.5) and D6 (ZDP6.2), also supplies 100mA of charging current when the shunt resistor (R6, 7Ω) is connected (see Fig.4).

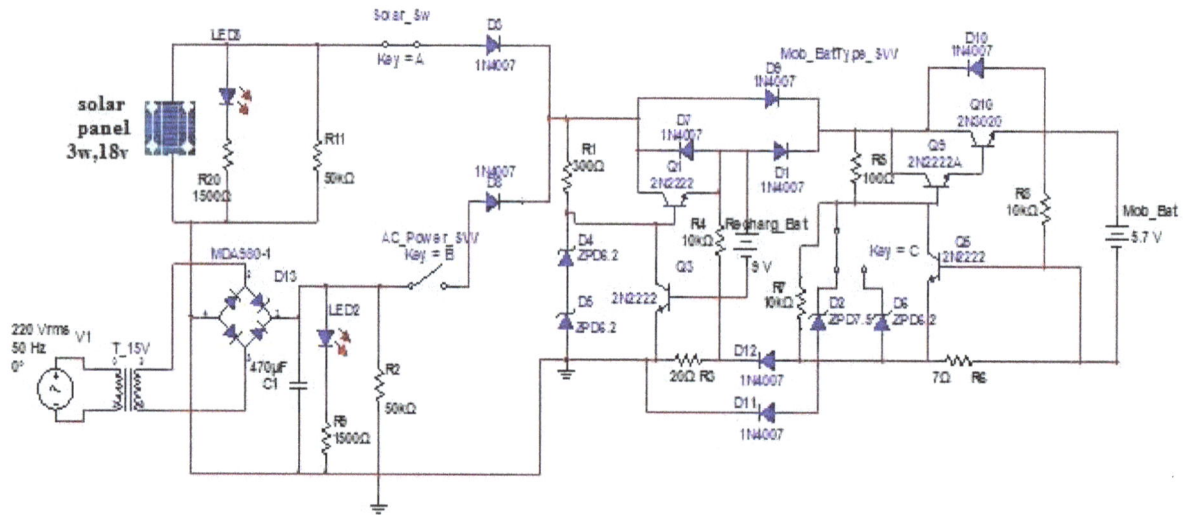

**Fig. 4.** *The Proposed Design for the Solar Charger with Current Limiter*

The current-limiting action is effected from measuring the current that passes through the shunt resistor. If it reaches the value lead to the voltage across the base and emitter equal 0.7V in will effect directly on the load voltage to make continuous current control on the load current (Charging Current), this action was done for 9V rechargeable battery during transistors Q1 and Q3. The same was done for cellular battery, with transistor Q4 and Darlington transistors Q9 and Q10. Fig. 3 is the proposed practical electronic circuit and all the distributed meters for the complete simulated measurements.

# 4. Simulation Results

The secondary coil of the stepdown transformer provided 15Vac, the load current was 124.79mA, the load resistance was 150Ω, and the dc load voltage was 18.7V; all these were measured by the third meter. The dc power supply delivered the required load currents in normal charging of rechargeable battery and cellular device.

• The Complete Charging Circuit with Current Limiter

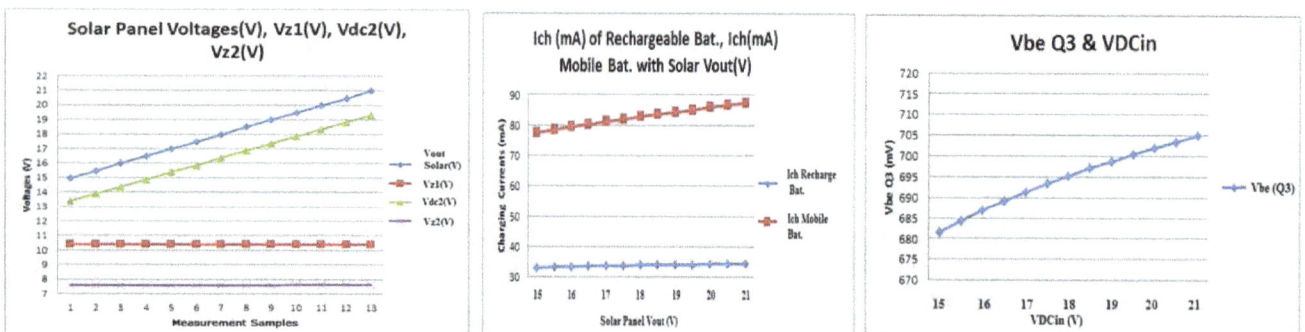

**Fig. 5.** *Simulation Data for the Charging Circuit, the Charging Current, and the Controlling Voltage*

Fig. 5 shows the complete simulations for the proposed charger. One is for rechargeable-battery charging current limited to 34mA (high-speed charging level), the other for cellular-battery charging current limited to 100mA (normal level). Calculations for the charging current levels were based on these: base emitter resistor R3=0.7V for transistor Q3 forward voltage (limiting to 34mA the rechargeable battery charging current). The value for a suitable base resistor will thus depend on the following: *R(be) = Vbe / I(pass through Rbe)*. The maximum value for Vbe was limited to 0.7V. After correct selection of the current to pass through the resistor (for rechargeable battery, we selected the current level to equal the high-speed charging limit of 34 mA), a suitable resistor value would be Rbe = Vbe / Ibe = (0.7V) / (34mA) = 20.5Ω.

Through the same procedure but for different levels of charging current, the resistor selected to limit the maximum charging current was 7Ω. Fig. 5 include all related records as a drawings data came from distributed multimeters, the reads cover different case with suitable range of dc input voltage which came from the meeting point of diodes connection of the switching supply and DC power supply, from data and the related drawing in the Fig. 4 that fixed zeners voltages for range of Vdc input, It explains the charging current level around the current value of 34mA for rechargeable battery, and It explains the controlling of the level of charging current came from the designed value of resistor and the effect of the base emitter voltage, by same principle the recorded data of the charging current (XMM8) in mobile devices battery not pass more than 100mA.

## 5. Implementing the Design

The design is a PV-based (3W, 18V) energy system for mobile applications. It contains a PV array, a circuit design model, an oscilloscope, and a 9V DC battery for charging (see Fig. 5). After full charging, the battery starts converting energy through the 9V DC battery (which is used when the solar source dries up or at night). Control of the battery charging involves maintaining the current level at the high-speed charging limit equaling 34mA.

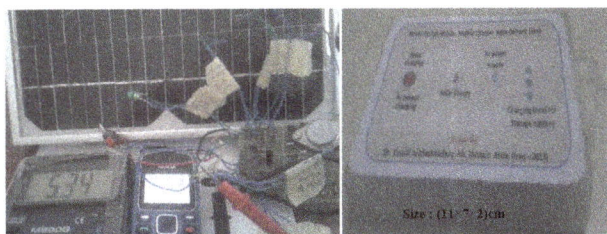

***Fig. 6.*** *Practical Implement of System Design and Final Product Form.*

Different levels of charging current are possible (the normal charging level is 100mA). The rechargeable battery was charged to 34mA and the results fully correspond with the simulation results. Fig. 6 also shows the final display of the mobile charger. The selection for the source type (either solar energy or AC) depends on the source available. The level of charging of the external battery also shows up on the panel.

## 6. Conclusion

The proposed design is novel. It is simple and cheap but high performance. It also functions on two sources. Its simulation and experiment results show:

- Above 95% charging efficiency (proving solar energy's feasibility in supplying energy to mobile phones).
- Its current limiter circuit extending battery life and it is safe even after full charging.
- Possible future work in increasing the solar panel efficacy and reducing the system size.

## References

[1] Attia H. A, Getu B, Ghadban H, Abu Mustafa K, Portable Solar Charger with Controlled Charging Current for Mobile Phone Devices, Int. J. of Thermal & Environmental Engineering, Vol. 7, No. 1, 2014 , pp.17-24

[2] Hild S, Leavey S, Sorazu B, Smart Charging Technologies for Portable Electronic Devices, Journal Of Latex Class Files, 2014.

[3] Geist T, Kamath H, Porter S, Designing Battery Charger Systems for Improved Energy Efficiency, Prepared for the California Energy Commission, Sep 28, 2006.

[4] F. Boico, B. Lehman, Multiple-input Maximum Power Point Tracking algorithm for solar panels with reduced sensing circuitry for portable applications, J. Solar Energy 86 (2012) 463–475

[5] P. Görbe , A. Magyar, K. M. Hangos, Reduction of power losses with smart grids fueled with renewable sources and applying EV batteries, J. Cleaner Production 34 (2012) 125-137

[6] P. Bajpai, V. Dash, Hybrid renewable energy systems for power generation in stand-alone applications: A review, J. Renewable and Sustainable Energy Reviews 16 (2012) 2926–2939

[7] B. ChittiBabu, et. al, Synchronous Buck Converter based PV Energy System for Portable Applications, Proceeding of the 2011 IEEE Students' Technology Symposium 14-16 Jan. (2011), 335 - 340

[8] A. Robion, et. al, Breakthrough in Energy Generation for Mobile or Portable Devices, 978-1-4244-1628-8/07/$25.00 ©2007 IEEE, (2007), 460 - 466

[9] M. H. Imtiaz, et. al, Design & Implementation Of An Intelligent Solar Hybrid Inverter In Grid Oriented System For Utilizing PV Energy, International Journal Of Engineering Science And Technology, Vol. 2(12), 2010, 7524-7530

[10] M. S. Varadarajan, Coin Based Universal Mobile Battery Charger, IOSR Journal of Engineering (IOSRJEN) ISSN: 2250-3021 Vol. 2, Issue 6 (June 2012), 1433-1438

[11] R. M. Akhimullah, Battery Charger with Alarm Application, Thesis of Bachelor Of Electrical Engineering (Power System), University Malaysia Pahang, November, 2008, pp: 1-24

[12] M. A. Baharin, Solar Bicycle, thesis MSc. Bachelor Of Electrical Engineering (Power Systems), University Malaysia Pahang, November, 2010, pp:1-24

[13] J. Tyner, et. al, The Design of a Portable and Deployable Solar Energy System for Deployed Military Applications, Proceedings of the 2011 IEEE Systems and Information Engineering Design Symposium, University of Virginia, Charlottesville, VA, USA, April 29, (2011), 50 - 53

[14] C. Li, et. al, Solar Cell Phone Charger Performance in Indoor Environment, 978-1-61284-8928-0/11/$26.00 ©2011 IEEE, (2011), pp: 1-2.

[15] Q. I. Ali, Design & Implementation of a Mobile Phone Charging System Based on Solar Energy Harvesting, Iraq J. Electrical and Electronic Engineering, Vol.7 No.1,( 2011), 69 - 72

[16] K. Ishaque, Z. Salam, A review of maximum power point tracking techniques of PV system for uniform insolation and partial shading condition, J. Renewable and Sustainable Energy Reviews, 19, (2013), 475–488.

[17] A. Higier, Design, development and deployment of a hybrid renewable energy powered mobile medical clinic with automated modular control system, J. Renewable Energy ,50, (2013), 847 - 857

[18] M.H. Taghvaee, et. al, A current and future study on non-isolated DC–DC converters for photo voltaic applications, J. Renewable and Sustainable Energy Reviews, 17, (2013) 216–227

[19] R. Komiyama, Analysis of Possible Introduction of PV Systems Considering Output Power Fluctuations and Battery Technology, Employing an Optimal Power Generation Mix Model, Electrical Engineering in Japan, Vol. 182, No. 2, (2013), pp. 1705–1714

# 14

# High-performance multilevel inverter drive of brushless DC motor

**Yousif Ismail Al Mashhadany**

Electrical Engineering Department, Engineering College, University of Anbar, Anbar, Iraq

**Email address:**

yousif_phd@hotmail.com

**Abstract:** The brushless DC (BLDC) motor has numerous applications in high-power systems; it is simple in construction, is cheap, requires less maintenance, has higher efficiency, and has high power in the output unit. The BLDC motor is driven by an inverter. This paper presents design and simulation for a three-phase three-level inverter to drive the BLDC motor. The multilevel inverter is driven by discrete three-phase pulse width modulation (DPWM) generator that forced-commuted the IGBT's three-level converters using three bridges to vectored outputs 12- pulses with three levels. Using DPWM with a three-level inverter solves the problem of harmonic distortions and low electromagnetic interference. This topology can attract attention in high-power and high-performance voltage applications. It provides a three-phase voltage source with amplitude, phase, and frequency that are controllable. The proposed model is used with the PID controller to follow the reference speed signal designed by variable steps. The system design is simulated by using Matlab/Simulink. Satisfactory results and high performance of the control with steady state and transient response are obtained. The results of the proposed model are compared with the variable DC-link control. The results of the proposed model are more stable and reliable.

**Keywords:** Brushless DC Motor, Multilevel Inverter, High-Performance Drive, Pulse Width Modulation (PWM), Maltlab, Simulink

## 1. Introduction

The brushless DC (BLDC) motor is a permanent-magnet synchronous machine. It is supplied by a six-transistor inverter whose on/off switching is determined by the rotor position of the motor. It has neither brush nor commutator. Its torque-speed characteristic is similar to that of a permanent-magnet conventional DC motor, minus possible brush/commutation failure. It is becoming more popular in high-performance variable-speed drives. It requires relatively little maintenance and has lower inertia, larger power-to-volume ratio, lesser friction, and lesser noise than a conventional permanent-magnet DC servo motor of similar output rating. However, these advantages are costly, and the controller of a BLDC motor is more complex than that of a conventional motor. Good armature current response is also necessary to drive a BLDC motor satisfactorily [1]–[6].

BLDC motors have higher power density than other motors (e.g., induction motors) because no loss of rotor copper and no commutation occur. The structure is compact and robust, which contributes to the popularity of BLDC motors in efficiency-critical applications or where commutation-induced spikes (which are unwanted) exist. Commutation necessitates using an inverter and a rotor position sensor. However, position sensor can add to drive cost and machine size and reduce reliability and noise immunity. Numerous studies reported on sensorless drives that can control position, speed, and/or torque without shaft-mounted position sensors

Four conventional sensorless control methods exist. One is open-phase current sensing, which detects the conducting interval of freewheeling diodes connected anti-parallel with power transistors. At low speeds, the synchronization is simple and the control is excellent. At high speeds, the resolution of the rotor position decreases. Detecting the freewheeling current requires supply of additional isolated power to a comparator. Another method detects the third harmonic of back Electro – Motive Force (EMF) [4, 5], removing all the fundamental and other polyphase components through simple summation of three-phase voltages. Not so much filtering is required for the integration function of a signal with a frequency three times the fundamental. The filter is much smaller than that in flux detection through back EMF. It is insensitive to filtering delays and performs well at many speeds. However, measuring phase voltages requires a neutral point not considered in the manufacture of the motor. At low speeds, the third harmonic is difficult to detect [12],[13].

Yet another method, integration of back EMF, uses the principle that integration is constant from zero crossing point (ZCP) to 30°. Operation of the main processor decreases because calculating an additional conversion point of the switching mode is unnecessary. This method does not synchronize the phase current with back EMF at the sensorless drive. Using a flux-weakening drive is also impossible. The most popular sensorless control method is open-phase voltage sensing, which indirectly estimates rotor position through ZCP detection of open-phase terminal voltage. However, its response deteriorates at the transient state, and to detect the terminal-voltage ZCP, its operational speed must be sufficiently high [14]–[16].

Multilevel inverter (MLI) topologies have been widely used in the motor drive industry to run induction machines for high-power and high-voltage configurations. Traditional multilevel converter topologies, such as neutral point clamped (NPC) MLI, flying capacitor (FC) MLI, and cascaded H-bridge (CHB) MLI, have catered to a wide variety of applications. The CHB MLI might be the only type of MLI where the energy sources (capacitors, batteries, etc.) can completely be the isolated DC sources. Induction motors have been traditionally used for mostly all types of commercial, industrial, and vehicular applications. However, studies in the last decade have shown that vehicular applications demand high performances that are delivered by certain special machines, which include BLDC machines, switched reluctance machines, and permanent magnet synchronous machines [17].

The obvious reasons for traditionally using induction motor is that the motor technology and control methodologies are understood by both the academia and the industry. The paradigm shift toward using permanent magnet synchronous machines and BLDC machines is the result of the increased demand in high performance, faster torque response, and enhanced speed and efficiency from vehicles [18].

BLDC motors offer numerous advantages including high efficiency, low maintenance, greater longevity, reduced weight, and more compact construction. They have been widely used for various industrial applications based on inherent advantages. They are the most suitable motors in application fields that require fast dynamic response of speed because they are highly efficient and can be easily controlled in a wide speed range. This paper consists of introduction to BLDC motor and MLI, mathematical model of BLDC motors, inverter topologies, design MLI-fed BLDC motor drive, simulation results, and conclusion.

## 2. Mathematical Model of BLDC Motors

BLDC motor is a rotating electric machine with a classic

three-phase stator similar to that of an induction motor. Its rotor mounts permanent magnets (see Fig. 1). The magnet rotates, and the conductors are stationary. This motor equals a reversed DC-commutator motor. Commutator and brushes alter the current polarity of a DC commutator motor, whereas a BLDC motor has its polarity reversed by power transistors synchronously switching with rotor position. Therefore, BLDC motors must often incorporate either internal or external position sensors that discern the actual rotor position or use sensorless detection [4],[8].

Fig. 1 shows the block diagram of the BLDC motor drive. Assuming the stator resistances of all the windings are equal and the self-inductance and mutual inductance are constant, the voltage equation of the three phases can be expressed as Equation (1), neglecting the magnets, the high-resistivity stainless-steel retaining sleeves, and the rotor-induced currents, and not modeling the damper windings [18].

$$\begin{bmatrix} v_a \\ v_b \\ v_c \end{bmatrix} = \begin{bmatrix} R_s & 0 & 0 \\ 0 & R_s & 0 \\ 0 & 0 & R_s \end{bmatrix} + \begin{bmatrix} L_s - M & 0 & 0 \\ 0 & L_s - M & 0 \\ 0 & 0 & L_s - M \end{bmatrix} \frac{d}{dt} \begin{bmatrix} i_a \\ i_b \\ i_c \end{bmatrix} + \begin{bmatrix} e_a \\ e_b \\ e_c \end{bmatrix} \quad (1)$$

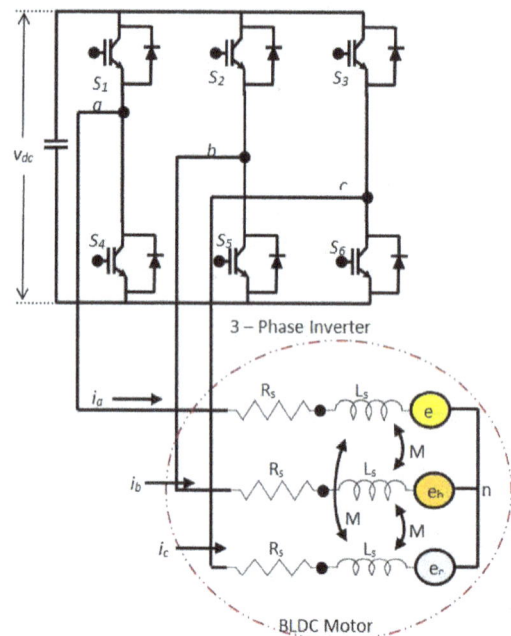

*Figure 1. Block diagram of a BLDC motor drive [18].*

va, vb, and vc denote the phase voltages; Rs the stator resistance; ia, ib, and ic the phase currents; Ls the stator inductance; M the mutual inductance; and L = Ls–M. The back EMFs of the phase are ea, eb, and ec, and the mechanical angular velocity is wm. Fig. 3 shows that injecting the square-wave phase current into the part that has the magnitude of the back EMFs fixed will reduce the torque ripple and stabilize control [19].

## 3. Inverter Topologies

The MLI has five basic types: isolated H-bridge, diode-clamped inverter, FC inverter, combinational multilevel topologies, and cascading fundamental topologies. Using MLI device in voltage sharing is automatic because of the independent DC supplies. No restriction is also observed on the switching pattern. With N devices (each capable of operating at voltage Vdc) per phase, the circuit can produce an output that varies between ± (N/2)*(Vdc/2). Very high voltage converters can be made by using a large number of H-bridges [17], [19].

The modular circuit is an advantage for manufacture and maintenance. The voltage stress on each of the switch also decreases. Using MLIs divides the main DC supply voltage into several DC sources that are used to synthesize an AC voltage into a stepped approximation of the desired sinusoidal waveform. The stepped approximation is also popularly known as the staircase model. An exhaustive literature survey was conducted to investigate the research previously performed in the area of MLIs [20].

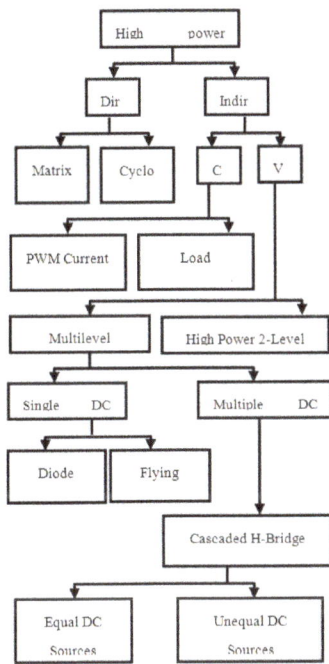

**Figure 2.** Classification of high-power converters.

The number of stages (cells or capacitors depending on the respective topology) helps decide the power capacity of the converter as a whole. Suitable connections in either series or shunt mode or both are performed to achieve higher voltage and/or current ratings. One of the biggest advantages of using an MLI is that the transformer can be eliminated, which helps enhance efficiency and cost effectiveness. The three popular topologies in MLI are as follows: NPC, FC, and CHB. Fig. 2 shows the classification of high power converters. Out of all power converters, cascaded bridge configuration is the most effective and popular. Cascaded bridge configuration is again classified into two types: half bridge and full bridge [21-22].

## 4. Design MLI-fed BLDC Motor

The proposed three-phase MLI fed to the BLDC motor is shown in Fig. 3. This model represents modeling a 50 kW, 380 V, 50 Hz, three-phase, three-level inverter. The IGBT inverter uses the discrete three-phase pulse width modulation (DPWM) technique (8 kHz carrier frequency) to convert DC power from a +/− Vdc source to V AC, 50 Hz. The inverter feeds a 50 kW resistive load through a three-phase transformer. L–C filters are used at the converter output to filter out harmonic frequencies generated mainly around multiples of 8 kHz switching frequency. The 12-inverter pulses required by the inverter are generated by the discrete three-phase PWM generator. The system operates in open loop at a constant modulation index. The inverter is built with individual IGBTs and diodes.

In a three-level voltage-sourced converter (VSC) using ideal switches, the two pairs of pulses sent to each arm could be complementary. For example, for phase A, IGBT1 is complementary of IGBT3 and IGBT2 is complementary of IGBT4. However, in practical VSCs, the turnoff of semiconductor switches is delayed because of the storage effect. Therefore, a time delay of a few microseconds (storage time + safety margin) is required to allow complete extinction of the IGBT that is switched off before switching on the other IGBT. Otherwise, a short circuit could result on the DC bus.

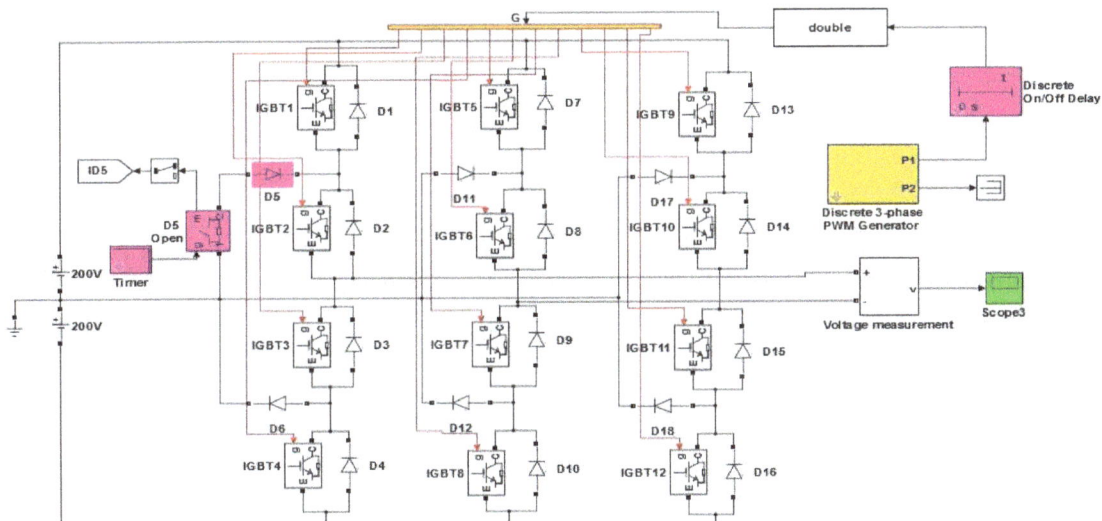

**Figure 3.** Three-phase three-level inverter with DPWM

## 5. Simulation Results

*Table 1. The numerical values for system design.*

| Parameter | Value | Parameter | Value |
|---|---|---|---|
| Stator phase resistance Rs (ohm): | 2.8750 | Torque Constant (N.m / A_peak) | 1.4 |
| Stator phase inductance Ls (H) : | 8.5e-3 | Back EMF flat area (degrees) | 120 |
| Flux linkage established by magnets (V.s) | 0.175 | Inertia, viscous damping, pole pairs, static friction [ J(kg.m^2) F(N.m.s) p() Tf(N.m)]: | [ 0.8, 1e-3 1e-3, 4] |
| Voltage Constant (V_peak L-L / krpm) | 146.6 | Initial conditions [ wm(rad/s) thetam(deg) ia,ib(A) ] | [0,0,0,0] |

The parameters of PID controller are: Proportional $K_P$= 20, Integral $k_I$=0.1, derivative $k_d$=1

The simulation of the system design in this study was conducted by employing Matlab/Simulink version 2013b. In this simulation used same numerical values for system design to make compression, table 1 presents the numerical values. The variable DC-link voltage controller technique was utilized to obtain the BLDC motor, and the results were compared with the proposed model. The technique of using a variable DC voltage source to control the applied voltage and consequently to control the motor phase currents is cheaper than a traditional PWM control, but the losses can be high at low voltage and high current conditions. However, at high speed, a linear power stage can be the best alternative when the switching losses and commutation delay of a pulsed power stage are significant. The variable DC-link voltage control technique is the only technique that does not cause high-frequency disturbances. Its performance was similar to that of the PWM method, but it produced much smoother torque because of the absence of high-frequency switching. In the frequency domain, the variable DC-link voltage control technique contains only harmonics caused by the current commutation. The full simulation for this technique is shown in Fig. 4. Its performance was similar to that of the traditional PWM method, but it produced much smoother torque because of the absence of high-frequency switching. In the frequency domain, the variable DC-link voltage control technique contains only harmonics caused by the current commutation.

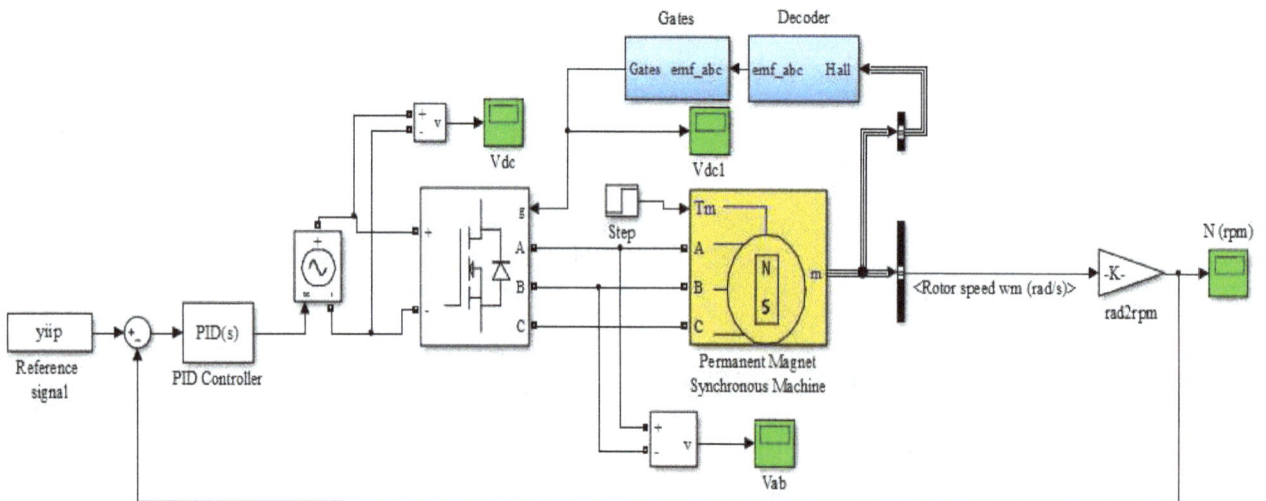

*Figure. 4. BLDC motor controlled by variable DC-link voltage controller.*

*Proposed MLI*

The proposed model of multilevel technique synthesizes the AC output terminal voltage with low harmonic distortion, thus reducing the filter requirements. In particular, MLIs are emerging as a visible alternative for high-power, medium-voltage applications. One of the significant advantages of multilevel configuration is the harmonic reduction in the output waveform without increasing the switching frequency or decreasing the inverter power output. The output voltage waveform of an MLI is composed of the number of levels of voltages, typically obtained from capacitor voltage sources starting from three levels, the number of levels can increase until the output is a pure sinusoidal. The output of the simulation for system design for MLI model described in Fig. 3 and used with BLDC motor is shown in Fig. 7. The output of this inverter based on DPWM with 12 pulses sequences, Fig. 6 explains the sample from the sequences used for operation MLI.

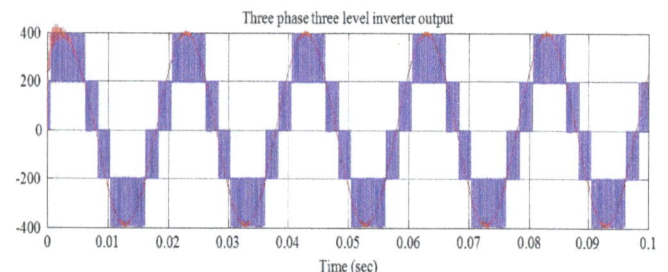

*Figure. 5. Output of three-phase three-level inverter with DPWM.*

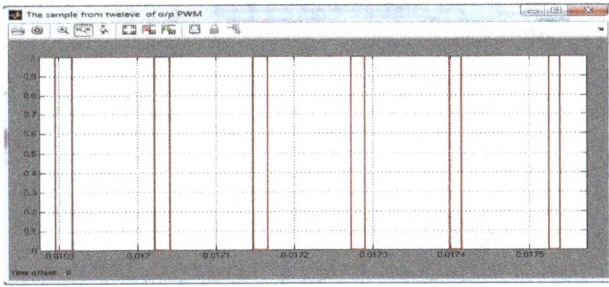

*Figure 6.* The sample from output of the DPWM

The full system design of speed control for BLDC motor based on MLI inverter driven by DPWM is shown in Fig. 7. A simple type of PID controller will be used with the proposed model of MLI-driven BLDC motor. A new DPWM strategy based on modulation that requires only a single carrier at 5 kHz and two reference signals is proposed, which is used to generate 12 PWM signals using the DPWM method, as shown in Fig. 7, with terminal G. If Vref1 exceeds the peak amplitude of the carrier signal Vcarrier, Vref2 is compared with the carrier signal until it reaches zero. At this point onward, Vref1 takes over the comparison process until it exceeds Vcarrier.

The simulation for control system design was described in Fig. 7, and the output is shown in Fig. 8. The test signal has a reference speed of 2000 rpm, and the desired high-performance input is achieved with small value of rising time (0.01 sec) and with accepted value of settling time (0.035 sec), with a steady-state error of less than 0.01% and maximum percent overshoot value of 3.48%; therefore, this response is reliability with high performance.

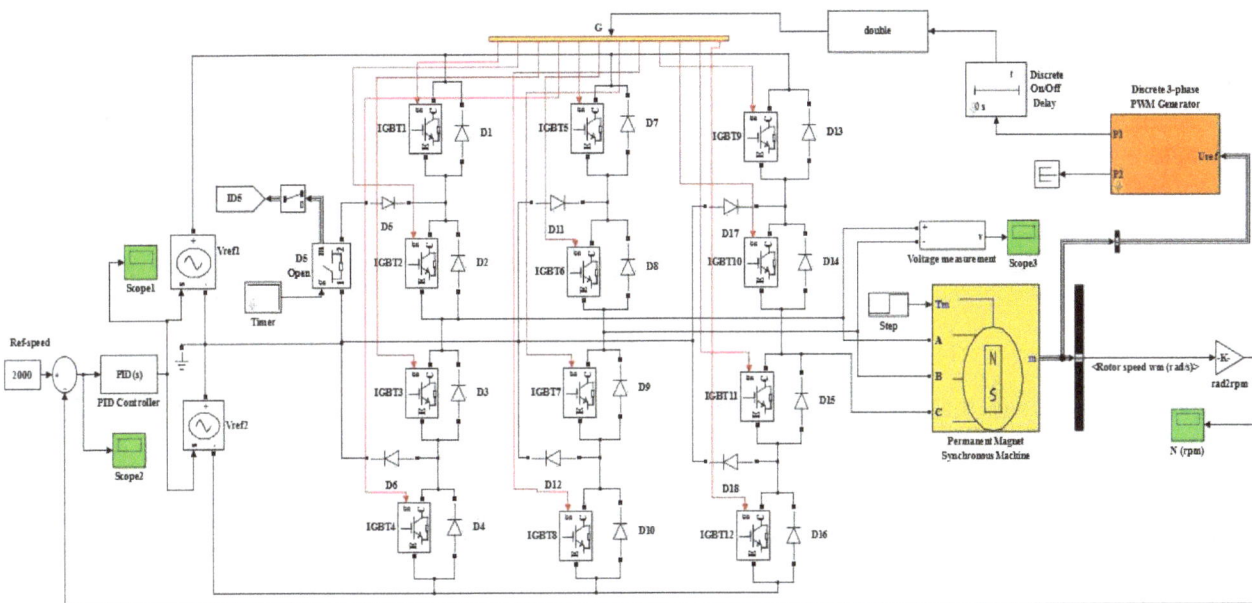

*Figure 7.* BLDC motor with MLI driven with PID controller.

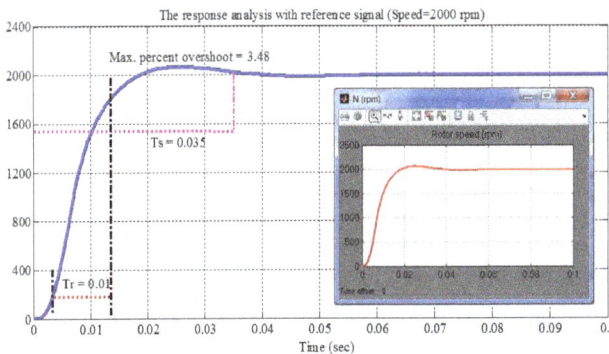

*Figure 8.* Analysis of response for the proposed MLI with PID controller of BLDC motor.

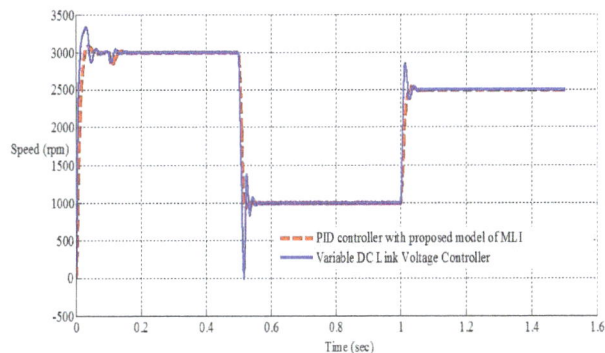

*Fig. 9.* Two outputs of controllers with proposed MLI and variable DC-link

Fig's. 8&9 show the compression response between two models. The proposed model was described in Fig. 6 and the model of variable DC-link voltage control in Fig. 4. The input of the two models was designed with hard variable test signal values of 3000, 1000, and 2500 rpm and with time (0.5 sec) for each step. Fig. 9 shows that the output of the proposed model responds better than the variable DC-link model and is more stable and reliable.

# 6. Conclusions

The proposed MLI performance analysis was successfully presented by using Matlab/Simulink software. The proposed topology can be easily extended to a higher-level inverter. The simulation results were sine waves and exhibited fewer ripples and low losses. This system would show its feasibility in practice. The vector control was described in adequate detail and was implemented with a three-level MLI. This method enabled the operation of the drive at zero direct axis stator current. Transient results were obtained when a DPWM was started from a standstill to a required speed. The performance of the vector control in achieving a fast reversal of PDPWM even at very high speed ranges is quite satisfactory. The performance of the proposed three-phase MLI was investigated and was found to be quite satisfactory. A comparison was made between the PID controller–based proposed model MLI and the controller with variable DC-link voltage. The results showed that the proposed model responded better in transient and steady states and was more reliability with high performance.

# Acknowledgements

Special thanks are due to University of Anbar – Iraq, Renewable Energy Research Centre for supporting me with the work with Grant No. RERC-PT26.

# References

[1]   P. D. Kiran, M. Ramachandra, "Two-Level and Five-Level Inverter Fed BLDC Motor Drives", International Journal of Electrical and Electronics Engineering Research, Vol. 3, Issue 3, pp 71-82, Aug 2013

[2]   N. Karthika, A. Sangari, R. Umamaheswari , "Performance Analysis of Multi Level Inverter with DC Link Switches for Renewable Energy Resources", International Journal of Innovative Technology and Exploring Engineering, Volume-2, Issue-6, pp 171-176, May 2013

[3]   A. Jalilvand R. Noroozian M. Darabian, "Modeling and Control Of Multi-Level Inverter for Three-Phase Grid-Connected Photovoltaic Sources", International Journal on Technical and Physical Problems of Engineering, Iss. 15, Vol. 5, No.2, pp 35-43, June 2013

[4]   P. Karuppanan, K. Mahapatra, "PI, PID and Fuzzy Logic Controlled Cascaded Voltage Source Inverter Based Active Filter For Power Line Conditioners", Wseas Transactions On Power Systems, Issue 4, Volume 6, pp 100-109, October 2011

[5]   D. Balakrishnan, D. Shanmugam, K.Indiradevi, "Modified Multilevel Inverter Topology for Grid Connected PV Systems", American Journal of Engineering Research, Vol. 02, Iss.10, pp-378-384, 2013

[6]   G.Nageswara, P. Sangameswara, K. Chandra, "Multilevel Inverter Based Active Power Filter for Harmonic Elimination", International Journal of Power Electronics and Drive System, Vol.3, No.3, pp. 271~278, September 2013

[7]   R. Pandey, S.P. Dubey, "Multilevel Inverter Fed Permanent Magnet Synchronous Motor Drive with Constant Torque Angle Control", Advance in Electronic and Electric Engineering, Vol. 3, No. 5, pp. 521-530, 2013

[8]   G. Su, D. Adams, "Multilevel DC Link Inverter for Brushless Permanent Magnet Motors with Very Low Inductance", IEEE IAS 2001 Annual Meeting, pp 1/6-6/6, October 2011.

[9]   A. Purna, Y.P. Obulesh, C.Babu, "High Performance Cascaded Multilevel Inverter Fed Brushless Dc Motor Drive", International Journal of Engineering Sciences & Emerging Technologies, Vol. 5, Iss. 2, pp 88-96, June 2013

[10]  S. S. Emani, Performance Evaluation Of A Cascaded H-Bridge Multi Level Inverter Fed Bldc Motor Drive In An Electric Vehicle, M. Eng. Thesis, Texas A&M University, India ,may 2010

[11]  Xx T. S. Kim, B. G. Park, D. M. Lee, J. S. Ryu, D. S. Hyun, " A New Approach to Sensorless Control Method for Brushless DC Motors", International Journal of Control, Automation, and Systems, vol. 6, no. 4, pp. 477-487, August 2008

[12]  S.M. Lanjewar, K. Ramsha, "Design of Control Scheme and Performance Improvement for Multilevel Dc Link Inverter Fed PMBLDC Motor Drive", International Journal of Electrical and Electronics Engineering, Vol.1, Iss.3,pp 79-85, 2012

[13]  C.Brahmaiah, M.Baba, K.Swathi, "A Series-Connected Multilevel Inverter Topology for Medium-Voltage BLDC Motor Drive Applications", International Journal of Electrical, Electronics and Data Communication, Vol. 1, Iss. 10, pp 55-60, Dec. 2013

[14]  M. P. Kumar, A. S. Hari, "A Five Level Inverter for Grid Connected PV System Employing Fuzzy Controller", International Journal of Modern Engineering Research,Vol.2, Iss. 5, pp-3730-3735, Sep. 2012

[15]  A. Mahendran, K. Muthulakshmi, D. Edison, L. J. Ganesan, "Design And Implementation Of A Fuzzy Logic Controller For Multilevel Inverter Topology", International Journal of Research in Computer Applications and Robotics, Vol.1 Iss.9, pp 109-116, December 2013

[16]  T. Chaudhuri, Cross Connected Multilevel Voltage Source Inverter Topologies for Medium Voltage Applications, 1st edition, 2008

[17]  P.Thirumuraugan, R.Preethi, " Closed Loop Control of Multilevel InverterUsing SVPWM", International Journal of Advanced Research in Electrical, Electronics and Instrumentation Engineering, Vol. 2, Issue 4,pp 1561-1572, April 2013

[18]  A. Yadav, J. Kumar, "Harmonic Reduction in Cascaded Multilevel Inverter", International Journal of Recent Technology and Engineering, Vol.2, Iss. 2, pp 147-149, May 2013

[19]  C. G. Real, E. V. Sánchez, J. G. Gil, "Position and Speed Control of Brushless DC Motors Using Sensorless Techniques and Application Trends", Sensors, pp 6901- 6964, 2010

[20]  M. Rajshekar, V. G. Swamy, T.A. Kumar, ""Modeling and \simulation of discontinues current mode inverter fee permanent magnet synchronous motor derive", Journal of Theoretical and Applied Information Technology, , pp. 64-94, 2011

[21]  V.M. Varatharaju, B. Mathur, U. Dhayakumar, "Adaptive Controllers for Permanent Magnet Brushless DC Motor Drive System using Adaptive-Network-based Fuzzy Interference System", American Journal of Applied Sciences, (8), pp. 810-815, 2011

[22] P. Tamilvani, K.R. Valluvan, "Hybrid Modulation Technique for Cascaded Multilevel Inverter for High Power and High Quality Applications in Renewable Energy Systems", International Journal of Electronic and Electrical Engineering, Vol. 5, No. 1, pp. 59-68, 2012.

# 15

# Simulation of solar off- grid photovoltaic system for residential unit

**Jasim Abdulateef**

Mechanical Engineering Department, Diyala University, Diyala, Iraq

**Email address:**

jmabdulateef@gmail.com (J. Abdulateef)

**Abstract:** The aim of this study is to design a solar off-grid PV system to supply the required electricity for a residential unit. A simulation model by MATLAB is used to size the PV system. The solar PV system is simulated with the case of maximum solar radiation on a sunny day. The results show that the average daily load requirement of the selected residential unit is 36 kWh/day. This load requirement can be meet by using an array of 44 solar panels. During the day time, the PV system supplies the desired 12.4 kWh of energy. During the night time, a battery storage system of 23.6 kWh (48V, 350 Ah) is used to meet the night load.

**Keywords:** Solar PV, Off-Grid, MPPT Controller, Isolated Places

## 1. Introduction

Photovoltaic system installation has played a big role in renewable energy because PV systems are pollution free, economically reliable for long-term operation and secure energy source. The major obstruction of PV technology is its high capital costs compared to conventional energy sources.

In isolated regions and because of the scarcity of means, it is necessary to optimize the solar off-grid (stand-alone) PV system in order to minimize the costs and to make the PV systems competitive with the other forms of renewable energies. Modeling and simulation techniques can be used to assess the performance of PV system components before installation in place hence reducing the overall system costs. Therefore, many research studies have focused on the optimization of PV systems [1-8].

This study presents an optimization procedure to design solar off-grid PV systems for a residential unit to find the effective way to use solar energy at the lowest cost possible. The proposed system may provide highly efficient, and clean solar powered electricity that can meet the daily load demands of the residential unit. Sizing of the off-grid PV systems based on the specific residential load requirement is also done. The area selected is a typical residential unit located in the campus of mechanical engineering department at Diyala University. The simulation is performed using the MATLAB software to validate the design results.

## 2. Residential Load Profile

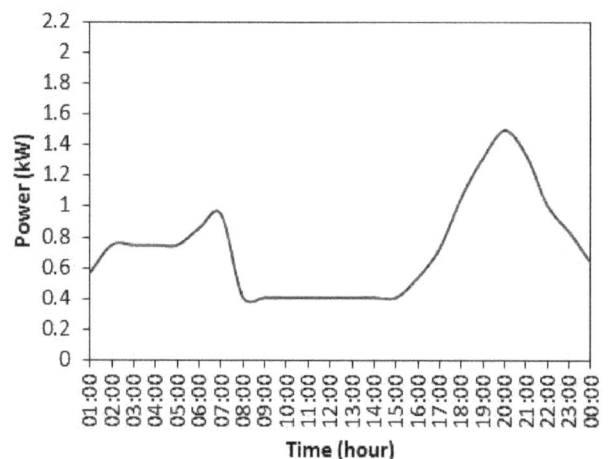

*Figure 1. Proposed residential load profile*

The remote area residential unit is simple and does not require large quantities of electrical energy used for lighting and electrical appliances. Figure 1 shows the proposed residential load profile. The Load profile was proposed considering the general hourly based load usage. At midnight hours, the power consumption for the residential unit comes down where only basic electrical appliances are consuming power. The load demand rises up during morning hours when everybody gets ready either to leave for schools or offices. Throughout the noon hours the load demand levels are minimum as most of the family members are outside. Again, during the evening hours when all the family members are present, the power consumption rises as everyone switches on various entertainment appliances.

The average energy consumption of electrical appliances of a typical residential unit is assumed 547 kWh/month, i.e. 17.64 kWh/day. According to the load profile shown in Fig. 1, the load requirement considered should be maximum hourly load consumption. Thus the proposed solar off-grid PV system should produce 36 kWh/day (1.5kW * 24h).

# 3. Meteorological Data of the Selected Site

Iraq is among the countries with remarkable potential in solar energy. The solar off-grid PV system of the interest area (Latitude 33.75°, Longitude 44.63°) that located in the campus of mechanical engineering department at Diyala University is simulated with average global solar radiation per year equal to 5.5 kWh/m$^2$/day [9]. Assuming that the solar panels will be placed on the roof with a possible inclination corresponding to the latitude of the selected area.

# 4. PV System Modeling

The proposed block diagram of a solar off-grid PV system that provides the required electricity for a residential unit is shown in Fig. 2. The main components of the system are namely PV array, controller, battery, inverter and load. The solar PV system is simulated such that the PV module charge the battery through the controller and battery also provides the power to the load when the solar radiation is insufficient. DC/AC inverter provides AC electricity to the required residential AC loads.

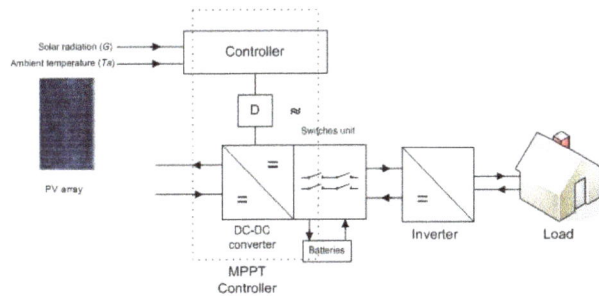

*Figure 2. Solar off- grid PV system configuration*

## 4.1. PV Panel Model

Table 1 shows the electrical specification of the PV module that was selected for a MATLAB simulation model [10]. The PV module is derived from the equivalent electric circuit of a solar cell [11]. The equation that describes the *I-V* relationship of the PV cell is written below [12].

$$I = I_{sc} - I_o \left[ e^{q\left(\frac{V+I.R_S}{nkT}\right)} - 1 \right] \tag{1}$$

where
$I$     Cell current
$I_o$    Reverse saturation current of diode
$I_{sc}$   Short-circuit current
$k$     Boltzman's constant (13,807 10 kJ)
$n$     Diode quality factor, ($n$=1-2)
$N_s$   Number of PV modules
$q$     Electronic charge (1.6022 10$^{-19}$ C)
$R_s$   Series resister
$T$     Cell temperature
$V$     Cell voltage = $V_m/N_s$
$V_m$   Voltage at maximum power ($P_{max}$)

*Table 1. BP SX 150 PV module specifications*

| Characteristic | Rating |
|---|---|
| Maximum power ($P_{max}$) | 150W |
| Voltage at $P_{max}$ ($V_{mp}$) | 34.5V |
| Current at $P_{max}$ ($I_{mp}$) | 4.35A |
| Open-circuit voltage ($V_{oc}$) | 43.5V |
| Short-circuit current ($I_{sc}$) | 4.75A |
| Maximum system voltage | 600V |
| Area | 1.2 m$^2$ |
| Efficiency | 15 % |

## 4.2. Battery Model

The battery model that used in the PV system was based on a lead-acid battery PSpice model [13]. The battery model has two modes of operation: charge and discharge. The battery is in charge mode when the current into the battery is positive, and discharge mode when the current is negative. The code of battery model was written in MATLAB and used to simulate the performance of solar PV system during charging and discharging.

The storage capacity of the battery was calculated using the following relation [14,15]:

$$Storage\ capacity = \frac{N_C E_{Load}}{DOD.\eta_b} \tag{2}$$

where
$DOD$    Maximum permissible depth of battery discharge
$E_{Load}$   Average energy consumed by the load
$N_C$      Largest number of continuous cloudy days of the interested area
$\eta_b$      Efficiency of the battery

The proposed off-grid PV system is intended to supply 1.5 kW/48 V for 24 hours (36 kWh). The largest number of continuous cloudy days $N_C$ in the selected site is about 1 day. Thus, for a maximum depth of discharge for the battery $DOD$ of 0.8 and battery efficiency 80%, the storage capacity using eq. (2) becomes 56.3 kWh. Since the selected DC bus voltage is 48 V, then the required ampere-hours of the battery =56.3 kWh/48 ≈ 1173 Ah. If a single battery of 12 V and

350 Ah is considered, then 4 batteries are connected in series; to give an overall number of 4 batteries required for the PV system.

### 4.3. Converter Model

The most basic DC-DC converter is based on the idea that the power is converted while altering the current and voltage. A DC-DC converter is used to increase the efficiency of the PV system by matching the voltage generated by PV array to the voltage required by the load. The output power ($P_{out}$) of DC-DC converter is given by:

$$P_{in} = P_{out} \qquad (3)$$

Assuming the efficiency factor of $x$:

$$P_{in}.x = P_{out} \qquad (4)$$

Substituting $V.I$ for $P$ results:

$$V_{in}.I_{in}.x = V_{out}.I_{out} \qquad (5)$$

For the DC-DC converter used in this model:
$V_{in}$= voltage across the PV array
$I_{in}$=current output of PV array
$x$=0.9 (assume 90% efficiency)
$V_{out}=V_b$= Battery voltage
$I_{out}$=current output from converter when all other values are known.

The output voltage is related to the input voltage as a function of duty cycle of the switch ($D$)[16]. For the Cuk converter [17], the relationship is expressed as:

$$\frac{V_{out}}{V_{in}} = \frac{D}{D-1} \qquad (6)$$

### 4.4. Controller Model

The MPPT controller is modeled in based on the DC-DC converter which is controlled by the MPPT algorithm in order to operate the PV array at its maximum power point. The MPPT algorithm has three inputs; PV module voltage ($V_{PV}$), ambient temperature ($T_a$) and solar radiation ($G$) to give two outputs which are the duty cycle and the optimum voltage at MPP. The block diagram of the MPPT controller model is shown in Fig. 3.

The number of controllers required for the off-grid PV system, [18] is calculated using:

$$Total\ max\ power\ of\ PV\ = P_{max} \cdot N_{PV} \qquad (7)$$

$$Controller\ max\ power\ = V_b \cdot I_{controller} \qquad (8)$$

$$Number\ of\ Controller\ required = \frac{Total\ max\ power\ of PV}{Controller\ max\ power} \qquad (9)$$

where
$I_{controller}$ Maximum current the controller which can handle from the PV system to the battery bank
$N_{PV}$ Total number of PV modules required to meet the residential load

Perturb and Observe algorithm is used for MPP tracking

because it has a simple feedback structure and fewer measured parameters.

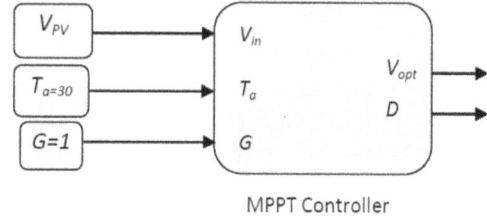

**Figure 3.** Block diagram of the controller

### 4.5. Inverter Model

The role of the inverter is to keep on the AC side the voltage constant at the rated voltage 230V and to convert the input power ($P_{in}$) into the output power ($P_{out}$) with the best possible efficiency. The inverter efficiency is thus expressed as:

$$\eta = \frac{P_{out}}{P_{in}} = \frac{V_{ac}I_{ac}cos\phi}{V_{dc}I_{dc}} \qquad (10)$$

where $I_{ac}$ is the output current by the inverter on the AC side. $I_{dc}$ is the current required by the inverter from the DC side (for example, from the controller) to be able to keep the rated voltage on the AC side (for example on the load). $V_{dc}$ is the input voltage for the inverter delivered by the DC side, for example by the controller.

### 4.6. Load Model

The load existing in a solar PV system is an AC load with an equivalent resistance given by:

$$R_{load} = \frac{V^2}{P} \qquad (11)$$

The load current can be modeled as:

$$I_{load} = \frac{V}{R_{load}} \qquad (12)$$

## 5. Results of the PV System Simulation

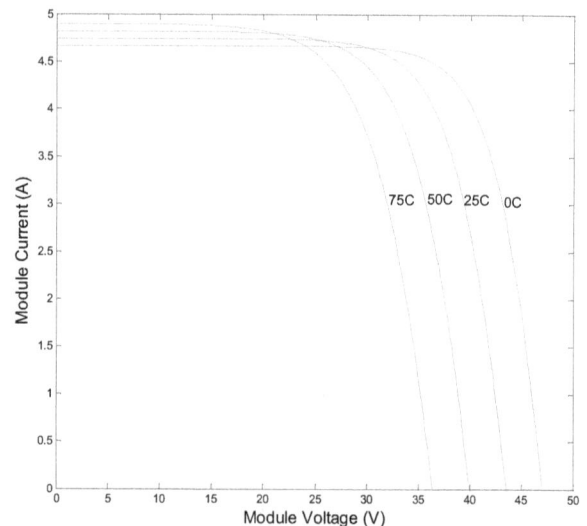

**Figure 4.** I-V curves of the PV module at different temperatures ($1kW/m^2$)

The simulation results of PV module using MATLAB are compared with the PV model data provided by the specification data sheet [10]. Figures 4 and 5 show the current-voltage characteristics of the PV module obtained from simulation. The results of the characteristics of the PV module obtained from simulation are almost identical to the PV specifications from data sheet [10].

The size of PV modules ($N_{PV}$) required to meet the load demand can be calculated by using the following equation:

$$N_{PV} = \frac{E_{load}}{E/m^2 \, A_m \eta_m} \cdot factor \; of \; safty \qquad (13)$$

where

$A_m$     PV module area

$E/m^2$   Average energy received by PV module on a horizontal mode during solar days

$\eta_m$     Efficiency of the PV module

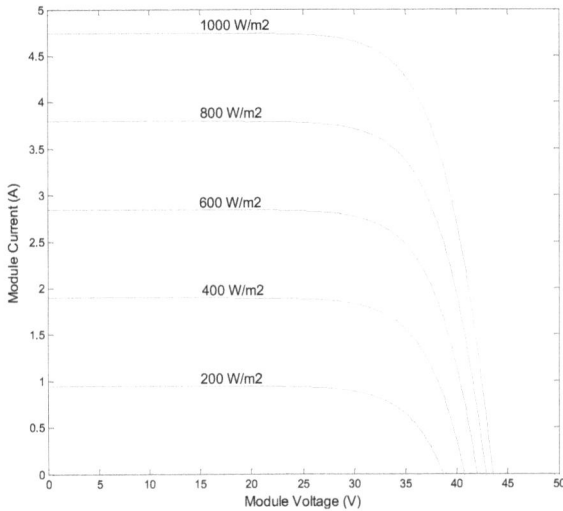

*Figure 5. I-V curves of the PV module at different irradiations (25°C)*

The safety factor used in eq. (13) has a value in the range of 1.2-1.3[19]. The PV system requires to supply1.5 kW/48 V for 24 hours (36 kWh). The solar radiation incident on one square meter in the selected site is considered equal to 5.5 kWh/m². By using eq. (13), an array configuration of 44 solar PV panels is required to meet the daily load demand of the residential unit. The required characteristics of the solar off-grid PV system used in this study are given in Table 2.

*Table 2. Solar off-grid PV system specifications*

| Solar system type | off-grid |
|---|---|
| Average solar radiation | 5.5 kWh/m²/day |
| Daily load requirement | 36 kWh/day |
| Number of PV modules needed | 44 |
| Power produced by PV panels | 6.6 kW |
| PV area | 52.8 m² |

It must be sure when selecting a controller; it has an output voltage rating equal to the nominal battery voltage. Also the maximum PV voltage should be less than the maximum controller voltage rating [20].

The controller which has been chosen has a maximum output current of 80 Amps and a maximum controller voltage of 150 V. By using eq. (9), the total number of controllers required for the proposed PV system that connected to a battery bank with a voltage of 48V is two. If we have 44 PV modules, one sub arrays of 22 PV modules and one sub-array of 22 PV modules should be configured and connected in parallel to each one of the controllers.

In order to study the state of charge of the battery (SOC %) for the given sunny day, we need to simulate our model for the whole day (24 hours) and compare it with the solar PV power and residential load profile. When the power generated by PV system exceeds the load power, the battery is charged with an increase in its SOC. If the solar power falls below the load power, the battery discharges with a decrease in its SOC. Figure 6 shows the battery performance on a sunny day in terms of the SOC.

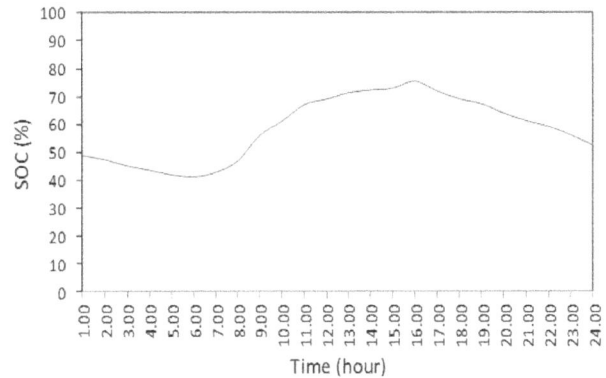

*Figure 6. SOC vs. Time on a sunny day*

During the day time, the PV system supplies the desired 12.4 kWh of energy. During the night time, battery storage system of 23.6 kWh (48V, 350 Ah) is employed to meet the night load.

It can be seen that during the daytime the power produced from the solar irradiation is used to meet the residential load requirements as well as charge the battery. Considering the whole 24 hours scenario, the SOC decreases as the battery discharges during the night hours, and the solar radiation tends to increase during the day while the SOC tends to follow it as well.

Finally, the charging of the battery is performed on a major scale during the afternoon to evening hours where the solar irradiation is at high level and the residential load requirement is less. Then it is clear that after reaching the peak, the SOC decreases with the increase in the residential load requirement and decrease in the sun radiation level during the evening to morning hours.

## 6. Conclusions

This study presents a simple but efficient off-grid photovoltaic system for a residential unit that can meet the residential daily load demands. The results show that the average daily load requirement of a residential unit of 36 kWh/day. In order to meet this load demand, an array of 44 solar panels. During the day time, the PV system supplies the desired 12.4 kWh of energy. During the night time, battery storage system of 23.6 kWh (48V, 350 Ah) is employed to meet the night load.

# References

[1]  J.K. Kaldellis," Optimum technoeconomic energy autonomous photovoltaic solution for remote consumers throughout Greece," Energy Conversion and Management, vol. 45, pp. 2745-2760, 2004.

[2]  N.D. Kaushika, N.K. Gautam, and K. Kaushik, "Simulation model for sizing of stand-alone solar PV system with interconnected array," Solar Energy Materials and Solar Cells, vol. 85, pp. 499-519, 2005.

[3]  T. Markvart, A. Fragaki, and J.N. Ross, "PV system sizing using observed time series of solar radiation," Solar Energy, vol. 80, pp. 46-50, 2006.

[4]  R. Posadillo, and R.L. Luque, "Approaches for developing a sizing method for stand-alone PV systems with variable demand," Renewable Energy, vol. 33, pp. 1037-1048, 2008.

[5]  P. Arun, and R. Banerjee, S. Bandyopadhyay, "Optimum sizing of photovoltaic battery systems incorporating uncertainty through design space approach," Solar Energy, vol. 83, pp. 1013-1025,2009.

[6]  T. Khatib, A. Mohamed, K. Sopian, and M. Mahmoud, "Optimal sizing of building integrated hybrid PV/diesel generator system for zero load rejection for Malaysia," Energy and Buildings, vol. 43, pp. 3430-3435, 2011.

[7]  J. Abdulateef, K. Sopian, W. Kader, B. Bais, R. Sirwan, B. Bakhtyar and O. Saadatian, "Economic analysis of a stand-alone PV system to electrify a residential home in Malaysia," in Proc. HTE'12, Istanbul, 2012.

[8]  H. A. Kazem, T. Khatib, and K. Sopian, "Sizing of a standalone photovoltaic/battery system at minimum cost for remote housing electrification in Sohar, Oman," Energy and Buildings, vol. 61, pp.108–115, 2013.

[9]  M. Sh. Salim, J. M. Najim, S. M. Salih, "Maximum power analysis of photovoltaic module in Ramadi city," International Journal of Energy and Environment, vol. 4 (6), pp.1013-1024, 2013.

[10]  bpsolar: bpsolar, www.bpsolar.com

[11]  R.A. Messenger and J. Ventre, Photovoltaic Systems Engineering, CRC Press, New York, 2004.

[12]  G.R. Walker, "Evaluating MPPT converter topologies using a MATLAB PV model," Australasian Universities Power Engineering Conference, AUPEC Brisbane, 2000.

[13]  L. Castaner, and S. Santiago, Modelling Photovoltaic Systems Using PSpice, John Wiley & Sons Ltd, 2002.

[14]  M.M. Mahmoud, I.H. Ibrik, "Techno-economic feasibility of energy supply of remote villages in Palestine by PV-systems, diesel generators and electric grid,"  Renewable Sustainable Energy Rev., vol. 10, pp. 128-138, 2006.

[15]  S.R. Wenham, M.A. Green, M.E. Watt, "Applied Photovoltaics," Center for Photovoltaic Devices and Systems: Australia, 1994.

[16]  Tyson Denherder, "Design and simulation of photovoltaic super system using Simulink," Senior Project, Faculty of California Polytechnic. State University, 2006.

[17]  Diong Bill, "Future energy challenge final report," University of Texas, EI Paso, (http://www.energychallenge.org/2001Report/UTEP.pdf), 2001.

[18]  PVDI 2007: Solar Energy International, "Photovoltaic Design and Installation Manual", New Society Publishers.

[19]  The German Solar Energy Society, "Planning control apparatus and method and power generating system using them," Patent US5, 654,883, 2005.

[20]  M. R. Rivera, "Small wind / photovoltaic hybrid renewable energy system optimization," Master's thesis, electrical engineering, University of Puerto Rico, 2008.

# 16

# Initial field testing of concentrating solar photovoltaic (CSPV) thermal hybrid solar energy generator utilizing large aperture parabolic trough and spectrum selective mirrors

**Jonathan Richard Raush[*], Terrence Lynn Chambers**

Department of Mechanical Engineering, University of Louisiana at Lafayette, Lafayette, U. S. A.

**Email address:**

jrr1239@louisiana.edu (J. R. Raush), tlchambers@louisiana.edu (T. L. Chambers)

**Abstract:** The University of Louisiana at Lafayette has completed initial field testing of a test unit of the MH Solar Concentrating Solar Photovoltaic (CSPV) system. The CSPV unit is a retrofit system for use with a parabolic trough type concentrating solar power (CSP) thermal solar collector which redirects a portion of the incident solar radiation spectrum to a PV module while allowing normal operation of the thermal system to continue. The system was tested at the UL Lafayette Solar Energy Laboratory utilizing the existing Large Aperture Trough (LAT) test field. The dichroic cold mirror reflected solar radiation of between 500 and 1000 nm to the MH Solar vertical multi junction (VMJ) silicon PV cells (known as the MIH VMJ cells) which provided high efficiency operation under a concentration ratio of 30. The testing produced a PV module efficiency of 30% across the portion of the spectrum which was redirected, while the thermal efficiency was reduced by only about 9 percentage points, resulting in an overall efficiency increase of the power plant. The total power output of the power plant could therefore be increased through utilization of the hybrid configuration.

**Keywords:** Solar Energy, Concentrating Solar Power, CSP, Photovoltaic, CPV-T, CSPV, Hybrid

## 1. Introduction

For solar energy technologies to reach full market penetration, continued improvements must be made in order to lower the levelized cost of electricity (LCOE) of primary solar energy technologies, including both photovoltaics (PV) and concentrating solar thermal power (CSP). According to the U.S. Department of Energy SunShot Initiative, which proposes a goal of $.06/kWh LCOE for solar energy technologies by the year 2020, a market share of 14% of total U.S. electric production would be reached by 2030 [1]. Current LCOE figures for utility scale projects range from about $0.11/kWh for utility scale PV to $0.13/kWh for utility scale CSP [2]. In an effort to improve overall plant efficiency, and thus drive down LCOE, significant interest has been directed toward the subject of hybrid solar collector systems, integrating multiple receivers and technologies in order to more fully leverage the breadth of the solar radiation spectrum. Differing techniques for hybrid systems have been investigated since as early as the middle of the 20th century, including those which combine PV and photo-thermal (PT)

energy conversion in parallel by incorporating spectral beam splitting technologies [3]. Since PT energy conversion processes tend to convert solar energy to thermal energy at efficiencies that maintain relatively constant over the solar spectrum, depending on the optical properties of the thermal receiver, beam-splitting allows the portion of the spectrum that is most advantageous for the PV cell, which is extremely wavelength dependent, to be directed to the cell, while the remaining spectrum can proceed to a thermal receiver for conversion to thermal energy, as in Figure 1 [4].

PV cells are most efficient when converting photons of energies close to the PV cell band-gap energy. Photons below this energy pass through the active area of the cell without being absorbed, and are dissipated as heat. Photons of energy larger than the band are partly utilized, with the remainder of their energy also dissipated as heat. Because of these limitations, a more optimal method of using solar cells would be to direct onto them only the part of the solar spectrum for which high conversion efficiency can be achieved, and to recover the radiation outside this range by diverting it to a second receiver, which could operate as a thermal, chemical, or different PV band-gap receiver. The concept of a PV-thermal solar hybrid system is that the incident beam is split into PV and thermal spectral components and directed to their respective receivers for conversion to electricity and thermal energy more efficiently.

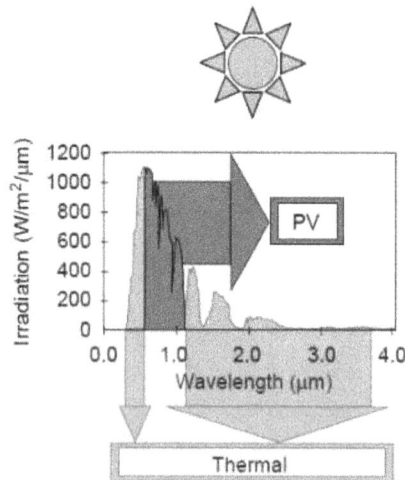

*Figure 1. Spectral filtering for hybrid solar energy production [4].*

Hybrid solar energy systems of various configurations have been proposed, through the inclusion or omitting of various forms of tracking, concentrating, beam splitting, and thermal receivers, in search for the optimal configuration to most completely leverage the given solar resource [5], [6], [7], [8],[9]. Tracking and concentrating PV-thermal hybrid systems are undergoing investigation due to the increased efficiency of some PV cells while under concentration and the cost savings of reducing the area of solar cell needed [10], [11],[12]. Much of this interest also evolves from the deleterious effect of increased temperature on the efficiency of CPV cells, creating a need for efficient thermal management through active heat removal or passive cooling of the cells [13]. Concentrating photovoltaic systems produce the advantage of high efficiency cells but also the complication of decreasing efficiencies with temperature increases. The spectral beam splitting alleviates much of this complication by allowing the longer wavelength radiation to travel directly to the PT device, decreasing the temperature gain and the need for active cooling [6]. Various configurations have been proposed in the literature for both commercial and residential applications, and producing a wide range of theoretical system efficiencies (solar-to-electric), ranging from 10% to upwards of 40%, although few economic estimates are available due to the lack of empirical testing of pilot scale systems.

Kosmadakis, Manolakos, and Papadakis proposed a hybrid system directly coupling a tracking parabolic concentrator with a silicon cell PV system with 10 suns concentration, and an organic Rankine cycle (ORC) thermal power generator. The full solar spectrum was directed to the PV cell with the thermal system acting as a heat sink. A numerical optimization concluded that the CPV–ORC combination improved the efficiency of CPV technology from 9.81% to 11.83% [14]. Liu, Hu, Zhang, and Chen proposed a hybrid system utilizing a two-axis tracking Fresnel concentrator with beam splitter and a secondary parabolic reflector and crystal silicone PV cells [15]. A numerically evaluated efficiency gain from 22.9% for the CPV system alone to 26.5% for the hybrid system was reported for cell operating temperature of 25 $^\circ$C and from 19.8% to 25.6% for cell operating temperature of 50 $^\circ$C. Ju, Wang, Flamant, Li, and Zhao proposed a hybrid system utilizing a Fresnel concentrator with beam splitter coupled with a gallium arsenide (GaAs) solar cell and a $CoSb_3$ thermo-electric generator. Analyzed numerically, the proposed system generated an optimal system efficiency of 26.62% and 27.49%, corresponding to heat sink heat transfer coefficients of 3000 $W/m^2K$ and 4500 $W/m^2K$, respectively[16]. The concentration ratio ranged from 550 to 770 suns.

Many of the recent investigations involving CSPV hybrid systems focus on the triple junction solar cell configuration, which when utilizing the full solar spectrum would operate at a higher theoretical efficiency than single crystalline silicon and other solar cell chemistries [12]. Chen et al, described a two-axis tracking concentrator focusing the full solar spectrum to a triple junction solar cell module coupled with a thermal energy collection system. The cell consisted of a top cell of InGaP, middle cell InGaAs and bottom cell Ge, connected in series. The system produced a solar cell efficiency of 26% while maintaining a thermal conversion efficiency of 52% [17].

## 2. Background and System Description

The University of Louisiana at Lafayette (UL Lafayette) has completed the initial field testing of a concentrating solar photovoltaic (CSPV) hybrid solar power system utilizing a vertical multi junction (VMJ) cell developed by MH Solar Co., Ltd. (MH Solar) - the MIH VMJ cell (VMJ cell). The parabolic trough CSPV system produces thermal energy by way of traditional concentrating solar power through the existing heat collection element (HCE) tubes, while producing electricity directly from a concentrating PV system operating in parallel. This is accomplished by splitting the solar radiation beam, filtering out the spectrum of wavelengths that are most efficiently converted to electricity by the silicon PV cells and redirecting the light onto the PV module, while leaving the remaining ultraviolet and infrared light to pass to the existing thermal receiver of the parabolic trough. This type of hybrid system, utilizing a single axis linear concentrator, a solar radiation beam splitter, and linear thermal receiver has been proposed previously [18]. However, the MH Solar system cell technology applies a novel VMJ cell which makes it uniquely suitable for this application.

The MIH VMJ cell is produced by stacking and bonding together a large number of P-N diffused silicon wafers, which is then cut vertically into thin MIH VMJ cells. This process has several distinct advantages to other CPV cell configurations. Because each P-N junction is in series, the MIH VMJ cell can generate a very high voltage in a small package. Each additional wafer added to the stack adds another 0.6V to the total device voltage, resulting in less current for a given solar cell area [19]. In addition the modules are easily scalable in size and output voltage. The parabolic trough CSPV application requires a solar cell that can perform well with moderate to high solar concentration, as well as under a partial solar spectrum. Similar to a triple junction solar cell, the MIH VMJ cell performance improves when in concentrated sunlight, and was shown to have linear performance out to 2500 suns concentration [15]. However, unlike triple junction cells which rely on the full solar spectrum for high efficiency operation, the MIH VMJ cell operates at peak efficiency within the spectral range most optimal for hybrid applications. According to Imenes and Mills, the band-pass region of the beam splitter for silicon

cells is from 590 to 1082 nm for the optimization of peak electrical output when combined with a thermal receiver [4]. Filtering this spectral band from the rest of the spectrum and directing it to silicon cells results in relatively high electrical conversion efficiencies such as 30% relative to the total solar energy of the spectral band [5]. The MIH VMJ cell displays peak solar conversion efficiencies within this range of the solar spectrum, while a triple-junction solar cell's performance will drop dramatically when illuminated by only a part of the solar spectrum, as shown in Figures 2 and 3 [20].

Additionally, the MIH VMJ cell's performance is reduced by only 3% for every 10 °C operation above its standard test conditions (Figure 4) [19]. This performance loss is not as dramatic as conventional silicon solar cells, which can lose 5% in performance for every 10 °C temperature rise [13].

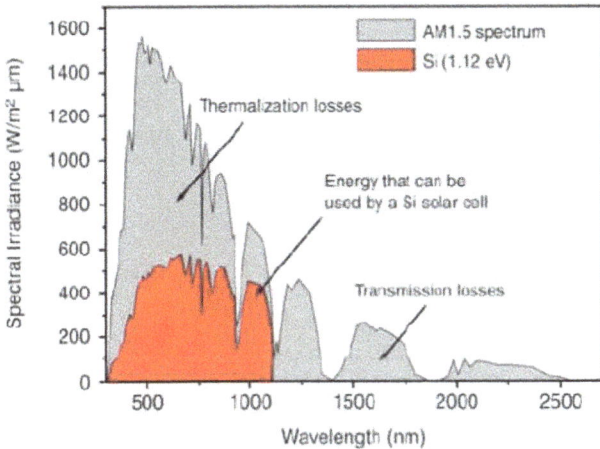

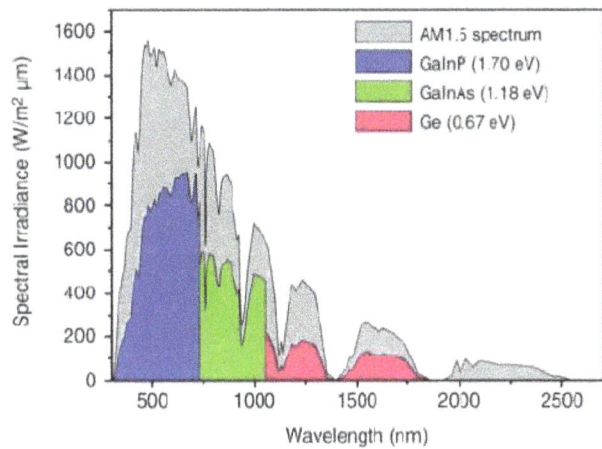

Figures 2 and 3. The AM1.5 solar spectrum and the parts of the spectrum that can be used by: Si solar cells (left figure) and certain triple junction solar cells (right figure) [20].

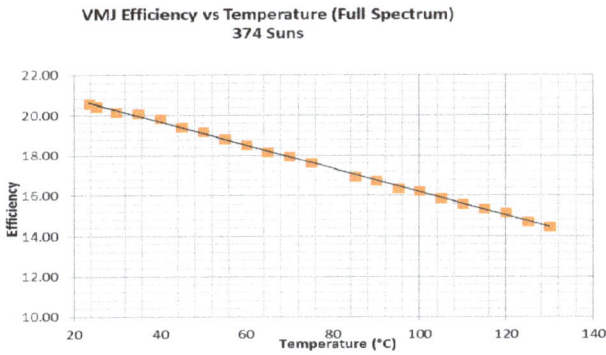

Figure 4. MIH VMJ cell efficiency vs. temperature. Courtesy of MH Solar.

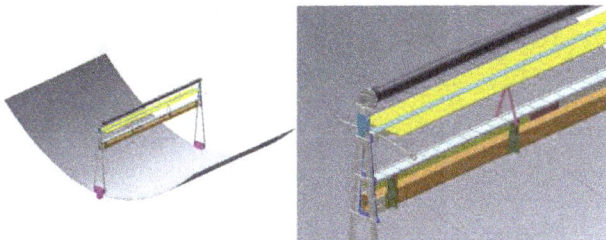

Figures 5 and 6. Depiction of CPV unit with cold mirrors and MIH VMJ cells. Courtesy of MH Solar.

## 3. Test Platform

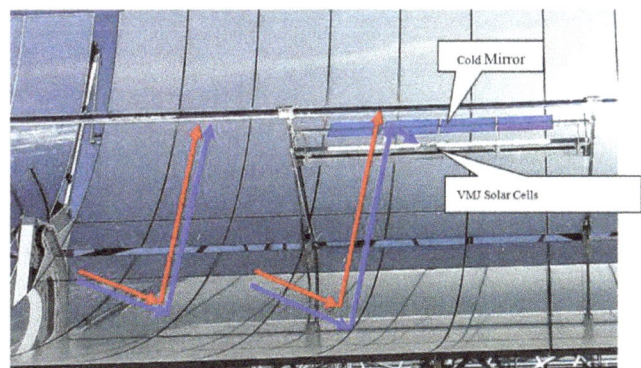

Figure 7. Schematic depiction of dichroic cold mirror operation. Red arrows represent infrared light; blue arrows represent visible light.

The VMJ cells were placed linearly in parallel with the HCE tube and with rotatable mirrors installed to either supply all of the incident solar radiation to the HCE tube while stowed, or to provide a selected spectrum of radiation to the cells while deployed. Depicted in Figures 5 and 6, the solar cells are passively cooled by use of a heat sink to ambient temperature. A depiction of the cold mirror operation in Figure 7 demonstrates the operation of the beam splitting "cold mirror", which reflects the visible portion of the spectrum (blue arrows) while allowing most infrared wavelengths (red arrows) to be transmitted efficiently. The cold mirror was a dichroic spectrally selective mirror with the dichroic film by Evaporated Metal Films Corporation (EMF). The cold mirror demonstrated excellent spectral reflectance in the band of 500 to 1000 nm as seen in Figure 8, while also demonstrating good transmittance outside of this band (Figure 9), presented here on a low iron glass substrate.

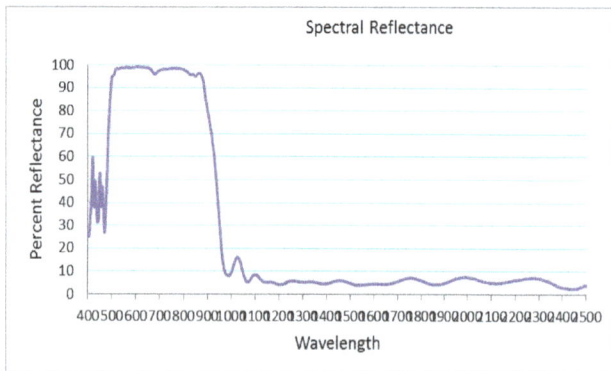

*Figure 8. Spectral reflectance of dichroic cold mirror.*

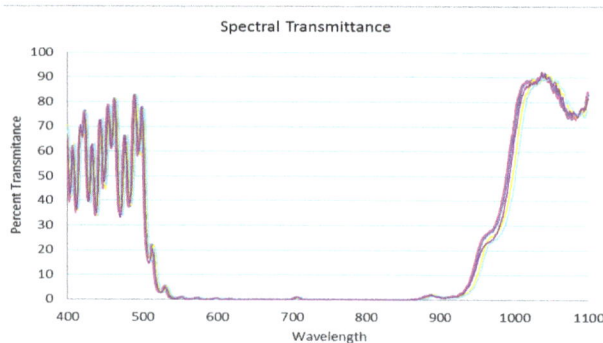

*Figure 9. Spectral transmittance of dichroic filter.*

Testing was conducted at the UL Lafayette Solar Technologies Application, Research and Testing (START) Center, where the only operating solar thermal power plant in Louisiana is located [21]. The CSPV module was installed on a Large Aperture Trough (LAT) from Gossamer Space Frames. The parabolic trough had an aperture of 7.3 meters and was oriented in a North-South configuration, while tracking from East to West. The thermal receiver was the PTR70, 70 mm heat collection element (HCE) tube from Schott. Previous testing completed by UL Lafayette has demonstrated direct normal (DNI) solar-to-thermal efficiencies of the solar collector in the range of 70 to 80 percent under optimum conditions [22]. Solar radiation measurements were taken onsite by the Kipp & Zonen SOLYS 2 Sun Tracker and CHP 1 pyrheliometer system. The LAT collector field consisted of 12 individual troughs, each 12 m in length arranged into two Solar Collector Assemblies (SCAs) (Figure 10). The collector field was coupled to an organic Rankine cycle (ORC) thermal power plant. Additional key metrics of the UL Lafayette solar thermal test facility are listed in Table 1.

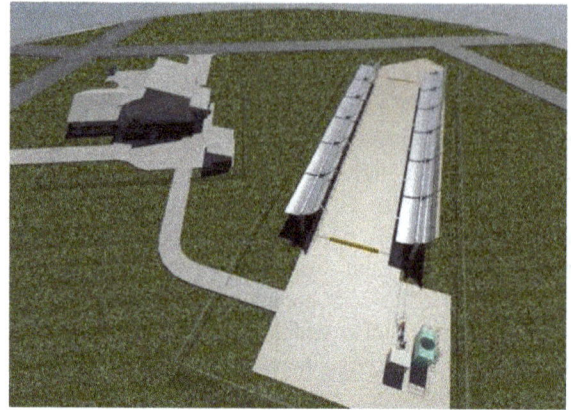

*Figure 10. Representation of UL Lafayette START Center.*

*Table 1. Plant characteristics.*

| Plant Location | Crowley, LA |
|---|---|
| Yearly Direct Normal Solar | 1590 kWh/m$^2$ |
| Plant Size (nominal) | 50 kWe |
| ORC Gross Output | 50 kWe |
| Solar Field Heat Transfer Fluid | Water |
| Inlet Temperature | 93°C |
| Outlet Temperature | 121°C |
| ORC Working Fluid | R245fa |
| ORC Design Point Efficiency | 8% |
| Solar Field Size | 1051 m$^2$ |
| Land Area | 4050 m$^2$ (1 acre) |
| Solar to Electric Design Point Efficiency | 6% |

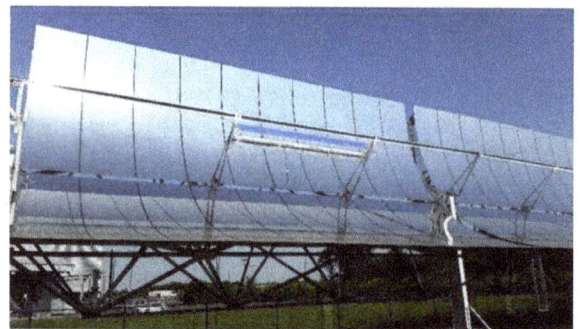

*Figure 11. MH Solar CSPV system installed at the UL Lafayette START Center.*

For the testing conducted at the UL Lafayette testing facility, one MH Solar module was installed in parallel with the HCE tube, four meters in length, on one parabolic trough Solar Collector Assembly (SCA) section (Figures 11 and 12). The focal length of the HCE tube was 2.0 m, and the cold mirrors, six inches in width, were placed at a distance 260 mm from the mirror surface, intercepting about 29 percent of the reflected light, according to the ray trace conducted by MH Solar. The mirror substrate was 3.2 mm thick glass. Both borosilicate float glass and low iron glass as the mirror substrate were tested. Additional key metrics of the PV module are listed in Table 2. The test plan included the operation of the CSPV system both with traditional full spectrum mirrors as well as the cold mirrors in both the deployed and stowed configuration.

*Figure 12. MH Solar CSPV system installed at the UL Lafayette START Center.*

*Table 2. PV cell module characteristics.*

| | |
|---|---|
| Cell width | 2 cm |
| Cell length | 5.5 cm |
| Area of the cell, $A_{cells}$ | 11 cm$^c$ |
| Cells/Module | 6 |
| How many cells wide (per module) | 3 cells |
| Packing Density (length) | 92% |
| Packing Density (width) | 97% |
| Overall packing density | 88.9% |
| Module Length | 12.0 cm |
| Module's / m | 8.4 |
| Module's / Trough | 100 |

# 4. Results

Experimental data was collected in an effort to quantify the effect of the CSPV module on the overall efficiency of the parabolic trough collector field. The solar radiation conversion efficiencies were calculated based on measured data for both the PV module and the thermal receiver to determine system performance.

### 4.1. PV Performance Data

Power and efficiency calculations of the MIH VMJ cells could be completed by use of the voltage and current measurements taken during operational testing of the cells under the concentrated and spectrally selective conditions described above. Open circuit voltage and short circuit current readings for two tests are given in Table 3. Here $I_{sc}$ (900) is the short-circuit current normalized to a DNI of a standard test condition of 900 W/m$^2$. The conversion efficiency $\eta_{PV}$ of the cell module is defined as:

$$\eta_{PV} = \frac{P_{PV}}{q_{PV}} \qquad (1)$$

where $P_{PV}$ is the power generated by the cell module and $q_{PV}$ is the solar energy incident on the cell module. The solar energy delivered to the cells can be expressed as:

$$q_{PV} = q_{in} * \eta_{CM} * \int_0^\infty \rho(\lambda)d\lambda * \eta_{T,cell} * \gamma_{cells} \qquad (2)$$

where $q_{in}$ is the incident solar radiation onto the trough, $q_{solar}$, multiplied by an adjustment factor due to the cosine

effect, $\cos(\theta)$ [23],

$$q_{in} = q_{solar*\cos(\theta)} . \qquad (3)$$

In (2), $\eta_{CM}$ is the efficiency of incident radiation reaching the cold mirror, $\rho(\lambda)$ is the spectral reflectivity of the cold mirror, $\eta_{T,cell}$ is the transmission efficiency of the cell module cover glass, and $\gamma_{cells}$ is the packing density of the cells in the module. In order to determine $q_{PV}$ a DNI of 900 W/m$^2$ was assumed as a standard test condition. The fraction of light reaching the cold mirror was determined from:

$$\eta_{CM} = A_{trough} * \eta_{opt} * \eta_{soiling} * \eta_{CM\ intercept} * \eta_{shading} \qquad (4)$$

where $A_{trough}$ is the area of the parabolic trough per unit PV module, $\eta_{opt}$ is the optical efficiency of the trough mirror, $\eta_{soiling}$ is the soiling factor of the trough mirror, $\eta_{CM\ intercept}$ is the fraction of trough aperture intercepted by the cold mirror, and $\eta_{shading}$ is the shading factor on the trough by the PV module, the latter two being determined by a ray trace conducted by MH Solar. The factors mentioned above relating to the testing conducted are listed in Table 3.

The power generated by the PV module, $P_{PV}$ can be expressed as [15]:

$$P_{PV} = A_{cells} * V_{OC} * I_{SC} * FF \qquad (5)$$

where $A_{cells}$ is the area of the PV cells, Voc is the open-circuit voltage, $I_{sc}$ is the short-circuit current density and FF is the fill factor. The open-circuit voltage and short-circuit current measurements for two tests are listed in Tables 4 and 5, along with the max power, normalized to a DNI of 900 W/m$^2$. Also listed are additional test conditions including the actual DNI at the time of measurement. Figure 13 depicts the fill factor measurements at a concentration of 374 suns versus temperature. An empirically measured fill factor of 0.75 is utilized for the efficiency analysis (Figure 14).

*Table 3. PV cell and cold mirror characteristics.*

| | |
|---|---|
| **Aperture Width** | **7.3 m** |
| **Module length** | **0.12 m** |
| $A_{trough}$ | 0.876 m$^2$ |
| $q_{in}$ | 900 W/m$^2$ |
| $\eta_{opt}$ | 85% |
| $\eta_{soiling}$ | 95% |
| $\eta_{CM\ intercept}$ | 29% |
| $\eta_{shading}$ | 96% |
| $\gamma_{cells}$ | 88.9% |
| $\rho$ (total) | 48.6% |
| $\eta_{T,cell}$ | 90% |

For the given results of Test 1 an average power produced per module was 20.05 Watts, with a       of 69 Watts, resulting in a PV efficiency of 29% under the spectrally selective and concentrated conditions. The second set of tests yielded a PV efficiency of 29.5%. This efficiency calculation is dependent on an accurate optical efficiency calculation not available at the time of the test and therefore Table 5 is given which illustrated the range of efficiencies based on potential optical efficiency corrections.

***Table 4.*** *Test 1 results.*

| Reading - 1 | | | | Reading - 2 | | | |
|---|---|---|---|---|---|---|---|
| $V_{oc}$ (V) | $I_{sc}$ (mA) | $I_{sc}$(900) | $P_{max}$(900)W | $V_{oc}$ (V) | $I_{sc}$ (mA) | $I_{sc}$(900) | $P_{max}$(900)W |
| 260.80 | 80.00 | 100.42 | 19.66 | 250.30 | 80.00 | 105.57 | 19.84 |
| 261.10 | 82.20 | 103.18 | 20.23 | 251.40 | 82.20 | 108.48 | 20.47 |
| Time | 3:58 | | | 4:01 | | | |
| DNI | 717 | | | 682 | | | |
| Ambient Temp. | 86.0 | | | 86.0 | | | |
| Mirrors Used | low iron, 3.2mm cold mirror | | | low iron, 3.2mm cold mirror | | | |
| cosine correction | 0.999 | | | 0.999 | | | |

***Table 5.*** *Test 2 results*

| Reading - 1 | | | | Reading - 2 | | | |
|---|---|---|---|---|---|---|---|
| Voc (V) | Isc (mA) | Isc(900) | Pm(900)W | Voc (V) | Isc (mA) | Isc(900) | Pm(900)W |
| 256.70 | 86.70 | 112.27 | 21.79 | 251.10 | 86.60 | 106.33 | 20.19 |
| 256.60 | 81.10 | 105.02 | 20.37 | 250.30 | 81.10 | 99.58 | 18.84 |
| Time | 4:20 | | | 4:23 | | | |
| DNI | 695 | | | 733 | | | |
| Ambient Temp. | 84.0 | | | 84.0 | | | |
| Mirrors used | boro float, 3.2mm cold mirror | | | boro float, 3.2mm cold mirror | | | |
| Cosine correction | 0.992 | | | 0.992 | | | |

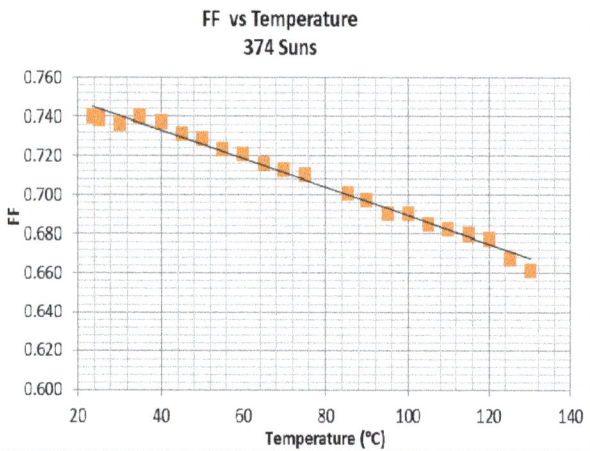

***Figure 13.*** *Fill factor vs. temperature for concentration ratio of 374.*

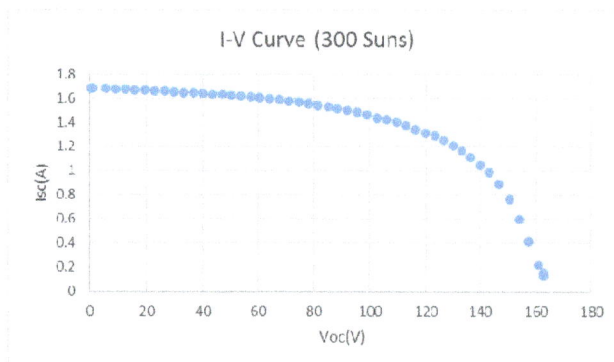

***Figure 14.*** *I-V curve for concentration ratio of 300.*

***Table 6.*** *Range of MIH VMJ cell efficiencies based on various optical efficiencies.*

| | Optical Efficiency | | | | | | | | |
|---|---|---|---|---|---|---|---|---|---|
| Cell | 78 | 79 | 80 | 81 | 82 | 83 | 84 | 85 | 86 |
| Efficiency | % | % | % | % | % | % | % | % | % |
| 29.0% | | | | | | | | | |
| 29.5% | | | | | | | | | |
| 30.0% | | | | | | | | | |
| 30.5% | | | | | | | | | |
| 31.0% | | | | | | | | | |
| 31.5% | | | | | | | | | |
| 32.0% | | | | | | | | | |
| 32.5% | | | | | | | | | |
| 33.0% | | | | | | | | | |
| 33.5% | | | | | | | | | |

### 4.2. Thermal Performance Data

The thermal efficiency, $\eta_T$, of the solar collector field can be expressed as:

$$\eta_T = \frac{P_T}{q_{in}} \qquad (6)$$

where $P_T$ is the thermal power generated by the solar field. The thermal power is calculated from the process flow data of the HTF through the collector field:

$$P_T = \dot{m} \int_{T_{in}}^{T_{out}} C_p \, dT \qquad (7)$$

where $\dot{m}$ is the mass flow rate of the heat transfer fluid through the solar collector field, $C_p$ is the specific heat of constant pressure, and $T_{in}$ and $T_{out}$ are the temperatures of the HTF leaving and entering the collector field, respectively. For the testing described here, the HTF was water with a 10 percent glycol mixture. Over the temperature ranges utilized in this study, the specific heat could be considered constant, resulting in:

$$P_T = \dot{m}C_p\Delta T \qquad (8)$$

where $\Delta T$ is the temperature gain through the collector field. The temperature gain through the portion of the parabolic trough which contained the cold mirrors, 4 meters in length, was examined. The $\Delta T$ was recorded for conditions without the CSPV mirrors installed, with the mirrors installed but stowed, and with the mirrors installed and deployed. A measurement was also recorded with full spectrum mirrors deployed in the place of the cold mirrors. In order to compare the effect on thermal performance with the CSPV cold mirrors in use, the thermal efficiency of the solar collector field was determined for each case and the difference taken. For the full mirrors, an average difference in thermal efficiency of 2.4 percentage points was observed during testing for deployed and stowed. Figure 15 depicts the results from one such test. The efficiency in both the stowed and deployed positions trended upward during the test day due to changes in tracking error which occur in normal operation. The tracking error typically maintains a 0.05 degree tolerance to the calculated sun angle. The absolute value of the tracking error for the test day is depicted in Figure 16. This fluctuation in tracking accuracy therefore accounts for some of the fluctuation seen in the efficiency calculations. Figure 17 displays the cosine corrected DNI during the CSPV test with the full mirrors in place.

In terms of efficiency, the deployed full mirror intercepted 29% of the aperture of a 4 meter linear distance, or 1/9th of the span where the efficiency calculation occurred. Therefore, considering the full 36 meter length (1/2 SCA) where the temperature measurements were taken, the expected measured reduction in efficiency across the span would be $\frac{1}{9} * 29\% = 3.2\%$ of the unobstructed thermal efficiency. During this testing, the unobstructed thermal efficiency was about 70.5%. A 3.2% reduction in efficiency would result in a drop of 2.25 percentage points, which is close agreement with the 2.4 point drop in thermal performance as measured in this testing. By applying the measured loss of efficiency across the full ½ SCA, the total reduction in efficiency of a fully equipped section (9 CSPV modules) would be 9*2.4 = 21.6 points while the full mirrors were deployed. This is in good agreement with the estimated 29%*70.5% = 20.5 percentage point reduction that would be expected.

The cold mirrors were tested utilizing a low iron glass substrate, resulting in an average difference in thermal efficiency of 1.03 percentage points, or 1.6 % reduction in efficiency. Figure 18 depicts results of the cold mirror testing. Comparing to the full mirror, this result would suggest that

about 45.5% of the radiation spectrum successfully passed through the cold mirror. This compares to the expected 48.6% of the spectrum to be reflected. Extending the thermal efficiency reduction from the cold mirror across the full length of the span (1/2 SCA) in which the temperature measurements were taken results in a 9.3 reduction in percentage points, or 14%, reduction in thermal efficiency. Figure 19 shows the cosine corrected DNI during the CSPV test with the cold mirrors.

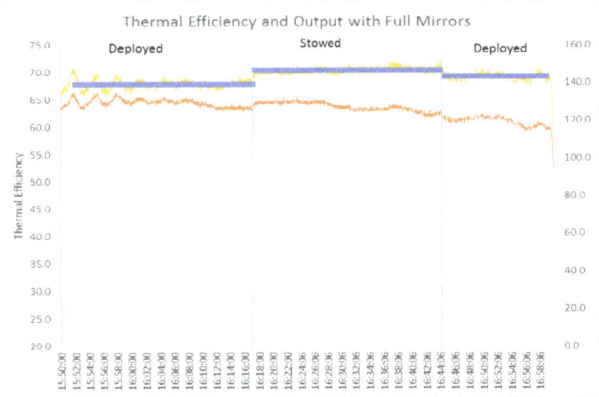

***Figure 15.*** *Thermal efficiency of CSPV system with full mirrors in deployed and stowed condition.*

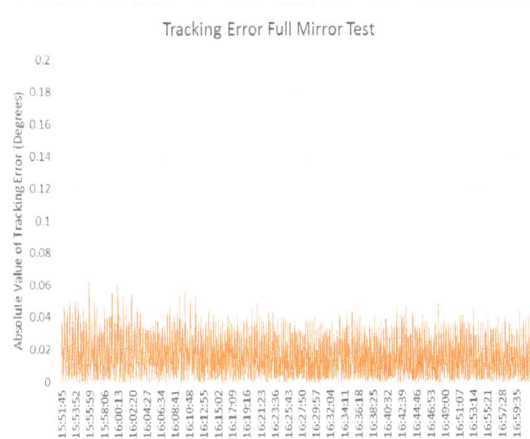

***Figure 16.*** *Absolute value of tracking error during CSPV test with full mirror.*

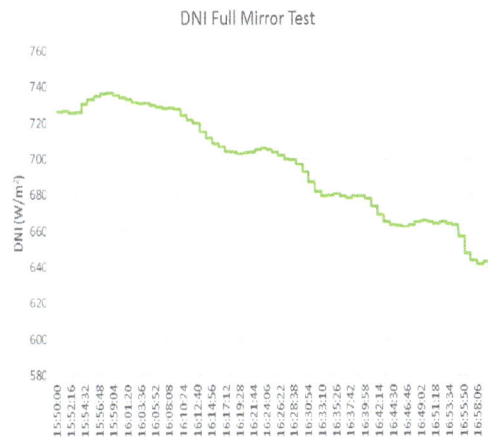

***Figure 17.*** *Cosine corrected DNI during CSPV test with full mirrors.*

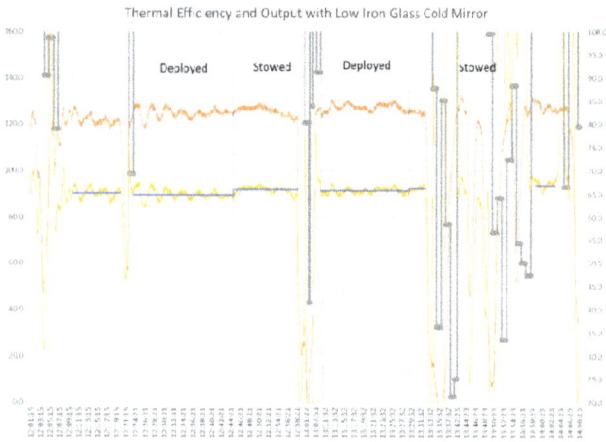

*Figure 18. Thermal efficiency of CSPV system with cold mirrors in stowed and deployed position.*

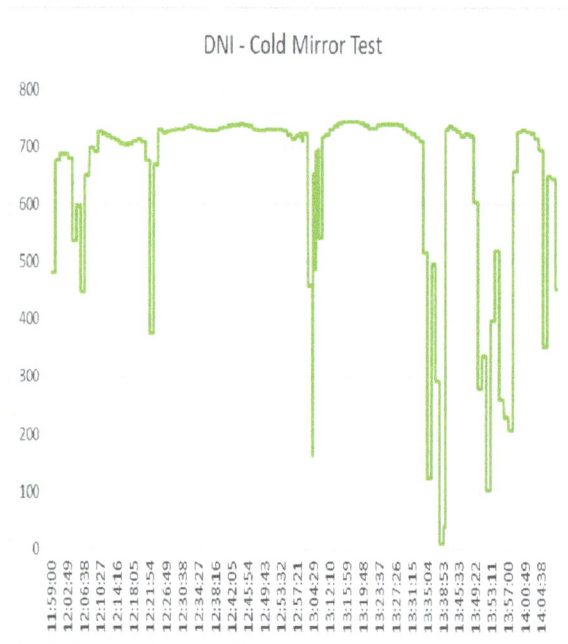

*Figure 19. Cosine corrected DNI curing CSPV test with cold mirrors.*

The system efficiency, $\eta_{system}$, with the PV and thermal systems operating in parallel can be express as

$$\eta_{system} = \frac{P_{system}}{q_{in}} \qquad (9)$$

where $P_{system}$ is the system power found from:

$$P_{system} = P_{PV} + P_{thermal} \qquad (10)$$

Using the results from the PV tests in Table 3, given a DNI of about 715 W/m$^2$, an additional 541 W of power are added to the system from the 4 m span. There are 9 four-meter spans in the area of the collector field under investigation, resulting in a total addition of 4,870 W of electric power. In order to determine the overall change in power output of the solar collector field, the thermal energy was assumed to be converted to electricity at an efficiency of between 33% and 35%, which is typical of installed parabolic trough solar thermal power plants, excluding any natural gas supplement [24]–[27]. Based on the assumed conversion rate and using the thermal data when the DNI was 715 W/m$^2$, a total of between 42,160 W and 44,420 W of electric power would be produced from the area of collector field under investigation (half of one SCA), 263 m$^2$ of aperture area. This compares to between 40,854 W and 43,330 W of electric power without the CSPV installation, resulting in an overall increase of 2.5% to 3.2% of power output from the solar collector field.

# 5. Discussion and Conclusions

Initial test results of a CSPV hybrid photovoltaic and thermal solar energy system have been presented. The combination of the CPV performance data and the thermal performance data show an overall increase in system efficiency from the base thermal only case. Based on the testing, an overall system solar-to-electric efficiency improvement from 21.7 percent to 22.5 percent was achieved. This increase in system efficiency was achieved through the employment of the MIH VMJ cell which has a high efficiency within the solar radiation spectrum bandwidth of 500 to 1000 nm. This bandwidth was selectively directed toward the PV portion of the spectrum at a concentration ratio of about 30. This allowed the PV cells to operate at high efficiency while also utilizing passive cooling. The remaining solar radiation spectrum was directed toward the parabolic trough HCE tube at a concentration ratio of 104, efficiently converting the remaining spectrum into thermal energy.

Future work involves conducting longer term additional optical efficiency measurements in order to more accurately determine the actual CPV cell efficiencies. Utilizing lower flow rates will increase the temperature gain through the field and provide higher resolution for data analysis. Additional analysis is needed to determine the optimal operational configuration for the CSPV system. Based on the operating scheme of a particular plant, the advantages built-in to operating the CPV system could be manifested in several ways. One, the operator could utilize the cold mirrors in the full collector field in order to capture the peak solar irradiation that is beyond the design rating of the field and would have been otherwise directed to storage. This could allow optimal generation of electrical power based on peak demand and value of the generated power. An additional scenario would involve using full mirrors to capture the solar irradiation in portions of the collector field that would otherwise have been dumped based on the rated capacity of the thermal power block.

# Acknowledgements

This work was supported and made possible by funding from Cleco Power LLC and the University of Louisiana at Lafayette.

# References

[1]   R. Margolis, C. Coggeshall, and J. Zuboy, "Sunshot Vision Study," *US Dept. Energy*, no. February, 2012.

[2]   "Progress Report: Advancing Solar Energy Across America | Department of Energy." [Online]. Available: http://energy.gov/articles/progress-report-advancing-solar-energy-across-america. [Accessed: 01-Aug-2014].

[3]   A. Mojiri, R. Taylor, E. Thomsen, and G. Rosengarten, "Spectral beam splitting for efficient conversion of solar energy—A review," *Renew. Sustain. Energy Rev.*, vol. 28, pp. 654–663, Dec. 2013.

[4]   A. G. Imenes and D. R. Mills, "Spectral beam splitting technology for increased conversion efficiency in solar concentrating systems: a review," *Sol. Energy Mater. Sol. Cells*, vol. 84, no. 1–4, pp. 19–69, Oct. 2004.

[5]   A. Mojiri, C. Stanley, and G. Rosengarten, "Spectrally Splitting Hybrid Photovoltaic/thermal Receiver Design for a Linear Concentrator," *Energy Procedia*, vol. 48, pp. 618–627, 2014.

[6]   Y. Li, S. Witharana, H. Cao, M. Lasfargues, Y. Huang, and Y. Ding, "Wide spectrum solar energy harvesting through an integrated photovoltaic and thermoelectric system," *Particuology*, vol. 15, pp. 39–44, Aug. 2014.

[7]   F. Shan, F. Tang, L. Cao, and G. Fang, "Performance evaluations and applications of photovoltaic–thermal collectors and systems," *Renew. Sustain. Energy Rev.*, vol. 33, pp. 467–483, May 2014.

[8]   V. V. Tyagi, S. C. Kaushik, and S. K. Tyagi, "Advancement in solar photovoltaic/thermal (PV/T) hybrid collector technology," *Renew. Sustain. Energy Rev.*, vol. 16, no. 3, pp. 1383–1398, Apr. 2012.

[9]   L. Tan, X. Ji, M. Li, C. Leng, X. Luo, and H. Li, "The experimental study of a two-stage photovoltaic thermal system based on solar trough concentration," *Energy Convers. Manag.*, vol. 86, pp. 410–417, Oct. 2014.

[10]  N. R. E. Wilson, Greg; Emergy, Keith; Laboratory, "Best Research-Cell Efficiencies." [Online]. Available: http://www.nrel.gov/ncpv/images/efficiency_chart.jpg. [Accessed: 13-Oct-2014].

[11]  E. F. Fernández, G. Siefer, F. Almonacid, A. J. G. Loureiro, and P. Pérez-Higueras, "A two subcell equivalent solar cell model for III–V triple junction solar cells under spectrum and temperature variations," *Sol. Energy*, vol. 92, pp. 221–229, Jun. 2013.

[12]  H. Helmers, M. Schachtner, and A. W. Bett, "Influence of temperature and irradiance on triple-junction solar subcells," *Sol. Energy Mater. Sol. Cells*, vol. 116, pp. 144–152, Sep. 2013.

[13]  E. Skoplaki and J. a. Palyvos, "On the temperature dependence of photovoltaic module electrical performance: A review of efficiency/power correlations," *Sol. Energy*, vol. 83, no. 5, pp. 614–624, May 2009.

[14]  G. Kosmadakis, D. Manolakos, and G. Papadakis, "Simulation and economic analysis of a CPV/thermal system coupled with an organic Rankine cycle for increased power generation," *Sol. Energy*, vol. 85, no. 2, pp. 308–324, Feb. 2011.

[15]  Y. Liu, P. Hu, Q. Zhang, and Z. Chen, "Thermodynamic and optical analysis for a CPV/T hybrid system with beam splitter and fully tracked linear Fresnel reflector concentrator utilizing sloped panels," *Sol. Energy*, vol. 103, pp. 191–199, May 2014.

[16]  X. Ju, Z. Wang, G. Flamant, P. Li, and W. Zhao, "Numerical analysis and optimization of a spectrum splitting concentration photovoltaic–thermoelectric hybrid system," *Sol. Energy*, vol. 86, no. 6, pp. 1941–1954, Jun. 2012.

[17]  H. Chen, J. Ji, Y. Wang, W. Sun, G. Pei, and Z. Yu, "Thermal analysis of a high concentration photovoltaic/thermal system," *Sol. Energy*, vol. 107, pp. 372–379, Sep. 2014.

[18]  S. Jiang, P. Hu, S. Mo, and Z. Chen, "Optical modeling for a two-stage parabolic trough concentrating photovoltaic/thermal system using spectral beam splitting technology," *Sol. Energy Mater. Sol. Cells*, vol. 94, no. 10, pp. 1686–1696, Oct. 2010.

[19]  B. Sater, M. Perales, J. Jackson, S. Gadkari, and T. Zahuranec, "Cost-effective high intensity concentrated photovoltaic system," *IEEE 2011 EnergyTech*, pp. 1–6, May 2011.

[20]  N. Yastrebova, "High efficiency multi-junction solar cells: current status and future potential," University of Ottawa SUNLAB, Ottawa, Canada. [Online]. Available: http://sunlab.eecs.uottawa.ca/?page_id=134. [Accessed 13-Oct-2014].

[21]  T. Chambers, J. Raush, and G. Massiha, "Pilot solar thermal power plant station in southwest Louisiana," *Int. J. Appl. Power Eng.*, vol. 2, no. 1, 2013.

[22]  J. Raush and T. Chambers, "Demonstration of Pilot Scale Large Aperture Parabolic Trough Organic Rankine Cycle Solar Thermal Power Plant in Louisiana," *J. Power Energy Eng.*, vol. 1, no. 7, pp. 29–39, 2013.

[23]  E. Leonardi and B. D'Aguanno, "CRS4-2: A numerical code for the calculation of the solar power collected in a central receiver system," *Energy*, vol. 36, no. 8, pp. 4828–4837, Aug. 2011.

[24]  "System Advisor Model ( SAM ) Case Study :," National Renewable Energy Laboratory, [Online]. Available: https://sam.nrel.gov/content/case-studies. [Accessed 13-Oct-2014].

[25]  R. Cable, "Solar Trough Generation - The California Experience," *ASES Forum 2001*, Washington D.C., 2001.

[26]  E. F. Camacho and A. J. Gallego, "Optimal operation in solar trough plants: A case study," *Sol. Energy*, vol. 95, pp. 106–117, Sep. 2013.

[27]  H. Price, "A Parabolic Trough Solar Power Plant Simulation Model Preprint," ISES 2003, International Solar Energy Conference. National Renewable Energy Laboratory, January, 2003.

# 17

# Present scenario of renewable and non-renewable resources in Bangladesh: A compact analysis

**Md. Niaz Murshed Chowdhury[1], Samim Uddin[2], Sumaiya Saleh[1]**

[1]Research Assistant, Department of Economics, South Dakota State University, Brookings, USA
[2]Department of Economics, University of Chittagong, Chittagong, Bangladesh

**Email address**

md.chowdhury@jacks.sdstate.edu (Chowdhury M. N. M.), shamimecocu@gmail.com (Uddin S.), sumaiya.saleh@jacks.sdstate.edu (Saleh S,)

**Abstract:** This research deliberately searches the present scenario of renewable and non-renewable resources in Bangladesh and also focuses on their effective management. Therefore, the research is unique in terms of focusing the present scenario of natural resources and the research highlights the present conditions of natural resources. Most of the study and research on NRs focus on superficial problems, poverty, gender, and scientific measure of resource degradation. This research gives special attention to find the present scenario and to find the actors who are responsible for NRs management. There are considerable opportunities of Bangladesh to boost the economic growth through renewable and nonrenewable resource. With the help of these resources Bangladesh can generate electricity and can meet the required demand in the future. Therefore, the Government and the Private sector should work hand to hand to emphasize more renewable energy sources to produce electricity to solve our power crisis problem. Renewable energy sources discussed above can help Bangladesh to produce more power in order to reduce Load-shedding problem. Time has come to look forward and work with these renewable energy fields to produce electricity rather than depending wholly on conventional method. In addition, we observed that Bangladesh has a huge amount of natural Gas and other mineral resources. Proper and corruption free management can be able to solve the problem of energy crisis.

**Keywords:** Natural Resource, Renewable, Non-Renewable, Public and Private Sector, NR Management

## 1. Introduction

Natural resources and their management are the most important for a country. There are many states which are straightly depends on natural resources. Bangladesh is a developing and a probabilities country, with small area. We have also some natural resources, some are renewable and some are nonrenewable. Renewable resources are natural resources that can be replenished in a short period of time. On the other hand, non-renewable resource is a natural that cannot be remade or re-grown at a scale of comparable to its consumption. In Bangladesh, there are many natural resources such as: renewable natural resources are Energy, Water, Fish, Forest etc. and Coal, petroleum, oil, natural gas Rock, Sand etc. are considered non-renewable natural resources. Energy is the prime ingredient for sustainable economic development of a country. Economic Development depends on effective management of one's natural resources. People all over the world have a large unsatisfied demand of

energy, which is growing rapidly in the span of time. Bangladesh has a massive potential for renewable energy and the natural availability of alternative energy that creates opportunities of Growth in Bangladesh. Technologies should be developed to produce energy in an environment friendly manner as well as enough importance should be given to conserve the energy in most efficient shape. In order to ensure energy security, the primary energy source of the country especially gas, coal and other mineral resources have been taken into consideration.

Majority rural poor of Bangladesh depend on Natural Resources (NR) for their livelihoods. Land, water, forests, and livestocks are the sources of livelihoods. The rural economy depends on productivity of the natural resources. Small trade and manufacturing process cannot replace dependency over agricultural and natural resources. The country lacks institutional framework in terms of Natural Resource Management (NRM), which has resulted chaos and conflict over NRs and eventually poor and marginal people do not have access to. People have been losing their entitlement to these resources. On the hand, degradation of land and other resources along with bio-diversity and eco-system are the prime concern for the entire population in Bangladesh. Bangladesh is experiencing an acute shortage of electric power that is likely to be worsening day by day that stresses the need for the deployment of renewable energy resources to extenuate this energy crisis. In Bangladesh, there are many natural resources such as coal, gas, petrol. The main source of energy in Bangladesh is Natural gas (24%) that is likely to be depleted by the year 2020. Then Bangladeshis people will be faced a problem. In this case renewable energy helps the people of Bangladesh. Bangladesh has a vast potential for renewable energy and the natural availability of alternative energy creates opportunities of Growth in power sector; the substantial availability of renewable energy sources in the form of solar, biomass, biogas, hydropower and wind energy can provide opportunities of sustainable energy based development. Bangladesh is one of the low energy consuming countries of the world. The national grid could so far cover only 35% of the total population, and only 3 per cent people are enjoying piped gas supply. About 72 per cent people of Bangladesh live in rural areas, where the situation is worse and renewable energy is considered to be the right choice for providing clean energy to these remote settlements. Bangladesh is endowed with rich and extensive fisheries resources. Due to natural conditions and geographical location, Bangladesh has huge fisheries resources having high potential of increasing fisheries production. The country's fisheries may be conveniently divided into inland and marine sectors, although the dividing line between salt and fresh water, and open sea and inland waterway is very nebulous. Fishermen to move seasonally from inland open waters to sea fishing so that any demarcation between the two fisheries must be arbitrary. Inland fisheries is further divided into two groups i.e. aquaculture and inland capture. An inland fishery occupies an area of 4.575 million ha and marine capture covers 1, 66,000 sq. km. The culture fisheries include ponds, ox-bow lakes and coastal shrimp farms. The flood plains and the beels, which cover an area of 29.5 lakh ha, offering tremendous scope and potential for augmenting fish production by the adopting aquaculture based enhancement techniques.

The objectives of this research are:

(i) To have an overview on natural resources: data and information about land, water, forest, minerals and fisheries could help to produce the overview.

(ii) To determine calamitous factors involved changing bio-diversity and destruction of ecology.

(iii) Figure out the leakage of natural resource management and suggest the appropriate policy laws.

# 2. Data and Methodology

The research has studied secondary data and information. This research carefully examines some existing reports and study on natural resources in Bangladesh. The literature review also includes policy papers, declaration, and conventions on natural resources, bio-diversity, and ecology. Information, data, and case studies are studied and compiled. Both Bengali and English daily newspapers are studied for seeking information and data. The daily news papers selected for this purpose are The Somokal, The New Age, The Financial express, The Independent, and The Daily Star. Information and data on natural resources are collected, compiled and analyzed for this research. Internet sources include research paper, reports, workshop outputs and information published in the web sites.

This research deliberately searches the present scenario of renewable and non-renewable resources in Bangladesh and also focuses on their effective management. Very few researches are available to focus on both things. Therefore, the research is unique in terms of focusing the present scenario of natural resources and the research highlights the present conditions of natural resources. Most of the study and research on NRs focus on superficial problems, poverty, gender, and scientific measure of resource degradation. This research gives special attention to find the present scenario and to find the actors who are responsible for NRs management.

# 3. An Overview of Renewable (Energy, Fish, Forest, Land, Water) Resources of Bangladesh

## 3.1. Energy Situation

### 3.1.1. Present Scenario of Electricity Production

Bangladesh, with its 160 million people in a land mass of 147,570 sq. km is an emerging economy of South Asia successfully maintaining sustained economic growth of least 6% since last decade resulted a considerable high electricity demand each year. A booming economic growth, rapid urbanization and increased industrialization and development have increased the country's demand for electricity. Presently, 62% of the total population (including renewable energy) has access to electricity and per capita generation is 321 kWH, which is very low compared to other developing countries. In Bangladesh, renewable energy is considered to be the right choice for providing clean energy to these remote settlements. Power generation in Bangladesh was mono-fuel dependent, i.e. indigenous natural gas since 2009 considering its apparent huge availability. About 89% of generated power comes from natural gas and the rest is from liquid fuel, coal and hydropower. The present share of renewable energy is only 0.5%. As per election manifesto of the present government electricity generation in the country would be 7000 MW by the year 2013, 8000 MW by 2015 and 20,000 MW by the year 2021. The government aims to generate additional 15,000 MW electricity, within 2016 under short, medium and long term plan. The government has further extended its vision targeting the upcoming years up to 2030 and prepared the Power System Master Plan 2010 (PSMP). This plan states that in 2030 the demand of power would be around 34,000 MW while the generation capacity would be about 39,000 MW. Presently, the generation capacity is nearly 9,713 MW (September 2013) which implies that much endeavor is required to achieve the goal.

### 3.1.2. Overview in Different Renewable Energy Sources

#### 3.1.2.1. Solar Power

Solar Energy can be a great source for solving power crisis in Bangladesh. Bangladesh is a south Asian country located in between latitude 20°34' and 26°39' north and longitude 80°00' and 90°41' east that is an ideal location for solar energy utilization. Bangladesh is a subtropical country, 70% of year sunlight is dropped in Bangladesh. For this reason, we can use solar panels to produce electricity largely. Solar radiation varies from season to season in this country and receives an average daily solar radiation of 4-6.5 kWh/m². In a recent study conducted by Renewable Energy Research Centre, it is found maximum amounts of radiation are available in the month of March-April and minimum in December-January.

The vision is targeted to achieve through a concerted effort of Bangladesh Government involving a number of government ministries and their affiliated agencies. In addition, there would be a strong involvement of private sector (more than 50%) in the project implementation.

Under this initiative there would be two types of projects:

1. Business type involving contribution from beneficiaries and private sector management.
2. Social Service type aiming to implement government's social commitment.

The snapshot of the program is as follows: Investments projects are 1. Installation of solar irrigation pumps. 2. Installation of mini grid solar system. 3. Solar park. 4. Roof-top solar power solution. Social sector projects are 1. Solar electrification at railway station. 2. Solar electrification at Union Information Services Centers. 3. Solar led street lighting. 4.Solar electrification in rural health center. 5.Installation of Solar Home System in Religious Establishments. 6.Solar Electrification in Remote Education Centers. 7. Installation of Solar Home System in Government / Semi-government offices.

#### 3.1.2.2. Biogas

Organic wastes such as dead plant and animal material. Animal dung, and Kitchen waste can be converted into a gaseous fuel called biogas. Biogas originates from biogenic material and is a type of biofuel. Major components of biogas are 40-70% methane ($CH_4$), 30-60% carbon dioxide ($CO_2$) and other gases (1-5%). It also contains several trace gases like Hydrogen sulfide ($H_2S$), Nitrogen ($N_2$), Ammonia ($NH_3$) and Carbon monoxide (CO). A biogas based electricity generation system consists of a digester, a biogas collection tank, a generator as well as the piping and controls required for successful operation. The biogas is produced in the anaerobic digester, where anaerobic fermentation takes place

which is provided every day with livestock manure in the form of cattle dung. Biogas production plays an important role in Bangladesh since the necessary resources are plentiful. The Government along with several NGOs are working together for development of power production from Biogas. Grameen Shakti is one of the most uttered NGO in field of biogas and has completed several works. They have completed 13,500 biogas plants. Recently Seed Bangla Foundation has proposed a 25 KW Biogas based Power plant in Rajshahi. IDCOL A Government owned Investment Company fixed a target to set up 37,669 biogas plants in Bangladesh by 2012, under its National Domestic Biogas and Manure Programmers (NDBMP). Bangladesh has a wonderful climate for biogas production. The ideal temperature for biogas is around 35 [deg. celsious] The temperature in Bangladesh usually varies from 6 deg. celsious to 40 deg. But the inside temperature of a biogas digester remains at 22 deg.-30 deg., which is very near to the optimum requirement.

#### 3.1.2.3. Biogas Based Electricity Generation Plants

IDCOL is financing setting up of three biogas based on electricity generation plants, one in Mymensingh and two in Gazipur, and one organic fertilizer plant in Gazipur by Paragon Agro Ltd. Electricity generated from these plants will be supplied to the adjacent poultry farms of Paragon Poultry Ltd. (PPL) at BDT 4 / kWh, while organic fertilizer will be sold in the market at BDT 15 per 1 Kg packet and BDT 400 per 40 Kg packet. Total project cost is BDT 149.40 million.

#### 3.1.2.4. National Domestic Biogas and Manure Program

Infrastructure Development Company Limited (IDCOL) is implementing National Domestic Biogas and Manure Programme (NDBMP) with support from GoB, SNV-Netherlands Development Organization. Under the project a total of 37,269 domestic sized biogas plants have been financed during the period 2006-2012. The overall objective of the NDBMP is to further develop and disseminate domestic biogas plants in rural areas with the ultimate goal to establish a sustainable and commercial biogas sector in Bangladesh.

#### 3.1.2.5. Wind

Bangladesh is in the midst of a severe energy and power supply crisis; one of the worst in South Asia. However, the government is now looking to explore the potential of wind energy, particularly along the country's 724 km long coastline. Wind energy can potentially generate more than 2000 -MW of electricity in the coastal regions. The growth of wind energy in the underdeveloped, coastal areas of the country holds hope for poor, isolated communities that are not connected to the national electricity grid and who are also unlikely to receive grid connection in the near future due to the high cost of establishing infrastructure and growing scarcity of traditional energy inputs. The Bangladesh Power Development Board has estimated that wind energy can contribute to 10% of the energy needs of the country. The

Board has also calculated the cost to generate one kWh from wind energy to be about half the cost of generating an equivalent unit of power from solar energy. The expansion of the potential of wind energy will be crucial in order for Bangladesh to achieve its national vision of providing electricity to all of its population by 2020.

### 3.1.2.6. Biomass

Bangladesh is an agricultural country and has strong potential for biomass gasification based on electricity. Cattle dung, agricultural residue, poultry dropping, water hyacinth, rice husk etc. used for biomass power generation are available in Bangladesh. More common biomass resources available in the country are rice husk, crop residue, wood, jute stick, animal waste, municipal waste, sugarcane bagasse

etc. Exploration of these resources for electricity generation is still at preliminary stage.

### 3.1.2.7. Micro Hydro

The Karnafuly Hydro Power Station is the only hydropower plant in the country with a capacity of 230 MW. It is operated by BPDB (Bangladesh Power Development Board). BPDB is considering increasing production up to 330MW. Micro hydro and mini hydro have limited potential in Bangladesh with exception of Chittagong Hill Tracts. Hydropower assessments have identified some possible sites from 10 kW to 5 MW implementation of which is still at large. Other renewable energy sources include bio-fuels, gasohol, geothermal, river current, wave and tidal energy. Potentialities of these resources are yet to be explored.

*Table 1. Potentialities of renewable resources.*

| Resources | Potential | Entities Involved |
|---|---|---|
| Solar | Enormous | Public and Private sector |
| Wind | Resource mapping required | Public sector / PPP |
| Hydro | Limited potential for micro or mini hydro max. (5 MW).Est. hydro potential: approx. 500 MW | Mainly public entities |
| Domestic Biogas System | 8.6 Million Cubic Meter of Biogas | Public and Private sector |
| Rice Husk based Biomass gasification Power Plant | 300 MW considering 2 kg of husk consumption per kWh | Mainly private sector |
| Cattle waste based Biogas power plants | 350 MW considering 0.752 m3 of biogas consumption per kWh. | Mainly private sector |

Other renewable energy sources include bio-fuels, gasohol, geothermal, river current, wave and tidal energy.

### 3.2. Power Sector

### 3.2.1. Market Overview

According to Bangladesh Power Development Board electricity demand in Bangladesh has been increasing by 200MW per year since 1996 .The total demand is projected to be more than 11497 MW by 2018 where contribution of public sector is 5933 MW, private sector is 5064 MW and 500 MW from import (Cross-border trading of 500 MW power with India has began in October 2013). Possibilities of trading of hydropower from Nepal, Bhutan and Myanmar are being explored. Highest generation so far was 6675 MW recorded on 12/07/2013 and it is increasing gradually.

Government has prepared the Power System Master Plan 2010 (PSMP 2010). The Bangladesh power sector master Plan indicates that to attain a 8% GDP. According to the PSMP 2010 the estimated demand for power would be about 19,000 MW in 2021 and 34,000 MW in 2030 (ref: Power Division). It has been estimated that power outage in this country results a loss of annual industrial output of $1 billion. Power is one of the prime reasons of slow GDP growth and the Government of Bangladesh (GOB) has recognized the power sector is a priority sector. GOB has decided to build more power projects through private sector and public private partnership

### 3.2.2. Generation Planning

*Table.2. Electricity capacity in private and public sector*

| Fiscal year | 2013(MW) | 2014(MW) | 2015(MW) | 2016(MW) | 2017(MW) | 2018(MW) | Total(MW) |
|---|---|---|---|---|---|---|---|
| Public | 763 | 889 | 1773 | 1285 | 450 | 1950 | 7110 |
| Private | 50 | 1864 | 1087 | 1098 | 2166 | 0 | 6265 |
| total | 813 | 2753 | 2860 | 2383 | 2616 | 1950 | 13375 |

### 3.2.3. An Assessment of Achievements in Power Sector

*Table 3. Bangladesh Power Sector at a Glance*

| Electricity Growth | 15% FY-2014 (Av. 7 % since 1990) |
|---|---|
| Derarted Capacity | 9675MW (February, 2014, source: BPDB) |
| Generation Capacity | 10241MW (February, 2014, source: BPDB) |
| Maximum Generation | 6060.00 MW ( 10th February 2014) |
| Maximum Generation in History | 6675.00 MW (12th July 2013) |
| Peak demand | 9268MW(2014) |
| Transmission Line | 9322 ckt. km |
| Distribution line | 2,90,000 km |
| Total Consumers | 14.2 Million |
| Newly connected people | 3.45 Million |

| Electricity Growth | 15% FY-2014 (Av. 7 % since 1990) |
|---|---|
| Present generation Capacity by Public sector | 58% |
| Present generation Capacity by Public sector | 42% |
| Reduction of System Loss(distribution) | 15.67% to 12.03%. |
| Per Capita Generation | 321 kWh (incl. RE) |
| Access to Electricity | 62% |

Regardless of financial constraints and gas supply shortages, the government deliberates a strategy to overcome the crisis and at the same time meet the ever-increasing demand for power. It launched immediate, short, medium and long term programs to increase power supply through introduction of fuel mix (gas, coal, liquid fuel, nuclear energy and renewable), demand side management, energy efficiency and conservation. After assessing the latest demand, the government has revised its targets for increasing power generation. The year-wise details of the additional power generation programs, both in public and private, are listed below 57 plants with a capacity of about 4,432 MW have been commissioned, 33 plants with a capacity of 6,569 MW are under construction. 19 projects with a capacity of about 3,974 MW are under tendering process and 9 plants with capacity of 3,542 MW are at initial stages.

### 3.2.4. Demand Supply Situation (February 2014)

- Generation: 9675 MW (Capacity- 10241 MW).
- Highest so far: 6060 MW (February 2014).
- Gas shortage causes 600 – 800 MW less Power Generation.
- Peak Demand: 9268 MW.
- Load shedding situation is in zero level(demand 5599MW, 12th February 2014). But, load shedding up to 1000 MW during hot summer days.
- Shortage and unreliable power supply has constrained economic growth.

Bangladesh has the fastest growing Solar Home System (SHS) in the world with over one million homes covered under the program being spear headed by IDCOL (a public infrastructure financing entity). Other projects include: 1 MW solar hybrid system along with 5 MW by diesel in Hatia island, 8 MW Solar PV plant in Kaptai, Solar Street lights in six City Corporation areas, replacement of diesel irrigation pump by Solar, 600 kW solar mini grid in a remote area Sullah, 11 KW solar power to the CHT area, nearly 230 W solar power in Angorpota and Dahagram Chitmahal area. Solar PV with capacity of 21.2 KW at the Bangladesh GOV has been installed as a demonstration project. Other line ministries have also undertaken projects on solar lighting. Bangladesh needs total US$22 billion investments in the power sector to minimize this demand-supply gap. The revised private sector power generation policy of Bangladesh provides a number of incentives for the foreign investment in the power sector.

### 3.2.5. Prospective Plan

The government of Bangladesh (GOB) has adopted Power System Master Plan (PSMP) 2010 as the basis for future projects to be undertaken in this sector. As per PSMP 2010,

electricity generation would reach to 34,000 MW by 2030. With new generation addition, the total generation capacity would be about 16,500 MW by FY 2018. By that time some power plants will be derated, contracts of some rental power plants will be over and the dependable capacity would be around 13,000 MW. Coal will be the dominating fuel in the future generation. Coal fire plants with capacity of 1320 MW will be set up in Khulna. The Khulna plant will be set up in joint venture with BPDB of Bangladesh and NTPC of India. Besides, other coal fired plants will be set up in different locations of Khulna, Chittagong, Matarbari and Moheshkhali; 100-200 MW power will be generated from wind. Along with wind mapping, a flag ship wind power project with capacity of 15 MW will be implemented within 2 years.

### 3.3. Fisheries

The country is crisscrossed with hundreds of rivers and it has established a credible record of sustained growth within a stable macro-economic framework where fisheries sector play an imperative and prospective involvement in agro-based pecuniary expansion, destitution easing, employment and delivering of animal protein and grossing the overseas exchange. Fish (including shrimp and prawn) is the second most valuable agricultural crop and its production contributes to the livelihoods and employment of millions. The key objectives of the sector are enhanced fisheries production; poverty alleviation through creating self-employment and improvement of socio-economic paradigm of the fishers; meet the demand for animal protein; achieve economic growth and earn foreign currency by exporting fish and fisheries products and maintain ecological balance; conserve biodiversity and improve public health. Bangladesh has achieved remarkable progress in the fisheries sector since its independence in 1971. Fisheries sector have been playing a very significant role and deserve potential for future development in the agrarian economy of Bangladesh. This sector contributes 4.39% to the national GDP and almost one fourth (22.76%) to the agricultural GDP (Bangladesh Economic Review 2012). In recent years, this sector performs the highest GDP growth rate in comparison to other agricultural sectors (crop, livestock and forestry). The growth rate of this sector over the last 10 years is almost steady and encouraging, varying from 4.76 to 7.32% with an average 5.61 percent. Whereas last four years average growth rate of this sector is 6.22 percent. The country's export earnings from this sector are 2.46% in 2011-12. The sector's contribution to the national economy is much higher than its 4.39% share in GDP, as it provides about 60% of the animal protein intake and more than 11% of the total population of the country is directly or indirectly involved in this sector for their livelihoods. There are 4.024 [million] ha open water bodies in our country. Among them 0.85 million ha are rivers and estuaries, 0.18 miilion ha sundarbans, 0.11 million ha *beel*, 2.832 million ha floodplains and 0.69 million ha Kaptai. *Beel* is one of the best natural habitats for the indigenous fishes of different food habits of Bangladesh. Most of the aquatic species specially the fish and prawn enter in the inundated areas of the *beel* from the adjoining rivers and canals to feed and grow during the monsoon months. The '*beel*' a Bengali term is used for large surface water body that accumulates surface runoff water through internal drainage channel. Bangladesh has thousands of *beels*, with the most common names being Chalan *Beel*, Gopalganj-Khulna *Beel*, Meda *Beel*, Aila *Beel*, Dekhar *beel*, Kuri *Beel*, Erali *beel* and Arial *Beel*. The average rate of production from *beel* is 714 kg/ha which can be increased manifold.

### 3.3.1. Inland Fisheries

The inland fishery resources of Bangladesh are considered to be unexcelled either in area or potential by any other inland fisheries of the world. Inland fisheries contribute nearly 90% to the total catch of fish in Bangladesh. Inland culture includes mainly pond/ditch, baor, shrimp/prawn farm, seasonal cultured water-body etc. covering an area of about 7.41 lakh ha and produces 17.26 lakh MT fish and shrimp in the 2011-12. Though the closed water area is only 15.55% of the total inland water-bodies, but 52.92% of the total yield comes from inland aquaculture. The inland water resources can be conveniently divided into the following categories:

a. Open inland waters which include the rivers and their tributaries; 'baors', 'haors' and 'beels' connected at least occasionally with rivers and streams; and the estuaries. The main river system in Bangladesh includes the Padma, the Meghna, The Brahmaputra and the Karnaphuli and their tributaries. The baors comprise ox-bow lakes and other forms of defunct rivers. There are many large and small baors in Jessore district and several in the districts of Kushtia and Faridpur. Among the larger baors in Jessore, Baluhar Baor (272ha), Joydia Baor (207), Sasta Baor (187), Morjad Baor (292), Bergobindapur Boar (214) and Jhampa Baor (183) represent only a few. The haors, synonymously called beels, are natural depressions used partially as agricultural lands, and seasonally or perennially filled from adjacent rivers or monsoon waters. Most of the larger haors/beels are located in Sylhet, Mymensingh and Faridpur. Kakaluki Haor (36,437ha), Tangua Haor (25,506) and Bardai Haor (3,239) in Sylhet district, and Beel Meskha (6,478) and Bengla Char Banda (6,073) in Mymensingh are among the big freshwater and salt water. The estuarine region of Bangladesh is interspersed with numerous distributaries of rivers, inlets of the sea and defunct streams that are interconnected with numerous channels. The deltaic area in Bangladesh is an archipelago.

b. Closed waters include ponds, dighis and tanks. For irrigation and other general purposes the feudal kings moharajas and land lords of the past had excavated tanks, dighis (larger tanks) and moats and thus set examples of their benevolent spirit for the good of their subjects. Simultaneously, well-to-do commoners also for their own requirements excavated quite a large number of tanks and dighis but of comparatively small size. As a result, water bodies of assorted shape and size are common almost everywhere in the country. In some places these ponds and dighis are widely scattered and, in some places in clusters or in groups. With political and social changes, the system of administration has also changed and the feudal system has since been abolished. All landed properties including the ponds owned by the Zamindars have been acquired by the Government. Due to continuous neglect most of these water bodies have turned into derelict waters. The number of such derelict ponds, dighis and canals are numerous

Rivers and estuaries are major sources of fish fry which are collected and used for fish culture in closed and semi-enclosed waters. The major carps species e.g., Rui, Catla, Mrigal etc., occur throughout the larger river systems in Bangladesh where the sexually mature brood fish spawn during monsoon when rivers are in flood. Major spawning grounds of these valuable fish species include:

i. Halda River - the lower reaches of the river from Sattaghat (near Gahira) to.
ii. its confluence with the Karnaphuli River.
iii. Arial Khan River - the region near Madaripur.
iv. Garai River - near the general area of Kushtia.
v. Ganges River - the area west of Rajshahi and estward in areas near Lalpur.
vi. Jamuna River - the regions adjacent to Sirajganj and near Fulchharighat.
vii. Old Brahmaputra River - the region north of Mymensingh.

### 3.3.2. Marine Fisheries

In addition to the above inland waters, the Republic has an extensive shelf area of which 37000 sq. km (is no deeper that 50 m. The entire shelf (down to 200 m) covers an area of 67,000 km. According to West, Bangladesh continental shelf covers an area of 27,000 mi (square miles) or 69,900 km. Beyond this continental shelf is the deep sea. The segment of marine water extending from the coast line into the sea up to the limit of 12 miles (19 km) constitutes the territorial water of Bangladesh. The Republic reserves the exclusive right for exploitation of fisheries from this territorial water mass measuring approximately 1 million ha. The declared economic zone extends 200 miles (320 km) out to the sea from the coast line. There are many indications that the continental-shelf waters of Bangladesh are rich in fish, shellfish and other biological resources potential of the Bay of Bengal.

### 3.3.3. Productions

The country has huge opportunities for the development of brackish water aquaculture boosting shrimp production and earning substantial amount of foreign currencies. Production of shrimp from culture and capture fisheries increased to a great extent in the beginning of 1980's. Since then, brackish water shrimp farming has been expanded to over 0.214 [million] ha of land by 2011 from 1.4 lakh ha in 1980. It is expected that with the introduction of improved scientific method of shrimp culture, the present production of shrimp will be increased substantially. The country has limited access to marine fisheries resources in the Bay of Bengal. Only demarsal fish and shrimp are being trapped from here. Other potential marine resources are yet to be exploited on commercial scale. Only 18% of total fish production comes from Marine capture fisheries and 82% from inland fisheries. The present democratic government has undertaken new policy for sustainable aquaculture production; provide need based aquaculture extension services, implements fish conservation activities which increase the national fisheries production as well as the growth rate in fisheries sector. Besides these, fisheries extension and conservation activities, AIGs and rehabilitation programs for poor fishers etc. were undertaken. Through the Execution of Fisheries Friendly Policy of the present government, total fish production has been increased from 2.7 million metric ton in 2008-09 to 3.062 million mt in 2010-11.

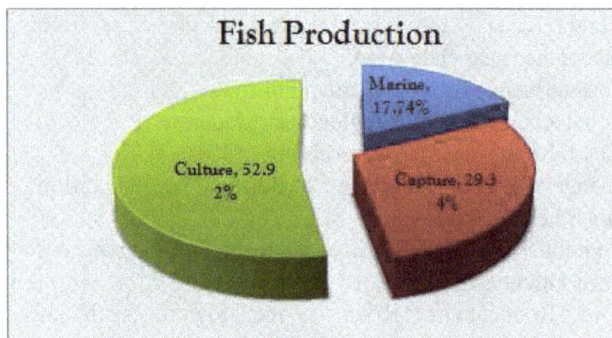

*Figure 1. Fish production is sector wise.*

### 3.3.4. Fish and Shrimp Culture

### 3.3.4.1. Fish Cultures

Pond aquaculture, Fish culture in paddy field, Fish culture in borrow-pit and *khal*, Fish culture in baor (Ox-bow lake), Cage culture, Pen culture, Integrated fish farming:

*Pond Aquaculture*

Currently pond aquaculture has been practiced in a total area of about 3.5 lakh ha which is 7.4 % of total inland water. Pond aquaculture is producing about 1199866 mt fish which contributing 47.70% of total inland production in 2010-11. The pond production involves composite culture produces an average 3430 kg/ha whereas there are records of 63 mt/ha production of pangas under intensive farming in Mymensingh region.

*Fish culture in paddy field*

Paddy fields and seasonal floodplains are promising and potential resources for aquaculture. It has been estimated that paddy fields cover an area of about 80 lakh ha of which 28.34 lakh ha floodplains which remain 4-5 months under

water.

*Fish culture in borrow-pit and khal*

Different types of waterbodies improved under Integrated Fisheries and Livestock Development Project in Flood Control, Drainage and Irrigation (FCDI) Project area and other waterbodies also included in the aquaculture systems.

*Fish culture in baor (Ox-bow lake)*

A total of about 600 baors having an area of 5,488 ha are situated in the south west part of the country. Different development projects have been implementing to increase the fish production from baor. The total water area of baors have been developed and brought under improved aquaculture through fingerling stocking and management practices. Six baors of Jessore district were under disposal of DoF till Feb/2009 and now these baors are under disposal of Department of Fisheries according to the MoU signed between Ministry of Land and Ministry of Livestock and Fisheries for next 6 years. Besides this, 30 baors are managing by OLP-2 project of DoF with the financial support of IFAD. These baors covered area of 1137 ha and fish production has increased from 80 kg to 750 kg/ha (DoF 2008). Local fisher communities are being involved in the baor management and improved their livelihood.

*Cage culture*

Cage aquaculture has been identified as a means of livelihoods for landless people. Northwest Fisheries Extension Project (NFEP) in Parbatipur, Dinajpur and Patuakhali Barguna Aquaculture Extension Project (PBAEP) demonstrated cage aquaculture as pilot basis. The production achieved through cage culture was encouraging and satisfactory but the activities were discontinued due to socio-economic condition of the farmers and some constrains. Cage culture of monosex tilapia is being practiced in Chandpur, Laxmipur Faridpur, Barishal, Mymensingh, Dhaka, Munsigonj, Gopalganj and other regions of Bangladesh. In 2011, about 6750 metric ton fish produced from 6000 cages.

*Pen culture*

Pen culture is also one of the potential means of producing fish from vast water body or water channel. In recent years, pens are made with different materials like bamboo, net, iron-meshed, wooden pillar etc. The area of pen also varies in size from half to few ha. The fish species reared in the pen are carp, tilapia, pangas etc.

*Integrated fish farming*

The integration of aquaculture with duck and chicken production was begun experimentally at the BFRI, Mymensingh producing some promising results. The project demonstrated that 500 khaki Campbell ducks can be profitably raised on a 1 ha carp pond while also producing 4.5 t/ha of fish without any additional need for supplementary feed or fertiliser for the fish. The most promising integrated farming in Bangladesh however, is rice fish culture, Ameen (1987) reported on the technique from many parts of Bangladesh. Traditionally one or more sump pond(s) are constructed at the lowest corner of the paddy field where fish accumulate as the water level reduces, thus fish are harvested from the sump without any additional stocking or management practices being required.

### 3.3.4.2. Shrimp Culture

The major shrimp producing districts are Bagerhat, Satkhira, Pirojpur, Khulan, Cox's Bazar and Chittagong, recently farmers especially in the Bagerhat and Pirojpur districts have begun shrimp farming in their paddy fields. Traditionally shrimp farming began by trapping tidal waters in nearby coastal enclosures known as 'gher' where no feed, fertilisers or other inputs were applied, with an increasing demand from both national and international markets farmers started to switch over into improved extensive and semi-intensive systems. With the expansion of aquaculture, environmental degradation, biodiversity and the control of the outbreak of disease especially in the coastal farming operations have become the major issues. Shrimp post larvae (PL) collectors are estimated to destroy nearly 100 other species of flora and fauna while collecting post larvae of Penaeus monodon , moreover, the destruction of mangrove forest for coastal shrimp culture and the introduction of viral diseases in semi-intensive farms has also become serious issues for concern. For inland aquaculture, habitat destruction, the use of insecticides and the introduction of diseases like epizootic ulcerative syndrome (EUS) have also become important issues. There are two types of culture

1 Shrimp (Bagda) Culture,
2 Shrimp (Golda) Culture.

### 3.4. Forest

Bangladesh is an independent and sovereign state since December 1971. It has about 157.22052 million people (2014) that is 2.19% of worlds total and growing at about 2.1% per annum with about 80% of them living in rural areas in 59, 990 villages having average household size of 5.3 persons. The overall literacy rate is 32.4% but the literacy rate of women is about 50% of men. The population density is very high and it lays on the active delta of three major rivers viz Padma, Meghna and Jamuna and their numerous tributaries. The country covers an area of 1,47,570 sq. km and bounded by India from the west, north and most of east. Myanmar lies on the south eastern edge and Bay of Bengal on the south. Forest Resources are renewable resources which can provide timber, pulp, pole, fuel wood, food, medicine, and habitat for wildlife and primary base for biodiversity A small tract of higher land occurs in Sylhet, Mymensingh, Chittagong, Cox's Bazar and Chittagong Hill Tracts (CHT) regions. The south-western region consists of a large number of dead and cut-off rivers. The coastal part of Bangladesh includes the famous Sundarbans Mangrove Forest. A number of depressed basins are found in the district of greater Mymensingh and Sylhet which are inundated by fresh water during the monsoon that gradually dry out during the dry winter season. These depressed basins are known as 'Haor'. Climate of Bangladesh is sub-tropical and monsoon rainfall varies from 1200-3500 mm. Rice is the major cereal crop while jute, sugarcane, and tea are the main cash crops. Other important

crops are wheat, tobacco, pulses, vegetable and tree fruits. Garments, raw and manufactured jute goods tea, fish and hides and skins are the chief exports. Bangladesh is noted for its estuarine environment, yet less than 10% of its total water flow originates from its own catchments and rest comes from India, Nepal and Bhutan. Normally, 20% of the country gets flooded during the monsoon period.

### 3.4.1. Land & Forest Areas

Of the total area of Bangladesh, agricultural land makes up 65% of its geographic surface, forest lands account for almost 17%, while urban areas are 8% of the area. Water and other land use account for the remaining 10%. The total forestland includes classified and unclassified state lands and homestead forests and tea/rubber gardens. In case of private forests, the data represent the tree-covered areas.

Of the 2.52 million hectare Forest Land, Forest Department manages 1.52 million ha which includes Reserved, Protected and Acquired forest and Mangrove forest on the newly accreted land in estuaries of major rivers. The remaining 0.73 million ha of land designated as Unclassed State Forest (USF) are under the control of Ministry of Land. Village forests (homestead land) form the most productive tree resource base in the country and accounts for 0.27 million ha.

### 3.4.2. Type of Forests in Bangladesh

Types of Forest: There are four types of forests, which have been managed, are as follows:

1 Mangrove Forests,
2 Tropical evergreen and semi-evergreen forests,
3 Tropical moist deciduous Forests,
4 Village Forest.

### 3.4.2.1. Mangrove Forest

#### i. Natural Mangrove Forests

The largest single tract of natural mangrove forest is the Sundarban. It consists of a total of 6,01,700 ha which is 4.07% of total land mass of the country and 40% of total forest land. Sundarban is a unique habitat for a number of wildlife. Among them some mammals are Bengal Tiger (Panthera tigris tigris), Gangetic Dolphin (Platanista gangetica), Monkey (Macaca mulatta), Indian Fishing cat (Felis viverrina), Indian Otter (Lutra perspicillata), Spotted Deer (Axis axis) etc. Reptiles like Estuarine Crocodile (Crocodylus porosus), Monitor Lizard (Varanus salvator), Rock Python (Python molurus) and Green Turtle (Chelonia mydas) etc. are found in the Sundarban.

#### ii. Mangrove Plantation

Mangrove afforestation along the entire southern coastal frontier is an innovation of foresters. During 1960-61, Government undertook afforestation programme along the shore land of coastal districts. This initiative got mementum from 1980-81 with the aid of development partners and afforestation programs are extended over foreshore islands, embankments and along the open coasts. Since 1965-66 up to 2012-2013, 1,96,000 ha of mangrove plantations have been raised under a number of coastal afforestation projects. The present net area of mangrove plantation is 132,000 ha after losing some area due to natural calamities.

### 3.4.2.2. Tropical Evergreen and Semi-Evergreen Forests

Tropical evergreen and semi evergreen forests are extended over Chittagong, Cox's Bazar, Chittagong Hill Tracts and Sylhet totaling an area of 6,70,000 ha which is 4.54% of total landmass of the country and 44% of national forest land. Depending on topography, soil and climate these area are categorized as i) Tropical wet evergreen forests and ii) Tropical semi-evergreen forests.

### 3.4.2.3. Tropical Moist Deciduous Forests

The Central and northern districts covering an area of 1,20,000 ha about 0.81% of total land mass of the country and 7.8% of the country's forest land are bestowed with Tropical Moist Deciduous Forests. This forest is intermingled with the neighboring settlements and fragmented into smaller patches. Sal (*Shorea robusta*) is the main species there with other associates like Koroi (*Albizzia procera*), Azuli (Dillenia *pentagyna*), Sonalu (*Cassia fistula*), Bohera (*Terminalia belerica*), Haritaki (*Terminalia* chebula), Kanchan (*Bauhinia acuminata*), Jarul (*Lagerstroemia speciosa*), Jam (*Syzygium spp*) etc.

### 3.4.2.4. Village Forests

Tree coverage in the village forests are 2,70,000 ha which acts as the source of a remarkable portion of national demand of forest produces. The latest inventory exhibits that a total of 54.7 million cubic meter [cum] forest products are available in this village forests.

### 3.4.3. Forest Products

NWFP (Non Wood Forest Products): Some of the important non-wood forest products are listed below:

Bamboo (Melocanna baccifera, Bambusa tulda etc, Sungrass (Imperata spp.), Cane (Calamus, viminalis,Calamus guruba):., Pati Pata / Murta (Clinogynae dichotoma, Gol-Pata (Nypa fruticans): Leaves, Bark & Fruits"Kurus pata", Honey,Shells, Conch-Shells, oysters etc.

### 3.4.4. Revenue Earnings from the Forest Sector

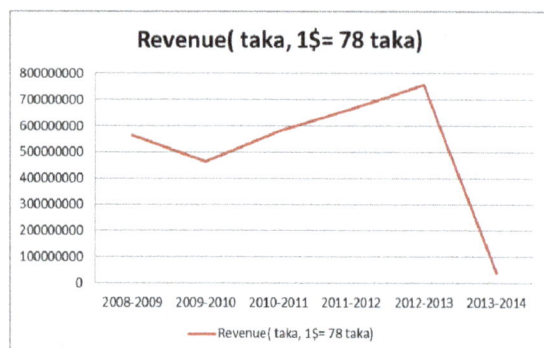

*Figure 2. Revenue earnings from Forest sector*

We observed an upward trend of revenue earning excluding 2013-14 fiscal year as it was partially calculated (till august 2013).

### 3.5. Water Resources

The country is bounded by India on the west, the north and the northeast; Myanmar on the southeast and the Bay of Bengal on the south. The area of the country is 147,570 sq.km. The country is the lowest riparian of the Ganges basin, the Brahmaputra basin and the Meghna basin. Most of its area is low lying floodplain formed by the alluvial soil deposited by three great rivers, namely the Brahmaputra/Jamuna, the Ganges and the Meghna. These rivers drain a catchment area of about 1.72 million km in India, Nepal, China, Bhutan and Bangladesh; only 8 percent of the catchment area lies within Bangladesh. These major rivers and their tributaries have their headwaters outside Bangladesh with about 90% of their annual flow originating outside the country. This flow has a huge annual variation, with the combined flow of the Ganges and the Brahmaputra typically increasing from less than 10,000 cubic meter per second [cumec] early in the year to a peak of 80,000 to 140,000 cumec by late August to early September. Shortage of water in the dry season in Bangladesh is exacerbated by the diversion of water at the Farakka Barrage, just upstream of where the Ganges enters Bangladesh. The country contains about 22155 km of river length for about 700 rivers. The Rivers and water bodies occupy about 5% of the land surface. The land topography is almost flat with little hilly areas in the southeastern part.

The country enjoys tropical monsoon climate with two prominent seasons; dry season (November-May) and wet season (June-October). Bangladesh is predominantly an agricultural country; about 54% of the lands are used for crop production. Up to 85% of the annual rainfall occurs between June and September. Mean annual rainfall ranges from about 1200 mm in the west to almost 6000 mm in the northeast. The average annual rainfall in the Himalayas and in the Meghalaya hills to the north of Bangladesh reaches about 10,000 mm. About 25% of the country is flooded to varying degrees each year during May through September when over 60% of the cereals is produced. Recurrent flooding severely restricts the farmers' choice of cropping to traditional low yielding broadcast variety of rice that can thrive in deep water and, in fact, the coverage is dominated by it. The real production potential is not harnessed due to flood depth. On the other hand, scarcity of irrigation water during March-April limits the cultivation of High Yielding Variety rice that accounts for about 36% of total rice production.

### 3.5.1. Sources of Water in Bangladesh

The sources of water in Bangladesh are surface water, groundwater and rainwater. The Ganges-Brahmaputra-Meghna river system discharges huge amount of surface water through Bangladesh, a part enters into ground to form groundwater. About 93% of the stream flow passing through the country originates from outside the Bangladesh (Khan, 1993). Rainfall within country contributes to the total water available in Bangladesh, a part of which infiltrates into ground to recharge existing groundwater and the remaining rainwater flows as surface run-off. These sources of water available for the development of water supplies have their relative advantages and disadvantages in Bangladesh context. The availability of water in terms of quantity and quality, present situation and problems associated with the sources have been discussed in the following sub-sections.

### 3.5.2. Surface Water

Surface water is abundant in the wet season in Bangladesh. An estimated 795,000 million cubic meter ($Mm^3$) of surface water is discharged through the Ganges-Brahmaputra system, in the downstream of the confluence of the Ganges and the Brahmaputra. This is equivalent to 5.52 m deep water over a land area of 147,570 sq.km. There are other rivers discharging surface water into the Bay of Bengal. An average annual rainfall of 2.40 m within the country partly replenishes surface water sources. Each year about one-third of Bangladesh is submerged in a normal flood, and the area submerged may increase to about two-thirds during severe floods. In the dry season water scarcity persists in many areas. In this period surface water is only available in part of the 22,155 km of major rivers, 1,922 $km^2$ major standing water bodies and about 1,475 $km^2$ of ponds in the country. Surface water irrigation systems in the country compete for this available water in the dry season. The perennial water bodies are decreasing with the use of more and more surface water.

### 3.5.3. Ground Water

The main source of ground water is the recharge from surface water. Most of the areas of Bangladesh have been formed from the sedimentary alluvial and deltaic deposits of three major rivers. These alluvial deposits have formed mainly an unconfined aquifer for most of the area of the country. Groundwater was supposed to be one of the major natural resources of the country except the safe drinking water supplies. But the presence of Arsenic in shallow aquifer has completely changed the situation. It is estimated that about 16% of present population of 123.15 million is exposed to arsenic contamination exceeding Bangladesh standard (0.05 mg/l). About 74452 sq.km. of groundwater use area (about 50% of the country) is unsuitable for use by hand tubewells (as a source of drinking water according to WHO standard) due to arsenic.

### 3.5.4. Water Demand and Supply

Demands arise from several factors such as, natural (evapotranspiration), water supply, irrigation, fisheries and livestock, industrial, navigation and the environment (demands for salinity control). Proportion of total water demands, as projected for 2025, is estimated to be: instream-56%, agriculture-32%, environment-9% and water supply-3%. So, consumptive use comes to be 44%. Environmental flow requirements according to IUCN as stated by Saleh (2003) should at least be 30% of the world's river flows so as to maintain a fair condition of freshwater ecosystems.

### 3.6. Land Resources

Total Geographical area of Bangladesh is about 56,000 square miles (143,998 square kilometer). Out of which about 9 million hectares are cultivable land. A government survey finds that total cropped land is nearly 14.1 million ha including single, double and triple cropping land. It is estimated that the growing population pressure will use up 50 per cent of the country's cultivable land by 2025[1].

Every person working in the agriculture sector now owns only an average of 0.12 ha of cropland. According to the classification of land, out of the total area, 63 per cent are being used for cultivation while 4.38 per cent for rural and urban housing and the rest includes forest & cultivable waste land[2]. There are two types of land in Bangladesh

1 Khas land.
2 Adivasi land.

### 3.6.1. Landlord and Land Mafias

More or less all state owned lands including 3.3 million acres of khas lands and new lands surfaced from the rivers & sea are occupied by the land grabbers, land mafias & terrorists under the patronage of former governments in power living in both rural & urban areas of Bangladesh.

### 3.6.2. Poor people's Access

According to government report [3] 57% people of Bangladesh are landless poor & they live below poverty level. But Non-Government sources say that the number of landless people in Bangladesh is more than 68%. They live in perpetual poverty, hunger, disease and deprivation. According to Dr. Mahboob Hossain & Prof. Abdul Byes, 45% of the landless & poor marginalized peasants in the rural areas own only 5% of the total cultivable land of Bangladesh and receives 10% needed credit from institutional sources.

### 3.6.3. Land Occupy by the Rich

22% of the rich & middle farmers of rural areas of Bangladesh own 71% of the total land & receive 31% institutional credit.

## 4. An Overview of Non-Renewable (Natural Gas, Coal, Oil, White Clay, Sand, Rock, Gold) Resources of Bangladesh

Geographically, Bangladesh occupies a bigger part of the Bengal basin and the country is roofed by Tertiary folded sedimentary rocks (12%) in the north, north eastern and eastern parts; uplifted Pleistocene residuum (8%) in the north western, mid northern and eastern parts; and Holocene deposits (80%) consisting of unconsolidated sand, silt and clay. The oldest exposed rock is the Tura Sandstone of Palaeocene age but older rocks like Mesozoic, Palaeozoic

---

[1] for detail see the report, "Needs for land and agrarian reform", prepared by ARBAN
[2] ibid.
[3] ARBAN

amid Precambrian basement have been encountered in the drill holes in the north western part of the country. The imperative mineral deposits of Bangladesh are natural gas, coal, limestone, hard rock, gravel, boulder, glass sand, construction sand, white clay, brick clay, peat, and beach sand heavy minerals because of its different biological environment. Tertiary Barailshales stirring within the oil and gas windows have generated natural gas and oil found in Bangladesh. Sustainable mineral resources are still playing a vital role in shaping the modern civilized industrial world. Modern urbanization, industrialization, transportation and communication systems are the achievements of worldwide sustainable mineral resource development and their proper utilization in various sectors. Richness in natural resource is the key indicator of socio-economic infrastructure for any country all over the world. Strong technological know how its ability to explore and exploit mineral resources, and its wisdom in utilizing those resources properly in the development activities of the nation.

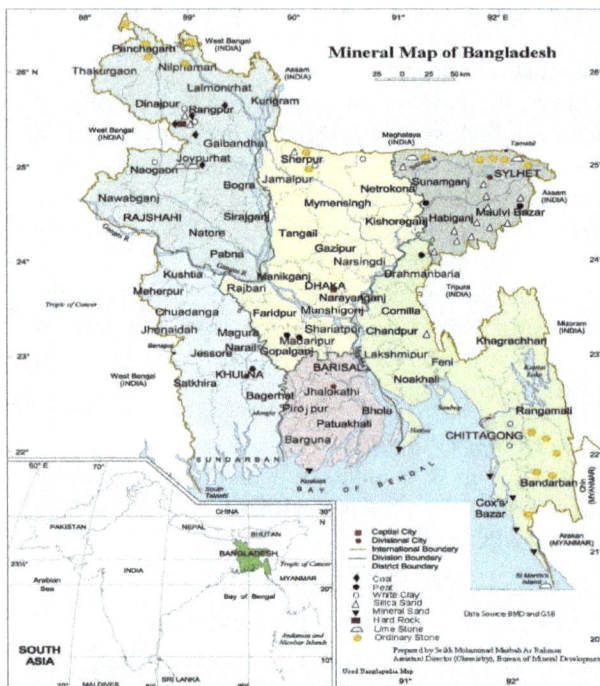

*Figure 3. Mineral Map of Bangladesh. (Source: Ministry of Mineral Resource)*

Developing world is generally far behind compared with developed world in the sense of development activities. At present, natural gas is the only mineral commodity significantly contributing to the national economy. Majority of the country's energy needs are met by gas that is about 90%. Energy is the key ingredient for socio-economic development of a country and economic development depends on the adequate energy supply that leads a country self reliant. Ministry of Power, Energy and Mineral Resources divided into two departments namely Energy and Mineral Resources Department and Power Department. Energy and Mineral Resources Department that entrusted to make all policies connected to natural gas, liquid petroleum

and mineral resources. EMRD is responsible to create policies and administrative control over Geological Survey of Bangladesh, Petrobangla, Bureau of Mineral Development and Department of Explosives. Hydrocarbon Unit and Bangladesh petroleum Institute is controlled and watches over by EMRD.

### 4.1. Natural Gas

Natural gas has a vital role to play in our socioeconomic development as a rudimentary source of energy. Its widespread use in power, fertilizer, industry and household has made it the energy of choice in Bangladesh that accounts for about 73% commercial energy of the country. It is pretty cheaper than other conventional sources of energy that we import from abroad. The use of indigenous natural gas has, indeed, helped to accelerate the pace of economic development and improve the quality of our life in the country. So far in Bangladesh 25 gas fields have been discovered with the rate of success ratio is 3.1:1 of which two of the gas fields are located in offshore area. Gas is produced from 20 gas fields (79 gas wells), 15 are state-owned and the remaining operated by international oil companies. Currently, Chevron contributes more than half of Bangladesh's total gas production. Bangladesh gas sector started its journey in early 60s, but its rapid expansion and integration stared to accelerate in 70s spurred by the raising oil price. Total recoverable proven and probable gas reserve from discovered 25 gas fields has been estimated as 27.04 trillion cubic feet [TCF] out of which estimated proven recoverable reserve (P1) is 20.70 TCF and recoverable probable reserve is 6.39 TCF. Up to December 2013 as much as 11.92 TCF gas has been produced leaving only 15.12 TCF recoverable gas. Currently 19 gas fields are in production and out of 104 wells located in 19 gas fields, 84 are in stream. A total of 600.86 billion cubic gas (BCF) was produced in FY 2007-2008, 653.57 BCF in 2008-2009, 703 BCF in 2009-2010, 708.92 BCF in 2010-11, 743.57 BCF in 2011-2012 and 805.67 BCF in 2012-13. The demand for gas has already surpassed 2700 million cubic feet [MMCF] per day whereas the peak supply of gas is nearly 2287 MMCF leaving shortfall of 413 MMCF per day. Against this backdrop, Petrobangla has drown time bound program to boost up gas production in the coming days, and with this end of view, short term, mid-term and long term (up to 2015) have been taken up for enhancing gas production to and additional amount of 1560 mmcfd with in the year 2015 as envisaged in the road map. Meanwhile, a volume of about 593 mmcfd of gas added to the natural grid.

Average daily gas production capacity is about 2000 mmcfd of which International Oil Companies (IOC) produce 1040 mmcfd and State Owned Companies (SOC) produce 960 mmcfd. The gas production recorded on 24 February, 2010 was 1996.7 mmcfd. At present the daily approximate projected gas demand throughout the country is 2500 mmcfd. The demand is increasing day by day. Energy and Mineral Resources Division (EMRD) has already undertaken an array of short, medium, fast track and long term plans to increase gas production to overcome prevailing gas shortage. According to this plan 188 mmcfd, 290 mmcfd, 995 mmcfd (including 500 mmcfd LNG), 500 mmcfd and 380 mmcfd gas will be added to the national gas grid by the year 2010, 2011, 2012, 2013 and 2015 respectively. After completion of these plans production capacity is expected to increase to about 2353 mmcfd gas by December 2015. To increase the gas production more programs will be taken in near future. Bangladesh's natural gas output increased only by 110 million cubic feet per day to 2.26 billion cubic feet per day until June 2013 from 2.15 BCF per day in June 2012, Petrobangla data indicated. Amid annual estimated demand growth of 10 percent, the country's entire recoverable gas reserves of 16.36 trillion cubic feet are expected to continue to 2025 and beyond, although in short supply. In a forecast of gas supply scene from the existing gas fields, production is expected to increase and reach its peak in 2016. As the demand for gas continues to grow, the gap between demand and supply will continue to widen as the production begins to decline after 2016. But if the natural gas consumption rate should exceed that 10 percent growth estimate, Bangladesh's reserves won't last more than a decade. The government's decision for over production raises questions, since it came without measuring the capability for over production of a gas field. In 2009, when the Awami League came to power, the country's gas supplies hovered around 1850-1900 mmcfd, a few hundred mmcfd short of the demand. Over the next five years, the supplies increased up to 2,250 mmcfd due to increased production by the national gas companies. Chevron alone provided 250 mmcfd. And now it is investing half a billion dollars to increase Bibiyana Gas Field's production by 300 mmcfd by early next year to address the country's gas crisis. Experts believe that this might cause Bibiyana to collapse just as Sangu did in the past. Besides over production, the lack of proper management and surveillance also poses some problem

### 4.2. Petroleum Product

Bangladesh imports annually about 1.3 million metric tons of crude oil. Besides these, another 2.7 million metric Tons (approx) of refined petroleum products per annum is imported. Condensate is mixed with crude oil. Major consumer of liquid fuel is transport sector followed by agriculture, industry and commercial sector which is mostly

met by imported liquid fuel. Eastern Refinery Limited (ERL), a subsidiary company of Bangladesh Petroleum Corporation (BPC), is capable of processing 1.3 million metric Tons of crude oil per year.

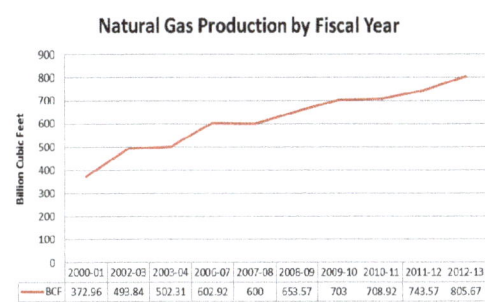

*Figure 4. Natural gas production by fiscal year.*

### 4.3. Oil

The only oilfield of the country has been discovered at Haripur in 1986 that is located in near Haripur in the eastern hilly district of Sylhet, but these have yet to be developed. The Haripur reserves are estimated at 40 million barrel [mbbl], with a recoverable reserve of about 6 million barrels and the total resource is likely to be much higher. For meeting the total requirement of commercial energy, Bangladesh imports yearly about 1.3 million metric Tons of crude oil.

### 4.4. Coal

As well natural gas, Bangladesh has significant coal reserve. Coal first discovered in the country by Geological Survey of Pakistan (GSP) in 1959 was at great depth. GEOLOGICAL SURVEY OF BANGLADESH (GSB) continued its efforts for exploration that resulted in the discovery of 4 coalfields. BHP Minerals, a US-Australian company, discovered a field in 1997 totalling 5 coalfields. Coal reserves of about 3.3 billion tons comprising 5 deposits at depths of 118-1158 meters have been discovered so far in the north-western part of Bangladesh. The name of these deposits are-Barapukuria, Phulbari and Dighipara coal field in Dinajpur district, Khalashpir in Rangpur district and Jamalganj in Joypurhat district.

*Table 4. Coalfields and coal quality*

| Coal Field | Depth (m) | Thickness (m) | Area (sqkm) | Reserve (m ton) | Fixed Carbon (%) | Volatile Matter (%) | Ash content (%) | Sulphur content (%) | Clorific Value BTU/lb |
|---|---|---|---|---|---|---|---|---|---|
| Jamalgonj | 640-1158 | 64 | 11.06 | 1053 | 47 (av) | 38 (av) | 22 (av) | 0.62 (av) | 11000 av |
| Barapukuria | 129-506 | 51 | 5.25 | 300 | 45.5-54.7 | 2.28-3.60 | 11.79-23.71 | 0.43-1.33 | 10547-12529 |
| Khalaspir | 257-483 | 50 | 12.56 | 143 | 32.0-80.8 | 2.93-30.47 | 7.6-50.51 | 0.24-3.15 | 7388-1388 |
| Dighipara | 328-407 | 61 | N.D | N.D | 51.3-65.6 | 25.29-38.23 | 2.64-20.05 | 0.51-1.02 | 10200-14775 |
| Phulbari | 151 | .... | .... | 386 | .... | .... | .... | .... | .... |

*Source*: Asian Mining Year Book (Seventh Edition), 2001: ND: Not determined.

### 4.5. White Clay

White Clay occurs in Sherpur, Netrokona, Dinajpur and

Chittagong district etc. and it is used to make crockery's, sanitary materials, insulator and tiles. In addition, it is also used in Paper, cement and sugar industries. There are surface to near surface deposits of white clay in Bijoypur and Gopalpur area of Netrokona district, Nalitabari of Sherpur district, Haidgaon of Chittagong district and BaitulIzzat of Satkaniaupazila, Chittagong district. Besides, there are subsurface deposits of white clay in Maddhyapara, Barapukuria, Dighipara of Dinajpur district and Patnitala of Naogaon district. The showing white clay is not good in quality that is used in the ceramic factories of Bangladesh after mixing with high quality imported clay.

### 4.6. Glass Sand

Important deposits of glass sand of the country are at Balijuri (0.64 million t), Shahjibazar (1.41 million t) and Chauddagram (0.285 million t) at or near the surface, Maddhyapara (17.25 million t) and Barapukuria (90.0 million t) below the surface. Glass sands consist of fine to medium, yellow to grey quartz. Total deposit is about 109.58 million t. Glass sand is used to make crockery's, lenses, glass sheet of windows and doors. It is further used to make quartz clock, frame of boat and aeroplane, foam glass and in various electronic equipments etc. Silicon chips are also made from silica, which is a main ingredient of glass.

There are two types of glass sand according to geological status.

1  Recent piedmont alluvium in the eastern zone.
2  Underground glass sand in the northern zone.

### 4.7. Limestone

It occurs in Sunamgonj, Takergat, Jahanpur, Paranagor, Joypurhat and St. Martin's Islands. Limestone is a very important mineral resource for Bangladesh. It is primarily used in cement industry. Limestone is used to make lime and cement and in paper, Ispat, sugar, glass industry. It is also used to decorate the building In Bangladesh limestone is found in Taker Ghat, Lalghat and Bangli Bazar of sylhet area, Jaypurhat and Saint-mertine of Cox's Bazar district. The lime stone of Takerghat limestone mining project is supplied to Chattak Cement Factory

### 4.8. Ilmenite, Garnet, Zircon, Kyanite, Magnetite, Rutile, Leucoxine, Monazite

All these are found in Cox's Bazar and Teknaf Beach, Kuakata Beach , Moheshkhali, Nijhum dwip, Kutubdia and Monpura Island. Ilmenite, Rutile and Leucoxine are used to make slag and in welding and in melting of metal. They are used as a dyeing subject and Refractory Brick. Ilmenite is used to make sand blasting and heavy mud as an alternative to Barite in drilling activities. Titanium metal, which is derived from this mineral, is used to make frame of aeroplane, missile, and in chemical reaction and salt removal process. Zircon is used to make foundry sand, Refractory brick and as a dyeing substance. Zirconium is used as a radioactive substance. Monazite is used to make catalyst, television tube, refractory substance, thermal insulator substance and in computer disk and line printer.

### 4.9. Peat

Deposits of peat occur at shallow depths in different low-lying areas of Bangladesh like Gopalganj, Madaripur, Khulna, Sylhet and Sunamganj district etc. the reserve of dry peat is about 170 million t. In 1953, a large scale of peat was discovered at Baghia-Chanda Bil in Faridpur and at Kola Monja in Khulna. The major deposits are in greater in the districts of Faridpur (150 million t), Khulna (8 million t). Peat requires drying before making briquettes for use as fuel. It is used as an alternative fuel to household, in brick and lime industries and in thermal power plant, and it requires drying before making briquettes for use as fuel. Petrobangla implemented a pilot project for extraction of peat and making briquettes but the result were discouraging and not economically feasible at current stage.

### 4.10. Hard Rock

It occurs at Maddyapara in Dinajpur district and has been discovered by GSB (Geological Survey of Bangladesh) at depth 132-160 m below the surface. The Rock Quality Designation (RQD) of fresh rock varies from 60% to 100%. The Maddhayapara project is extended over a wide range of area, from Baborgonj and Mithapukur of Rangpur district to Fulbari and Parbatipur of Dinajpur district and its capacity to recover 1.65 million t hardrock every year. It is used in construction such as road, highway, and railway track, regulator dam, river training, and river bank erosion. It is also used as construction material and as mosaic stone.

### 4.11. Gravel Deposit

Deposits of gravel are found along the piedmont area of Himalyas in the northern boundary of Bangladesh. It occurs in Lalmonirhat, Panchagar, Sylhet district, Greater Chittagong and Chittagong Hill Tracts. These river borne gravels come from the upstream during the rainy season. It is used to construct buildings, road, railway, bridge and in river training and flood control. A total reserve of gravel is about 10 million cubic meters.

### 4.12. Metallic Minerals

GSB has carried out investigation for mineral deposits and succeeded in locating a few potential zones. Relatively high content of metallic minerals like chalcopyrite, bornite, chalcocite, covelline, galena, sphaleriteetc have been found in the core samples from the north-western region of the country.

### 4.13. Construction Sand

It is very much available in the river beds throughout the country. Sand consists mostly of quartz of medium to large grains. It is extensively used as construction materials for buildings, bridge, roads all over the country.

### 4.14. Beach Sand

This may be potential source in the future. Deposits of

beach sand have been identified in the coastal belt, and in the coastal island in Bangladesh. Deposits of beach sand have been identified in the coastal belt and in the coastal islands of Bangladesh. Different heavy minerals and their reserves (in ton) are: Zircon (158,117), Rutile (70,274), Ilmenite (1,025,558), Leucoxene (96,709), Kyanite (90,745), Garnet (222,761), Magnetite (80,599) and Monazite (17,352). An Australian company has applied for the permission to carry out the feasibility study for exploitation.

### 4.15. Brick Clay

In Bangladesh the mineralogical, chemical and engineering properties of Pleistocene and Holocene brick clays of Dhaka, Narayanganj and Narsingdi districts are well documented. The bulk chemistry and engineering properties of the Holocene and Pleistocene samples have been found satisfactory for manufacturing good quality bricks. These are being exploited and widely used in the country.

### 4.16. Black Gold

Black gold is a very valuable mineral resource. Among it, the main are zircon, monazite, riotile etc. It is discovered in Cox's Bazar.

*Table. 5. Estimated reserve of Mineral resources.*

| Name of the Mineral | Place | Estimated Reserve (Probable) (million ton) | District | Remarks |
|---|---|---|---|---|
| Coal | #Jamalganj | 1053 | Joypurhat | #Development of Barapukuria Coal field is going on. Discovered in July/, 1995. |
| | #Barapukuria | 300* | Dinajpur | |
| | #Khalashpir | 143 | Rangpur | |
| | Dighipara | 150 | Dinajpur | |
| | Baggie Chanda | 150 | | |
| Peat | Kolamouza | 8 | Gopalganj | Instead of fuel wood may be used as fuel. |
| | Chatalbil | 6.21 | Khulna | |
| | Paula, Sunamganj | 3.50 | Sunamganj | |
| | Moulavibazar | 3.00 | | |
| Limestone | Jaypurhat | 100 | Jaypurhat | #612371 t of limestone from Takerghat have been exploited During 1972-93 |
| | Bagalibazar | 17 | Sunamganj | |
| | #Takerghat | 12.9 | Sunamganj | |
| | Lalghat | 12.9 | Sunamganj | |
| | Naogaon | -- | Naogaon | |
| White clay | Barapukuria | 25 | Dinajpur | #109541 t of whit clay from Bijoypur have Been exploited during 1972-93 |
| | #Bijoypur | 25 | Netrakona | |
| | Maddyapara | ** | Dinajpur | |
| | Dighipara | ** | Dinajpur | |
| | Patnitala | | Naogaon | |
| Glass sand | Barapukuria | 90 | Dinajpur | #94773 t of glass sand have been exploited during 1975-93. |
| | Maddyapara Bhatera | 17.25 | Dinajpur | |
| | Shahajibzar & | 8 | Moulvibazar | |
| | Bahubal | .30 | Habiganj | |
| | #Chaddagram | .17 | Comilla | |
| | Baljiuri Dighipara | ** | Sherpur | |
| | | | Dinajpur | |
| Hard rock | Maddyapara | 115 (Exploited) | Dinajpur | Mine development activities is going on. |
| Gravel deposit | Bholaganj | 4 | Sunamgaj | Gravel deposits are being exploited from different places of the country. |
| | Tetulia | 2.5 | Pachagarh | |
| | Patgram | 2.5 | Lalmonirhat | |
| | Chittagong Hill Tract | 1.00 | Chittagong | |
| Mineral sand | Sea beach of Cox's Bazar, Moheshkali, Kutubdia and Kuakatha. | | | |

*Reserves are in million tones except that of gravel that is in million cubic metre.
**Reserves have not yet been estimated.

# 5. Concluding Remarks and Recommendation

The summary of this paper exhibits that there is a considerable opportunity of Bangladesh to boost the economic growth through renewable and non-renewable resource. I am trying to show that solar power is the emerging sector of Bangladesh that meets the majority portion of the energy and power demand in this country. Besides these forests, fishery, land, water and other nonrenewable resources are abundant in Bangladesh. These non-renewable resources can help to boost the GDP growth in Bangladesh if government of Bangladesh comes forward to take healthy steps to reduce the corruption from natural resources sector from the country. Bangladesh, a country with a very low per capita GDP, is suffering from mounted pressure of huge population and ever-increasing budget deficits. Effective management of natural resources can help to reduce this problem and save the country from budget deficit. Conservation of forest is needed to stop the deforestation and enhance the environmental quality. Government has to produce strong laws and order and reformulate the existing laws for forest conservation. Different environment programs have to be undertaken to plant eco-friendly tress forestation. The rural and marginal poor should guarantee access to water bodies such as beel, haors, and baors. Right to safe drinking water has to be ensured for the citizens. I discussed previously, rain water is abundant in Bangladesh and proper utilization of rain water could be developed to stop overwhelming demand on ground water. Government should take necessary steps with regard to National Fisheries Policy (1998) importance of conserving fish breeding grounds and habitats, especially in relation to water management infrastructure such as flood control, irrigation and drainage projects. Finally, government has to introduce advanced technology to extract mineral resources and appropriate laws should be enacted to get the best outcome from natural resources sector. Government should take immediate steps for the conservation of ecology and bio-diversity. Environmental Conservation Act of February 1995 along with other policies that are concern about ecology should be modified to conserve the environment and should identify the local and foreign actors who violate rights and livelihood of the marginal people. In addition, government needs to ensure enabling atmosphere for marginal people to access natural resources. The following issues require to be addressed in future for fisheries:

1. To bring all available water bodies under modern fish culture regimes.

2. To generate increased employment opportunities in fisheries and allied industries.

3. To conserve fisheries resources and species biodiversity.

4. To develop fish landing and marketing systems.

5. To establish institutional frameworks to ensure research findings are made available to the relevant people.

6. To provide an adequate provision of financial assistance to fish and shrimp farmers.

7. To increase and sustain fish production for both domestic consumption and export.

# References

[1] Annual Report 2012-13, National Board of revenue

[2] A.K.M. Sadrul Islam, M. Islam and Tazmilur Rahman, "Effective renewable energy activities in Bangladesh," Renewable Energy, vol. 31, no. 5, pp. 677-688, Apr. 2006.

[3] Bangladesh Gazette: Renewable Energy Policy of Bangladesh 2008", published in November 06, 2008. http://lib.pmo.gov.bd/.pdf

[4] Bangladesh Economic Review 2012, Ministry of Finance, available at www.mof.gov org

[5] Bangladesh Statistical Pocket Book 2012, Bangladesh Bureau of Statistics

[6] Bangladesh Bank, http://www.bangladesh-bank.org/

[7] Banglapedia, National Encyclopedia of Bangladesh

[8] http://en.wikipedia.org/wiki/Bangladesh.

[9] http://www.powerdivision.gov.bd/index.php?page_i d=263

[10] Ministry of Power, Energy and Mineral Resources, Govt. of the People's Republic of Bangladesh, Renewable energy policy of Bangladesh, Dhaka, Bangladesh, Nov. 2008

[11] Ministry of Power Energy and natural resources http://www.powerdivision.gov.bd/user/brec/67/66

[12] Ministry of land Bangladesh, http://www.minland.gov.bd/

[13] Ministry of Fishery and Livestock, http://www.mofl.gov.bd/

[14] M.A.R. Sarkar, M. Ehsan and M.A. Islam, "Issues relating to energy conservation and renewable energy in Bangladesh," Energy for Sustainable Development, vol. 7, no. 2, pp. 77-87,Jun. 2003.

[15] National Bureau of revenue http://www.nbr-bd.org/

[16] Power Development Board, http://www.bpdb.gov.bd/bpdb/

[17] Reform in revenue Administration(RIRA) project website

[18] Rahma,M.S. Saha,S.K Khan, M. R.H Habiba U. & Chowdhury, S.M 2013Present Situation of Renewable Energy in Bangladesh: Renewable Energy Resources Existing in Bangladesh, Global Journal of Researches in Engineering Electrical and Electronics Engineering Volume 13 Issue 5 Version 1.0

[19] Water Development Board, http://www.bwdb.gov.bd/

# 18

# The Current World Energy Situation and Suggested Future Energy Scenarios to Meet the Energy Challenges by 2050 in the UK

**Ahmed Fakhri[1], Waleed Al-sallami[2], Nawar H. Imran[3]**

[1]School of Electronic and Electrical Engineering, University of Leeds, Leeds, UK
[2]School of Mechanical Engineering, University of Leeds, Leeds, UK
[3]School of Chemical and Process Engineering, University of Leeds, Leeds, UK

**Email address:**
e.ahmedalheety@gmail.com (A. Fakhri)

**Abstract:** Currently, energy demand is ever increasing along with the high levels of population world-wide. The global dependence on fossil fuels is very high and the need for reducing our energy consumption in line with mitigating the greenhouse gasses emissions is compelling. With the current global reduction of oil prices, companies or even governments tend to import more energy due to economic reasons. For instance, recently, DHL which is a famous company providing international express delivery, introduced a helicopter express delivery in London. Such development gives a real indication that as people/agencies consume more fossil fuels, in fact, the world become closer to the reserves ending point. Accordingly, this makes renewables deployment and hence reducing energy cost is quite difficult. This paper gives an overview of the current world energy situation along with three energy scenarios for the UK to achieve the official announced targets by 2050. Finally, looking for liquid metal battery advantages to secure our future energy needs.

**Keywords:** Liquid Metal Battery, Greenhouse Gasses, Global Energy Consumption, Energy Scenarios for the UK by 2050, DECC

## 1. Introduction

Global primary energy consumption is increased at an alarming rate during the last three decades. For instance, Electricity generation consumption raises from around 4000/MTOE in 1971 to ~8000/MTOE as shown in Fig.1 with expected demand growth by 37% by 2040[1, 2]. With the current continues demand growing, it is questionable if the current energy system could withstand the expected challenges: fossil fuel domination, climate change concerns, hindering the environmentally benign resources deployment along with volatile industry market prices [3]. Hence, it is worthwhile trying to predict the future energy scenarios even though the energy system itself is highly complex. As people, nowadays, take the energy for granted in line with less awareness about future energy issues, it is mandatory to publicise or predict some future energy scenarios depending on the current statistics. Therefore, in this study, three energy scenarios with respect to the UK energy system by 2050 will

be presented along with assessment with regard to each energy trilemma triangle leg: energy security, cost, and carbon emissions. Then, looking to some advanced technological solutions or inventions that help to alleviate some energy challenges. Ultimately, a conclusion will be given.

## 2. Future Energy Scenarios

Generally, there have been three energy scenarios along the history of the human beings [3]. In the ancient ages, the population was small, and the human or animal power was used to generate the energy. Then, with the industrial revolution, the use of heat engines facilitates the energy delivery to the demands. Afterwards, with the advanced of induction motors and power electronic converters, the electrical age has begun and still continues. Accordingly, nowadays, with the current higher standard of living, the unprecedented energy consumption, along with the world population inflation, graver concerns start looming such as global warming or reserves ending points [4,5]. Hence, in this section, three scenarios for the UK will be investigated based on the following world current energy assumptions:

- The global energy generation is mostly dominated by fossil fuels with more than 83% as shown in Fig.2
- Most of the oil which is imported from the Middle Eastern countries with highly likely being in turmoil and increasingly affected by geopolitics factors, thus, in turn makes the oil prices more volatile [6,7]. For instance, in the previous months, oil prices were suddenly climbed around 6% in one day from 55$/barrel to 59$/barrel as Saudi Arabia (Top World Producer) launched air strikes upon Yemen [8].
- With the current rate of consumption, the coal is expected to last for 200 years, the oil is projected to last for 100 years while the gas is expected to last 150 years as depicted in Fig.3 [9].
- The temperature rise due to greenhouse gasses (mostly CO2), as the big eight countries generally agreed, should not exceed 2C° with (1100 Gigatonnes) carbon emissions reduction from 2011 to 2050 [10].

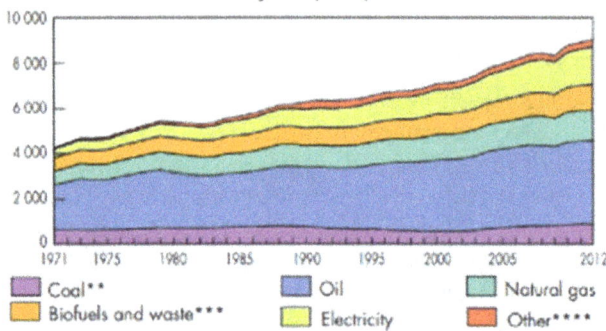

**Fig. 1.** *World Primary Energy Consumption since 1971 until 2012 per MTOE[1].*

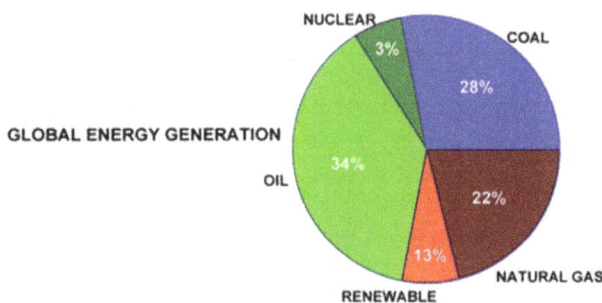

**Fig. 2.** *Current Global Energy Generation mix [7].*

Since it is quite difficult to predict the overall world scenarios due to different regional priorities, all the data and the analysis will be presented according to the UK energy system using '*The UK2050 pathway calculator*' which is a tool available online from Department of Energy and Climate change (DECC) to increase public awareness about climate change and energy security along with help policy makers and energy industry [11]. Yet, for reference guidance, the analysis has been done by taking into account the followings; transport system electrification, demand side management or energy

efficiency, renewables deployment with others, and growth in industry (see Appendix A). Hence, three predicted scenarios for 2050 are shown below depending on the way of ensuring affordable cost, secure supply and abated emissions.

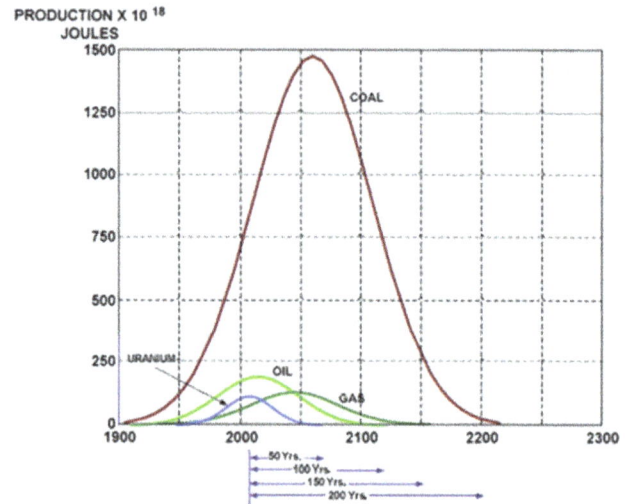

**Fig. 3.** *The depletion curves of fossil fuels sources with Uranium (Nuclear energy) [9].*

### 2.1. First Energy Scenario

In this scenario, a try to ensure energy supply security, low cost along with low carbon emissions, according to the UK official commitments by 2050. These commitments are to ensure most of the energy consumption is met from renewables in line with a cost reduction during time and 80% carbon reduction compared to 1990 levels [12,13]. Hence, the choices will mainly be based to solve the energy trilemma shown in Fig.4 by trying to adequately meet all the three targets despite the extreme difficulty.

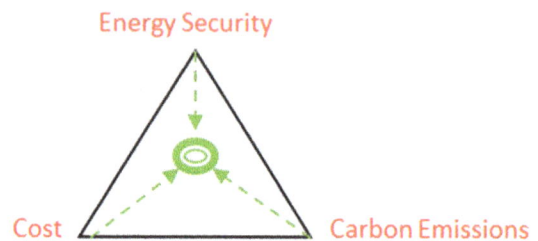

**Fig. 4.** *Energy Trilemma represents the difficulty of meeting all three objects at the same time.*

However, the constraints of approaching the three optimum targets comes from the fact of Kaya Identity [14], as follows:

$$C = P \ x \ \frac{E}{GDP} \ x \ \frac{GDP}{P} \ x \ \frac{C}{E} \qquad (1)$$

Where C is the carbon emissions, P is the population, $\frac{E}{GDP}$ is the energy consumption, $\frac{GDP}{P}$ is the Gross Domestic product and $\frac{C}{E}$ is the carbon intensity of energy supply.

At first glance, reduce any of the four aforementioned factors consequently leads to reduce carbon emissions which in turn means increase renewables share and decrease fossil fuels. But reducing (P) and ($\frac{GDP}{P}$) is undesirable due to obvious reasons. Thus, reducing either ($\frac{E}{GDP}$) or ($\frac{C}{E}$) represents attractive solutions before any increase of other factors. However, firstly, reducing ($\frac{E}{GDP}$) means reducing the energy taken per GDP or 'energy efficiency' but that will not likely solve the problem entirely as it is still on the demand side management. Secondly, reducing ($\frac{C}{E}$) means decarbonizing the energy supply by using renewable energy, nuclear or carbon capture and storage (CCS). So far, this may be a good solution to alleviate the problem of energy trilemma.

However, in this scenario and with respect to the previous analysis, the main focus will be for ensuring the carbon emissions not below 80% as required and reflect the results upon the cost and supply security. Thus, the choices to effectively meet this target can be seen in (Appendix B) while the results and the analysis are shown below:

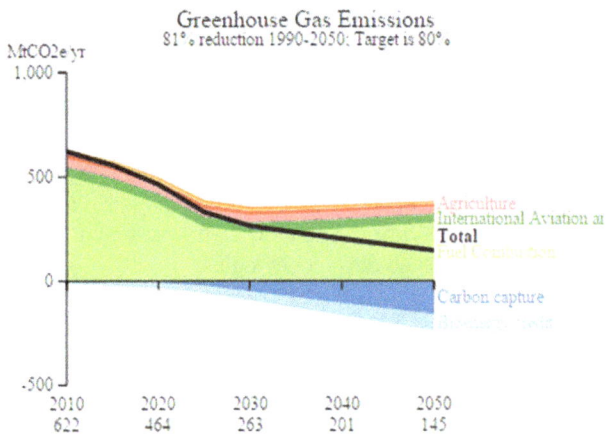

Fig. 5. *GHG emissions for UK 2050, 1ˢᵗ Scenario.*

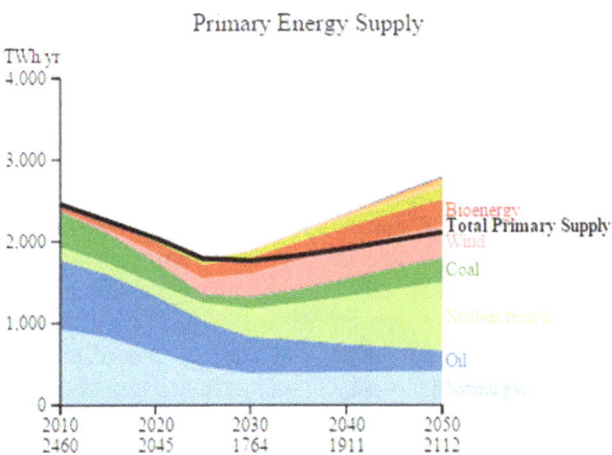

Fig. 6. *Primary Energy Supply for UK 2050, 1ˢᵗ Scenario.*

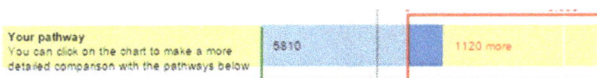

Fig. 7. *The cost required in (£/year/person), 2050, 1ˢᵗ S.*

It could be clearly seen from Fig.5 that the total GHG was

reduced from 622 MTCO₂ in 2010 levels to 145 MTCO₂ by 2050 with 81% reduction. While Fig.6 shows the total primary supply experienced a low decrease but Bioenergy and wind are dominant instead of fossil fuels particularly oil. However, as the new technologies associated with higher cost such as offshore and Wave, the customer should pay as much more as (1120£/year/person) than they paying now as shown in Fig.7 which is more expensive.

In conclusion for this scenario, the energy trilemma is not completely solved albeit the CO₂ emissions target and supply security were achieved, due to renewables high investment cost. However, it should be noted that maintenance and operation cost of renewables is much less than fossil fuels as follows:

$$C = \frac{I}{aE} + \frac{M}{E} \qquad (2)$$

Where C is the total cost of energy, I is the investment cost, E is the annual energy production, a is the discounting factor, and M is the operating cost.

### 2.2. Second Energy Scenario

In this scenario, the main focus is to achieve acceptable cost by increasing the energy efficiency particularly at the demand side. Hence, the choices are shown in (Appendix B) while the results and the analysis are shown below:

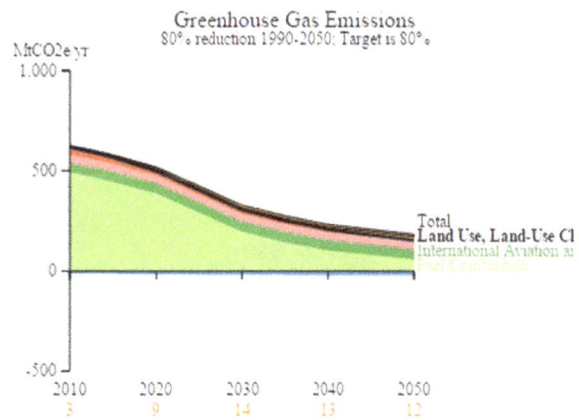

Fig. 8. *GHG emissions for UK 2050, 2ⁿᵈ Scenario.*

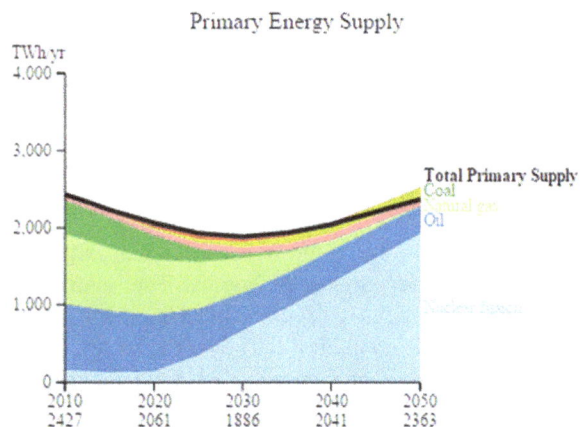

Fig. 9. *Primary Energy Supply for UK 2050, 2nd Scenario.*

*Fig. 10. The cost required in (£/year/person), 2050, 2$^{nd}$ S.*

As could be seen from Fig.8, the GHGs emissions goal once again was achieved thanks to the huge efforts made on demands side management along with the tremendous industry growth. Fig.10 shows that the added energy price per person per year is £567 which is superior to the previous scenario. However, the big debate is with Fig.9 when the supply security left mostly to the nuclear power along with Coal, Oil, and Gas. Accordingly, nuclear power, indeed, is cost effective, low carbon and secure thus it solved all the energy trilemma triangle. But after Fukushima nuclear plant accident in 2011, all countries have been investigated their plants and some others already started to shut down some of these plants if not all of them [15]. Nevertheless, there are many problems associated with nuclear power to be globally scalable such as the nuclear waste, the proliferation, the Uranium resource, the accident rate and the land area [16].

In conclusion for this scenario, in our opinion, this could be the worst one as the predicted energy supply cannot withstand the expected future energy challenges. And we believe that the (DECC) made a mistake when they assigned the GHGs emissions as the ultimate goal. Because this can be achieved even if nuclear and fossil fuels being used as seen here.

### 2.3. Third Energy Scenario

In this scenario, the choices were selected very carefully as shown in (Appendix B) to solve the energy trilemma as adequate as possible to ensure secure, affordable and environmentally benign energy sector. The results and analysis are discerned below:

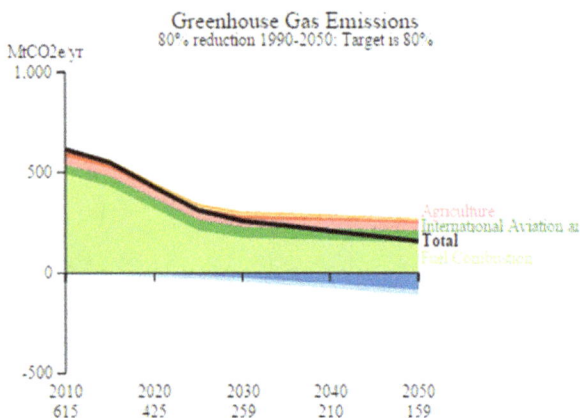

*Fig. 11. GHG emissions for UK 2050, 3$^{rd}$ Scenario.*

It could be seen from Fig.11, the GHGs emissions target is achieved same like the previous scenarios. The supply is dominated mainly by wind power and nuclear. Then with fossil fuels mix as depicted in Fig.12. The cost is far less than the previous scenarios with an increase of £335 per person per year as shown in Fig.13.

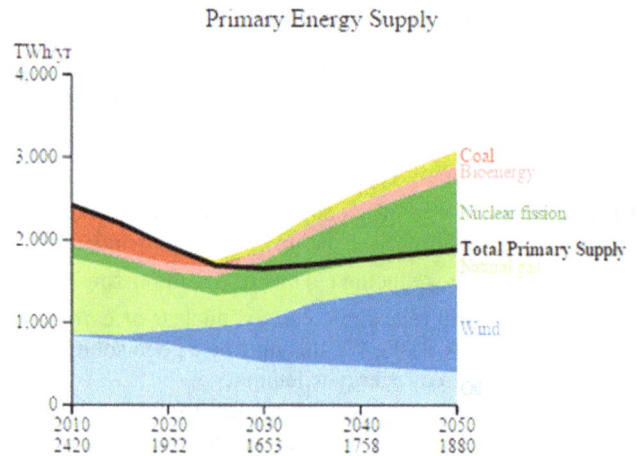

*Fig. 12. Primary Energy Supply for UK 2050, 3$^{rd}$ Scenario.*

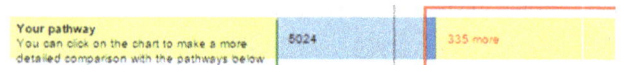

*Fig. 13. The cost required in (£/year/person), 2050, 3$^{rd}$ S.*

The reduced cost of energy in this path is mainly due to three reasons. Firstly, maximizing demand side management choices. Secondly, electrifying the transport system is costly because, for instance, neither electric vehicles is deployed nor it's charging infrastructure. Ultimately, reducing the cost without fossil fuels mix is difficult, at least currently, due to high investment cost related with renewables and new technologies.

## 3. Recent Invention: Liquid Metal Battery

One of the main constraints confronting renewable energy deployment is the intermittency of these resources. Hence, it is essential, for instance, to effectively store the energy when the sun does not shine or the wind dose not blow. Moreover, for instance, Lithium –ion battery, which is regarded the best battery nowadays is not suitable for grid-level storage as well as its high cost [17]. However, with the recent invention from MIT, liquid metal battery could play a crucial role to adequately mitigate the problem of intermittency. The battery is suitable for grid level storage with low cost, low fade rate, emissions-free, silent, made from earth abundant elements (Antimony Sb, Lithium Li, Magnesium Mg), and more effective if compared to the current battery technologies as shown in Fig.14 which represents the daily cycling of these batteries [18,19,20]. Thus, in our opinion, the use of this invention is essential for the UK as well as other nations to secure the future energy needs.

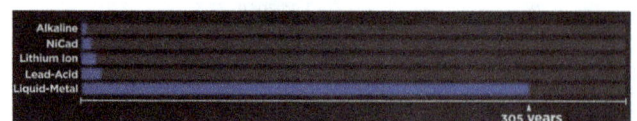

*Fig. 14. The daily cycling of number of batteries showing liquid metal battery is capable of maintaining 80% of its initial efficiency during 305 years' period [20].*

# 4. Conclusion

In conclusion for this paper, all global nations should agree to move by leaps and bounds to secure our energy future by meeting the demand, mitigate GHGs emissions and support renewables deployment since the dependence on fossil fuels, particularly oil is uncertain. For the UK energy future scenarios till 2050, the following points can be summarised:

- The 80% reduction of GHGs is achievable for all scenarios even if nuclear with fossil fuels are used without renewables. But this is very risky.
- The demand side management and energy efficiency is highly needed in all scenarios to achieve the future targets. Thus, increasing public awareness is necessary.
- Electrifying the transport system is also essential and the need for electric vehicles infrastructure with low cost is compelling.
- Currently, the investment cost for renewables technologies is very high. Therefore, government policies support is more likely needed along with 'angle investors', market competition to reduce the cost.
- Solving energy trilemma is difficult unless all players (Industry, public awareness, government support, global policies) agree to move in the same direction.

Ultimately, liquid metal battery is a promising battery technology for securing our energy future.

# Acknowledgements

The authors would like to thank the Higher Committee for Education Development in Iraq (HCED) for providing the fund for their master's study period at University of Leeds/UK. Ahmed would like also to thank his parents (Dhaferah and Maher) for their immense support.

# Appendix (A)

Refer to the website of The UK 2050 Calculator for better understanding and high quality view:
https://www.gov.uk/2050-pathways-analysis
Please note the following:

- The transition from (1) to (4) or from (A) to (D) as shown below represents the transition from the lowest case to the highest case. For example, level (1) means there is no effort to reduce emissions or save the energy while (4) means extremely ambitious target was chosen.
- For more details about the choices, because they are extremely versatile and there is no space to cover them all, please refer to the website above.
- When choosing one of any factors shown below, the results will appear immediately at the website showing the effect of the chosen indicator for supply security, CO2 emissions and the cost.

# Appendix (B)

## 1-FIRST SCENARIO CHOICES
Choices Criteria and Analysis:

1. There is an urgent need to electrify the transport system as it is much oil based in the UK, nearly 95% of the transport system is oil based. In addition, domestic transport behavior should change to a very ambitious level as well as International shipping and aviation. Hence, the choices are shown in the first energy scenario figure below.

2. For energy efficiency, certain measures are needed to reduce the cost of energy bills. For instance, homes insulation is as high as 18m.
3. For industry choices, the main points are to fully support the growth in industry and the commercial demand of heating and cooling should be dropped.
4. For Renewables share and others, offshore, onshore wind, and CCS power stations should substantially increase to meet the demand and secure the supply along with moderate choices associated with other technologies as shown.

Domestic transport behaviour
Shift to zero emission transport
Choice of fuel cells or batteries
Domestic freight
International aviation
International shipping
Average temperature of homes
Home insulation
Home heating electrification
Home heating that isn't electric
Home lighting & appliances
Electrification of home cooking
Growth in industry
Energy intensity of industry
Commercial demand for heating and cooling
Commercial heating electrification
Commercial heating that isn't electric
Commercial lighting & appliances
Electrification of commercial cooking
Nuclear power stations
CCS power stations
CCS power station fuel mix
Offshore wind
Onshore wind
Wave
Tidal Stream
Tidal Range
Biomass power stations
Solar panels for electricity
Solar panels for hot water
Geothermal electricity
Hydroelectric power stations
Small-scale wind
Electricity imports
Land dedicated to bioenergy
Livestock and their management
Volume of waste and recycling
Marine algae
Type of fuels from biomass
Bioenergy imports

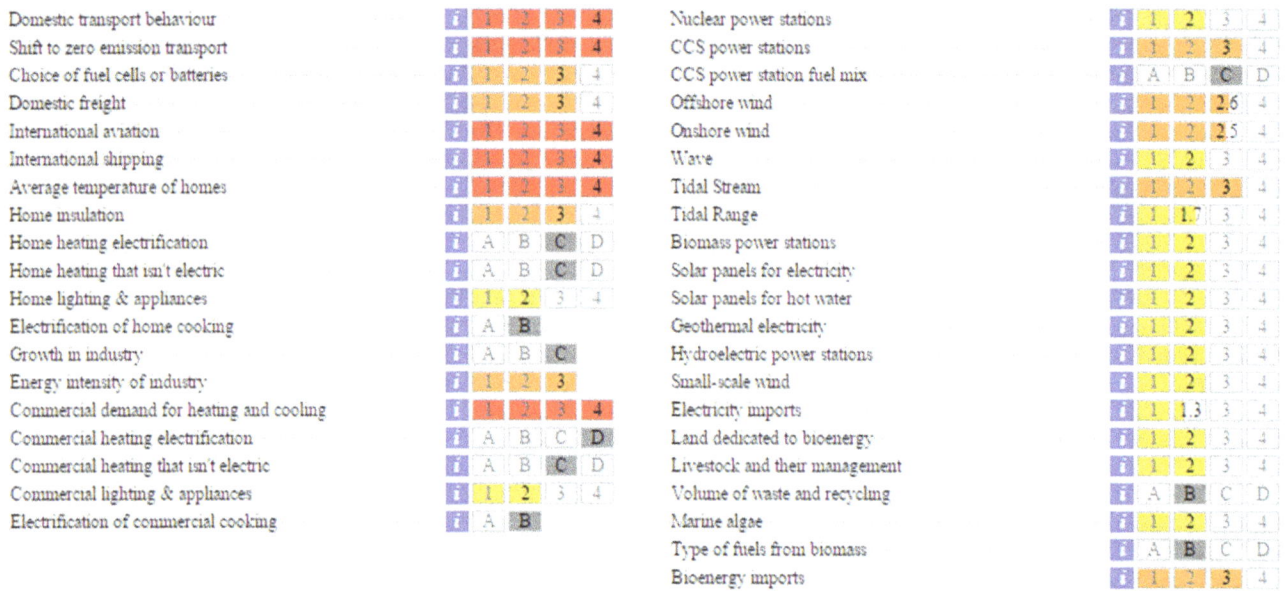

*First Energy Scenario Choices Figure*

*2- SECOND SCENARIO CHOICES*

Choices Criteria and Analysis:

1. Electrifying or decarbonizing the transport system was kept nearly the same as Scenario 1 as that is a prerequisite to achieve our future energy targets.
2. For energy efficiency, or demand side management, maximizing all the choices to the highest ambitious levels indicating full public awareness was achieved.
3. For industry choices, same like the energy efficiency, all were put to maximum.
4. Surprisingly, all renewables were set to the current limits Without any alteration. Except Nuclear power station were moved to level 3 which means more nuclear power plants need to be built.

Domestic transport behaviour
Shift to zero emission transport
Choice of fuel cells or batteries
Domestic freight
International aviation
International shipping
Average temperature of homes
Home insulation
Home heating electrification
Home heating that isn't electric
Home lighting & appliances
Electrification of home cooking
Growth in industry
Energy intensity of industry
Commercial demand for heating and cooling
Commercial heating electrification
Commercial heating that isn't electric
Commercial lighting & appliances
Electrification of commercial cooking
Nuclear power stations
CCS power stations
CCS power station fuel mix
Offshore wind
Onshore wind
Wave
Tidal Stream
Tidal Range
Biomass power stations
Solar panels for electricity
Solar panels for hot water
Geothermal electricity
Hydroelectric power stations
Small-scale wind
Electricity imports
Land dedicated to bioenergy
Livestock and their management
Volume of waste and recycling
Marine algae
Type of fuels from biomass
Bioenergy imports

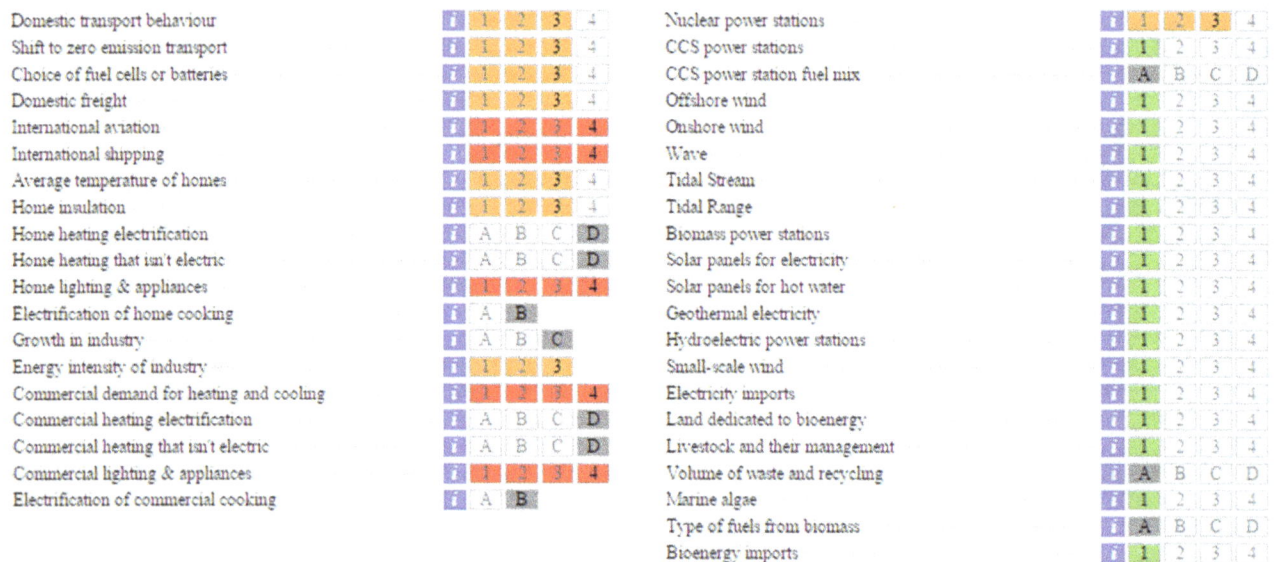

*Second Energy Scenario Choices Figure*

*3- THIRD SCENARIO CHOICES*

Choices Criteria and Analysis:

1. For the transport system, the choices were chosen to be nearly less ambitious than the second scenario due to high cost associated with deploying electric vehicles and charging infrastructure, for instance.
2. For energy efficiency, or demand side management, maximizing all the choices to the highest ambitious levels indicating full public awareness was achieved (same like 2)
3. For industry choices, same like the energy efficiency, all were put to maximum.
4. For Renewables and others, maximizing offshore wind, onshore wind, CCS, with one added nuclear plant, along

with the use of Tidal Stream and Range. This is primarily because these are promising technologies, clean, effective, albeit currently expensive. Solar was not chosen due to high cost mainly attributed to power electronic converters involved. And finally, any imports were not chosen because this can directly affect supply security.

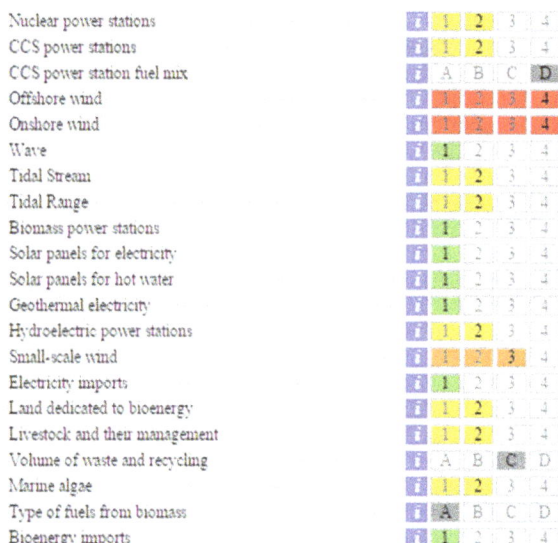

Domestic transport behaviour
Shift to zero emission transport
Choice of fuel cells or batteries
Domestic freight
International aviation
International shipping
Average temperature of homes
Home insulation
Home heating electrification
Home heating that isn't electric
Home lighting & appliances
Electrification of home cooking
Growth in industry
Energy intensity of industry
Commercial demand for heating and cooling
Commercial heating electrification
Commercial heating that isn't electric
Commercial lighting & appliances
Electrification of commercial cooking

Nuclear power stations
CCS power stations
CCS power station fuel mix
Offshore wind
Onshore wind
Wave
Tidal Stream
Tidal Range
Biomass power stations
Solar panels for electricity
Solar panels for hot water
Geothermal electricity
Hydroelectric power stations
Small-scale wind
Electricity imports
Land dedicated to bioenergy
Livestock and their management
Volume of waste and recycling
Marine algae
Type of fuels from biomass
Bioenergy imports

*Third Energy Scenario Choices Figure*

# References

[1] BP Statistical Review of World Energy. [Online]. [Accessed on 16/02/2014].Available from: http://www.bp.com/en/global/corporate/about-bp/energy-economics/statistical-review-of-world-energy.html

[2] International Energy Agency (IEA). [Online]. [Accessed on 16/02/2014]. Available from: http://www.iea.org/Textbase/npsum/WEO2014SUM.pdf

[3] M. Geidl, G. Koeppel, P. Favre-Perrod, B. Klockl, G. Andersson, and K. Frohlich, "Energy hubs for the future," *Power and Energy Magazine, IEEE,* vol. 5, pp. 24-30, 2007.

[4] W. F. Pickard and D. Abbott, "Addressing the intermittency challenge: Massive energy storage in a sustainable future," *Proceedings of the IEEE,* vol. 100, p. 317, 2012.

[5] B. K. Bose, "Global warming,"*IEEE Ind. Electron.* Mag., vol. 4, no. 1, pp. 1–[2]17, Mar. 2010.

[6] World Energy Outlook.[Online].[Accessed on 18/02/2015].Available from: http://www.worldenergyoutlook.org/pressmedia/quotes/6/

[7] B. K. Bose, "Global Energy Scenario and Impact of Power Electronics in 21st Century," *Industrial Electronics, IEEE Transactions on,* vol. 60, pp. 2638-2651, 2013.

[8] BBC New (26 March 2015). [Online]. [Accessed on 30/03/2015].Available from: http://www.bbc.co.uk/news/business-32062571

[9] J. R. Roth, *Long Term Global Energy Issues in Industrial Plasma Engineering*, vol. 1. Philadelphia, PA: Institute of Physics, 1995.

[10] C. McGlade and P. Ekins, "The geographical distribution of fossil fuels unused when limiting global warming to 2 [deg] C," *Nature,* vol. 517, pp. 187-190, 01/08/print 2015.

[11] 2050 Pathways analysis report. DECC. Available from: https://www.gov.uk/2050-pathways-analysis#creating-your-pathway

[12] UK government. 2050 Pathways Analysis. 2010. [Online]. Available from: https://www.gov.uk//government/uploads/system/uploads/attachment_data/file/42562/216-2050-pathways-analysis-report.pdf

[13] P. Allen and T. Chatterton, "Carbon reduction scenarios for 2050: An explorative analysis of public preferences," *Energy Policy*, vol. 63, pp. 796-808, 12// 2013.

[14] Y. Kaya and K. Yokobori, Environment, Energy and Economy; Strategies for Sustainability: United Nations University Press, 1997.

[15] H. Altomonte, "Japan's Nuclear Disaster: Its Impact on Electric Power Generation Worldwide [In My View]," Power and Energy Magazine, IEEE, vol. 10, pp. 96-94, 2012.

[16] D. Abbott, "Is Nuclear Power Globally Scalable? [Point of View]," Proceedings of the IEEE, vol. 99, pp. 1611-1617, 2011.

[17] D. Larcher and J. M. Tarascon, "Towards greener and more sustainable batteries for electrical energy storage," Nature Chemistry, vol. 7, pp. 19-29, 01//print 2015.

[18] Bradwell, D.J., et al., Magnesium–Antimony Liquid Metal Battery for Stationary Energy Storage. Journal of the American Chemical Society, 2012. 134(4):p. 1895-1897

[19] Wang, K., et al., Lithium-antimony-lead liquid metal battery for grid-level energy storage. Nature, 2014.

[20] Energy 2064 Video with Professor Donald R. Sadoway. [Online]. OCT 2014. [Accessed 07/04/2015].

# A newly isolated green alga, *Pediastrum duplex* Meyen, from Thailand with efficient hydrogen production

**Ramaraj Rameshprabu[1], Rungthip Kawaree[2], Yuwalee Unpaprom[2, *]**

[1]School of Renewable Energy, Maejo University, Sansai, Chiang Mai-50290, Thailand
[2]Program in Biotechnology, Faculty of Science, Maejo University, Sansai, Chiang Mai-50290, Thailand

**Email address:**

yuwalee@mju.ac.th (Unpaprom Y.), rrameshprabu@gmail.com (Ramaraj R.), rameshprabu@mju.ac.th (Ramaraj R.)

**Abstract:** Biofuels are gaining attention worldwide as a way to reduce the dependence on fossil fuels. Biological Hydrogen ($H_2$) production is considered the most environmentally friendly route of producing $H_2$, fulfilling the goals of recycling renewable resources and producing clean energy. It has attracted global attention because of its potential to become an inexhaustible, low cost, renewable source of clean energy and appears as an alternative fuel. $H_2$ production processes offer a technique through which renewable energy sources like biomass can be utilized for the generation of the cleanest energy carrier for the use of mankind. This paper presents laboratory results of biological production of hydrogen by green alga was isolated from fresh water fish pond in Sansai, Chiang Mai province, Thailand. Under light microscope, this green alga was identified as belonging to the genus *Pediastrum* and species *P. duplex* Meyen. The successful culture was established and grown in poultry litter effluent medium (PLEM) under a light intensity of 37.5 $\mu mol^{-1} m^2 sec^{-1}$ and a temperature of 25°C. The nutrient requirements and process conditions that encourage the growth of dense and healthy algal cultures were explored. The highest $H_2$ was produced when cultivated cells in PLEM for 21 hours under light and then incubated under anaerobic adaptation for 4 hours.

**Keywords:** Freshwater Algae, *Pediastrum Duplex* Meyen, Poultry Litter Effluent, Biohydrogen

## 1. Introduction

Today environmental pollution is a great concern to the world, mainly due to rapid industrialization and urbanization. And the utilization of fossil fuels is the main contributing factor to global climate change, mainly due to the emission of pollutant, especially carbon dioxide ($CO_2$) released into the atmosphere upon their combustion. Algae, the major biomass of living organisms in marine and freshwater [1], are the most important $CO_2$ fixer, the primary producer in aquatic ecosystem through the photosynthesis into biomass to fulfill the responsible duty of $CO_2$ [2, 3], and play a crucial niche of $CO_2$ bio-fixation of the ecosystem [3].

At present algae is broadly recognized as one of the best strategies of $CO_2$ fixation and biomass production [4]. There are several reasons for this approach: (i) the best growth rate among the plants, (ii) low impacts on world's food supply, (iii) specificity for $CO_2$ sequestration without gas separation to save over 70% of total cost, (iv) excellent treatment for combustion gas exhausted with $NO_x$ and $SO_x$, (v) high value of algae biomass including of feed, food, nutrition,

pharmaceutical chemicals, fertilizer, aquaculture, biofuel, etc [5–7].

The current boom in microalgae biotechnology has led to a further strong increase in the expectation that the production of biofuel from microalgae will be sustainable both energetically and financially. As efficient photosynthetic organisms, microalgae have unique advantages in capturing solar energy to generate reducing equivalents and converting atmospheric $CO_2$ to organic molecules [9]. Microalgae also have special advantages in ability to adapt to various stressful environments, non-requirement of agricultural land and so on [4]. Microalgae's high ability to use inorganic nutrients (nitrogen and phosphorous) from wastewater makes them a useful bioremediation tool in waste water treatment process. Many green microalgae species are commonly used in the wastewater treatment system due to their high tolerance to soluble organic compounds [10–12]. Algae can be successfully cultivated in wastewaters. It can be potentially a sustainable growth medium for the algal feed stock [12, 13].

Use of microalgae in treatment and recycling of waste water has attracted a great deal of interest because of excessive biomass generation at cheaper cost without extra input of nutrients such as inorganic fertilizer (chemical medium) [14]. The selection of carbon source is important for microalgae cultivation also. Microalgae can use inorganic carbon ($CO_2$) and organic carbon source (glucose, mannitol, acetate, sucrose). Cultivation of microalgae in swine wastes, dairy manure, and other animal residues has been reported by several authors [14–16].

More importantly, there were no bacteria and pathogens found on aerated manure when algae are grown. The algae culture carried out in different types of reactors, flasks and plastic bags are common practice with mineral medium as nutrient sources. Kumar et al. [17] stated digested piggery effluent could be an alternative nutrient source for mass algal production since anaerobically treated animal waste contains nutrients such as phosphorous, nitrogen species including, ammonium nitrogen, nitrate, nitrite and which are suitable for growing algae. In addition, pathogens in the effluent were eliminated by anaerobic digestion particular after two-stage (thermophilic and mesophilic) digestion [17]. Therefore, digested effluent as medium of algal culture could be an alternative to using costly mineral medium which is not environmentally sustainable.

In addition, microalgae grown under heterotrophic condition (using organic carbon) could have more potential to produce hydrogen than in autotrophic condition (inorganic carbon). In heterotrophic condition, high biomass is achieved. Moreover, in heterotrophic mode of cultivation organic wastes, as cheaper carbon sources can be used. Unlike autotrophic condition, light is also not required in heterotrophic cultivation [18]. Therefore, heterotrophic cultivation can be cheaper than autotrophic cultivation. One of the advantages of green algae for producing hydrogen is their ability to grow under photoautotrophic and photoheterotrophic condition. There is little information available for undertaking intensive algal production by using digested poultry effluent as a nutrient re-source. The present article reports simultaneous algae growth, biomass production and waste recycling with the green microalga P. duplex Meyen from the poultry litter waste water.

At present, the potential of microalgae as a source of renewable energy has received considerable interested in biofuel such as biodiesel, bioethanol, biogas, biohydrogen and bio-oils [17, 18]. Hydrogen gas ($H_2$) is a valuable energy carrier, an important feedstock to the chemical industry, and useful in detoxifying a wide range of water pollutants. As an energy carrier, it is especially attractive due to its potential to be used to power chemical fuel cells. It is considered an ideal energy carrier for the future. Compared to fossil fuels as traditional energy sources, hydrogen is a promising candidate as a clean energy carrier in the future because of its higher heating values 141.6 MJ $kg^{-1}$, or 12.6 MJ $m^{-3}$ [8,9]. Heat and water are the only products of combustion of $H_2$ with no releasing of $CO_2$ into the atmosphere [19, 20]. Thus, hydrogen is a clean, renewable, and non-polluting fuel. Previous reports showed that photosynthetic microorganisms, *cyanobacteria* and *green algae*, can generate hydrogen from solar energy and water [17–20].

Green algae are act as the pioneer photosynthetic organism or producer in the world of ecosystem. The genus *Pediastrum* Mayen (Chlorophyceae) is a free floating, coenobial, green algae occurs commonly in natural freshwater lentic environments like ponds, lakes, reservoirs etc [21]. In this study, we examined a newly isolated species of green microalga *P. duplex* Meyen for biohydrogen production applied with inexpensive poultry litter effluent medium (PLEM).

## 2. Materials and Methods

### 2.1. Isolation and Identification of Microalgae

Microalgae were collected by plankton net (20-µm pore size) from freshwater fish pond (18° 55′4.2″N; 99° 0′41.1″E) at a location near Maejo University, Sansai, Thailand. The collected samples were samples of about 5 ml were inoculated into 5-ml autoclaved Bold Basal Medium (BBM) [22] in 20-ml test tubes and cultured at room temperature (25 °C) under 37.5 $\mu mol^{-1}m^2$ $sec^{-1}$ intensity with 16:8 h photoperiod for 10 days. After incubation, individual colonies were picked and transferred to the same media for purification in 250 mL conical flask. The culture broth was shaken manually for five to six times a day. The pre-cultured samples were streaked on BBM medium-enriched agar plates and cultured for another 10 days with cool white fluorescent light using the same light intensity.

The single colonies on agar were picked up and cultured in liquid BBM medium, and the streaking and inoculation procedure was repeated until pure cultures were obtained. The purity of the culture was monitored by regular observation under microscope. The isolated microalgae were identified microscopically using light microscope with standard manual for algae [23,24].

### 2.2. Inoculums Preparation

Isolated and purified microalgae (*P. duplex* Meyen) were inoculated in 250-ml Erlenmeyer flasks containing 125 ml culture medium (BBM). Flasks were placed on a reciprocating shaker at 120 rpm for 7 d at room temperature of 25±1 °C. Light was provided by cool white fluorescent lamps at an intensity of 37.5 $\mu mol^{-1}m^2$ $sec^{-1}$. The algae culture was then transferred to 500-ml Erlenmeyer flasks containing 450 ml.

### 2.3. Preparation of Poultry Litter Effluent Medium

The anaerobically digested poultry litter effluent (PLEM) raw was collected from the chicken farm at Mae Fak Mai, Chiang Mai province, Thailand (18°97'20.01"N; 98° 97'99.41"E) and transported to the Maejo University laboratories in 10 L plastic containers and stored in a cold room maintained at 4°C. Pre-treatment was carried out by sedimentation and filtration with a filter cloth to remove large, non-soluble particulate solids. After filtration the substrate was autoclaved for 20 min at 121°C, after which the liquid was stored at 4 °C for 2 days for settling any visible particulate solids and the supernatant was used for microalgae growth studies.

### 2.4. Growth Conditions and Measurements

The isolated green alga was grown in 450 ml Erlenmeyer flasks each containing 450 ml of poultry litter effluent medium (PLEM) for the heterotrophic growth condition. It was cultivated at room temperature for 3 days under fluorescent light illumination for 18 hours per day with a shaking speed of 120 rpm. All experiments were carried out in triplicate. After obtaining the optimal growth medium, cultures were selected with an initial optical density of 0.01, *P. duplex* Meyen in PLEM under continuous light illumination at various light intensities and temperatures (same as before mentioned). Optical density at 730 nm was measured using a spectrophotometer (HACH, DR/4000U) every 3 days interval for 18 days.

The cell density and growth was determined by measuring the chlorophyll concentration; the algal culture was measured every 3 hours of cultivation by optical density measurement at wavelength 750 nm. The chlorophyll of cell suspension was extracted with 90% methanol and the total chlorophyll concentration was calculated by the method of Kosourov et al. [25].

### 2.5. Analytical Methods

All the indices including pH, chemical oxygen demand (COD), total nitrogen (TN), total phosphorous (TP) were continuously monitored throughout the study, following the standard protocols of APHA [20].

### 2.6. Measurement of $H_2$ Production

One hundred ml of cell culture was harvested by centrifugation at 5000xg for 10 min at 4°C. The cell pellet was washed with PLEM medium and the cell resuspension was transferred into a 10 ml glass vessel and sealed with a rubber stopper. The anaerobic adaptation was performed by purging Argon gas to the cell suspension for 5 min under dark condition and the cells were incubated at room temperature for 2h. Hydrogen was determined by analyzing gas phase by a gas chromatography (Agilent 6890 gas chromatography coupled to electron impact, EI, 70 eV with HP 5973 mass selective detector and a molecular sieve 5A 60/80 mesh packed column using a thermal conductivity detector). The injector and detector temperatures were kept at 100°C whereas the oven temperature was maintained at 50°C. Argon gas was used as a carrier gas during hydrogen analysis.

Hydrogen production was calculated as a term of hydrogen evolved per chlorophyll content per time ($nmolH_2/\mu g$ chl/h). Hydrogen evolution of cells at 6, 12, 18, 24 and 36 hours of cultivation was measured under light and dark conditions. In addition, algae cells were harvested and incubated under anaerobic condition for 2, 4, 6, 8 and 24 hours in darkness before measuring hydrogen evolution.

### 2.7. Statistical Analysis

Data are reported as mean ± SE from triplicate observations. Significant differences between means were analyzed. All statistical analyses were performed using SPSS Version 20.0.

## 3. Results and Discussion

### 3.1. Species Identification

*Pediastrum* is green algae occur frequently in lentic environment like pond, puddles, lakes, mostly in warm and humid terai region [21]. The green alga isolated from a fresh water fish pond, was identified as *P. duplex* Meyen. Under light microscope, single cells, four cells surrounded by transparent sheath, and truncated transparent sheaths can be observed. Figure 1 shows the morphology of *P. duplex* Meyen observed under a light microscope.

Systematic classification:
Phylum: Chlorophyta
Subphylum: Tetraphytina
Class: Chlorophyceae
Order: Sphaeropleales
Family: Hydrodictyaceae
Genus: Pediastrum
Species: *Pediastrum duplex* Meyen

Colonies free floating, disc-shaped to stellate, flat, monostromatic with 4-8-16-32-64 or more polygonal cells, compact or perforate; cells coenocytic, smooth or rough walls, marginal cells with or without process and usually differently shaped than interior cells; chloroplast parietal, disc shaped, in later stages filling entire cell, with 1-4 pyrenoids; reproduction by formation of zoospores, aplanospores, isogametes and zygotes [21].

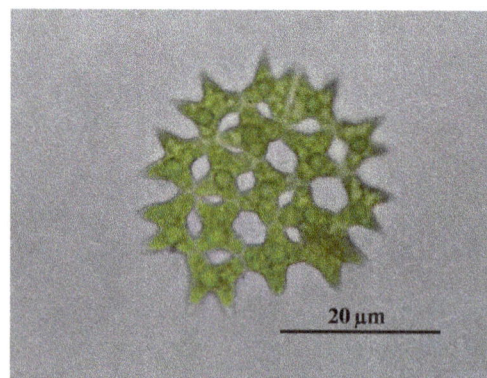

**Figure 1.** *Cell morphology of cells of P. duplex Meyen observed under a Light Microscope*

### 3.2. Algae Growth Medium, Biomass Production and Characterization

Algal growth was monitored by utilizing the optical properties of the culture to measure either its chlorophyll content / optical density. Growth of *P. duplex* Meyen was examined when grown in PLM medium, *P. duplex* Meyen showed a fast growth. PLM medium was identified as the most suitable for *P. duplex* Meyen since high cells density. The nutritious characteristics of PLEM presented in Table 1. The PLEM had high concentrations of nitrogen (1890 mg L$^{-1}$), phosphorus (187 mg L$^{-1}$), potassium (11749 mg L$^{-1}$) and other micronutrients (Table 1), which were significantly higher than recommended for algae cultivation. Algae growth was presented in Figure 2. It was found that the cells stayed in the lag phase period for the first 9 hours of cultivation. After that cells grew rapidly and entered the log phase period until reaching the stationary phase at about 15-27 hours of growth. Consequently, anaerobically digested poultry litter effluent medium was revealed that feasible algae biomass production. And the biomass is projected as a virtually eternal raw material for H$_2$ production.

*Table 1. Elemental composition of PLEM (after removing TSS).*

| Elemental analysis | PLEM (mg$^{L-1}$) |
|---|---|
| Aluminum (Al) | 10.55 |
| Boron (B) | 2.33 |
| Cadmium (Cd) | <0.1 |
| Calcium (Ca) | 210.2 |
| Chromium (Cr) | <0.1 |
| Copper (Cu) | 14.7 |
| Iron (Fe) | 28.71 |
| Lead (Pb) | 1.05 |
| Magnesium (Mg) | 57.35 |
| Manganese (Mn) | 5.99 |
| Molybdenum (Mo) | 1.16 |
| Nickel (Ni) | 1.02 |
| Phosphorus (P) | 187 |
| Potassium (K) | 1749 |
| Silicon (Si) | 48.07 |
| Sodium (Na) | 359 |
| Sulfur (S) | 124 |
| Zinc (Zn) | 10.27 |
| Nitrate nitrogen (NO$_3$-N) | 4.811 |
| Ammonia nitrogen (NH$_4$-N) | 1384 |
| Total nitrogen | 1890 |
| Total organic carbon | 881 |

energy intensive as compared to thermochemical and electrochemical processes. Furthermore, biohydrogen produced from biological processes has the potential for renewable biofuel to replace current unsustainable hydrogen production technologies, which rely on nonrenewable fossil fuels through thermochemical processes [26]. The basic advantages of biological hydrogen production over other "green" energy sources are that it does not compete for agricultural land use, and it does not pollute, as water is the only by-product of the combustion. These characteristics make hydrogen a suitable fuel for the future. Among several biotechnological approaches, photobiological hydrogen production carried out by green microalgae has been intensively investigated in recent years. One of the promising biohydrogen production approaches is conversion from microalgae, which is abundant, clean, and renewable [26, 27].

The ability of green algae to photosynthetically generate molecular H$_2$ has captivated the fascination and interest of the scientific community because of the fundamental and practical importance of the process. Below is an itemized list of the properties and promise of photosynthetic H$_2$-production [28]:

- Photosynthesis in green algae can operate with a photon conversion efficiency of > 80%.
- Microalgae can evolve H$_2$ photosynthetically, with a photon conversion efficiency of > 80%.
- Molecular O$_2$ acts as a powerful and effective switch by which the H$_2$-production activity is turned off.

Figure 3 exhibited the H$_2$ evaluation which was measured in cells at different time of cultivation period. The result showed that H$_2$ production was highest with 1.82 nmolH$_2$ /µg chl/h in cells grown for 21 hours (late-log phase cells) in PLEM. After 21 hours of cultivation, hydrogen production of cells decreased very sharply.

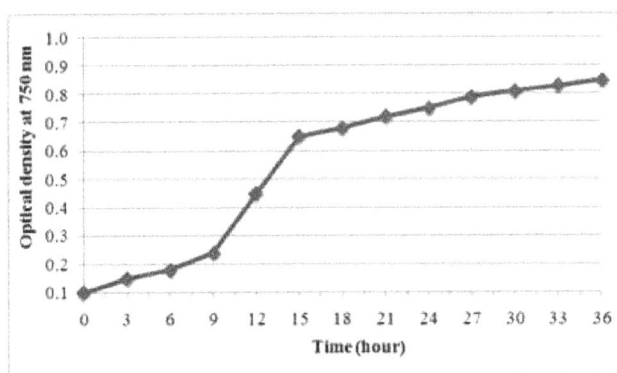

*Figure 2. Algae growth under photoheterotrophic condition*

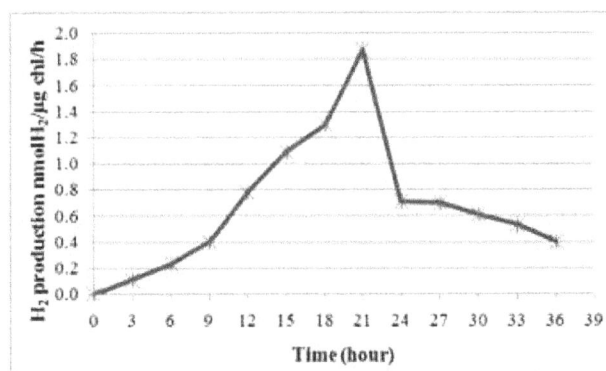

*Figure 3. Hydrogen production during cultivation periods*

### 3.3. Hydrogen Production P. Duplex Meyen in Photoheterotrophic Condition

Biohydrogen is a renewable biofuel produced from biorenewable feedstocks by a variety of methods, including chemical, thermochemical, biological, biochemical, and biophotolytical methods [26]. Biological hydrogen production processes are found to be more environment friendly and less

It was found that the late-log phase cells or 21-hours old cells could produce the highest hydrogen yield due to the enough accumulation of glycogen from the fermentation process in the PLEM. In the lag-phase cells or 18-hours old cells, under photoheterotrophic condition they need acetic acid for generating energy utilized in the cellular metabolism and for dividing cells. The generated energy and reducing powers are necessarily used for cell growth instead of producing hydrogen. In case of stationary phase cells (24-and 36-hours old cells), they were not fit and began to die because of carbon source starvation. Hence, the results from P. *duplex* Meyen verified that 18 to 21 hours are suitable time for $H_2$ production using biological method.

### 3.4. Hydrogen Production P. Duplex Meyen in Anaerobic Adaptation Time, under Light and Dark Condition

P. *duplex* Meyen was heterotrophically grown in PLEM at room temperature with same the shaking speed and time as before mentioned. Algae cells were incubated under anaerobic condition for 2 to 36 hours in darkness before hydrogen production measurement. It was found that hydrogen production was highest, with 0.137 nmol$H_2$/μg chl/h, in cells incubated under anaerobic adaptation for 4 hours (Figure 4). After that hydrogen production of cells was obviously decreased. It might be explained that during anaerobic adaptation, oxygen, an inhibitor of hydrogenase enzyme, was decreased resulting in an increase of hydrogen production in the first 4 hours after anaerobic adaptation. Incubation under anaerobic condition for more than 4 hours did not promote the higher hydrogen production because of the decrease of electron and proton donors in cells as well as the limitation of hydrogenase enzyme.

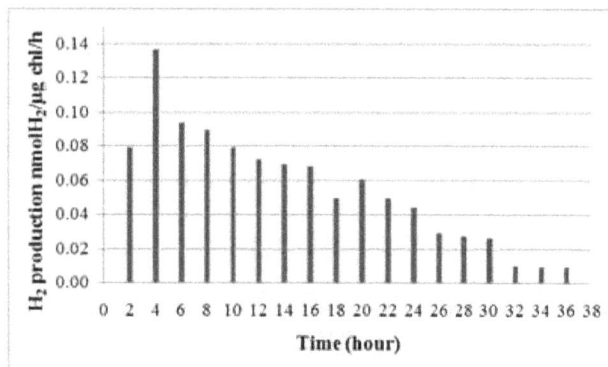

**Figure 4.** *Hydrogen production under different anaerobic adaptation time*

P. *duplex* Meyen cells were separately grown in PLEM either providing light illumination of 37.5 μmol$^{-1}$m$^2$ sec$^{-1}$ intensity or under dark condition. The result showed that cells grown under dark condition have higher optical density than those under light condition (data not mentioned). It might be explained that cells grown under dark condition used only acetic acid as carbon source for growing and dividing cells whereas cells grown under light condition could fix $CO_2$ from the atmosphere via photosynthetic process, therefore requiring more time for the initial growth.

The result showed those cells grown under light produced hydrogen about 4 times higher than cells grown under dark condition (1.82 and 0.37 nmol$H_2$/μg chl/h, respectively); the results presented in Figure 5. It was suggested that under light condition the energy in form of ATP (adenosine triphosphate) and the reducing powers NADPH (nicotinamide adenine dinucleotide phosphate) or NADH (nicotinamide adenine dinucleotide) were obtained from the light reaction of

photosynthesis, giving their electrons to excess protons for hydrogen production. Under dark condition cells produced less ATP and reducing powers, resulting in less hydrogen production. Consequently, the isolated green alga P. *duplex* Meyen produced the highest hydrogen when cultivated cells for 21 hours in PLEM under light and then incubated cells under anaerobic adaptation for 4 hours.

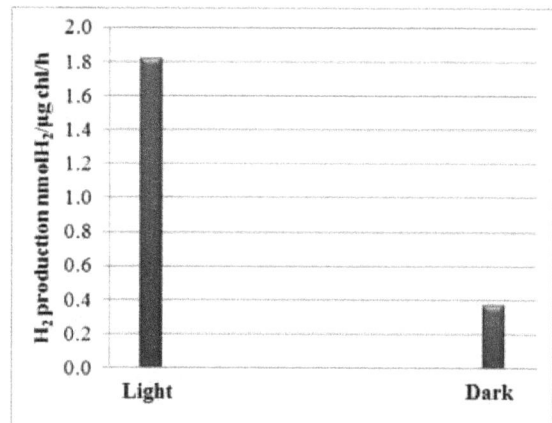

**Figure 5.** *Hydrogen production under light and dark condition*

## 4. Conclusions

Biological hydrogen production from biomass is considered one of the most promising alternatives for sustainable green energy production. One of the promising hydrogen production approaches is conversion from biomass, which is abundant, clean and renewable. We identified and isolated a new green alga, *Pediastrum duplex* Meyen, from freshwater fish pond at a location near Maejo University, Sansai, Thailand with a capacity of efficient biohydrogen production *using with* poultry litter medium (PLM). P. *duplex* Meyen can be produced by using anerobically digested poultry effluent without any chemical supplementation. This method may reduce the cost of commercial algal production and biofuel applications. P. *duplex* Meyen produced the highest hydrogen at 21 hours in PLEM under light and then incubated cells under anaerobic adaptation process. This study results revealed that biological dark fermentation is also a promising hydrogen production method for commercial use in the future.

## Acknowledgements

We thank Mr. Warunsiri Sujinda (Head of Strategic Planning and Assessment Sector, Maejo University, Sansai, Chiang Mai 50290, Thailand) for offered chicken manure, poultry forming support & Poultry litter effluent transportation.

## References

[1]    R. Ramaraj, D. D-W. Tsai, P. H. Chen, "Algae Growth in Natural Water Resources", Journal of Soil and Water Conservation, 2010, 42: 439–450.

[2] R. Ramaraj, D. D-W. Tsai, P. H. Chen, "Chlorophyll is not accurate measurement for algal biomass", Chiang Mai Journal of Science, 2013, 40: 547–555.

[3] R. Ramaraj, D. D-W. Tsai, P. H. Chen, "An exploration of the relationships between microalgae biomass growth and related environmental variables", Journal of Photochemistry and Photobiology B: Biology, 2014, 135: 44–47.

[4] R. Ramaraj, D. D-W. Tsai, P. H. Chen, "Freshwater microalgae niche of air carbon dioxide mitigation", Ecological Engineering, 2014; 68: 47–52.

[5] A. Demirbas, "Production of biodiesel from algae oils", Energy Conversation Management, 2009, 50: 4–34.

[6] DOE, "Recovery and Sequestration of $CO_2$ from Stationary Combustion Systems by Photosynthesis of Microalgae", Office of Fossil Energy National Energy Technology Laboratory, U.S. Department of Energy, 2006.

[7] H. J. Ryu, K. K. Oh, Y. S. Kim, "Optimization of the influential factors for the improvement of $CO_2$ utilization efficiency and $CO_2$ mass transfer rate", Journal of Industrial and Engineering Chemistry, 2009, 15: 471–475.

[8] C. N. Dasgupta, J. Gilbert, P. Lindblad, T. Heidorn, S. A. Borgvang, K. Skjånes, D. Das, "Recent trends on the development of photobiological processes for the improvement of hydrogen production", International Journal of Hydrogen Energy, 2010, 35: 10218–10238.

[9] R. Ramaraj, "Freshwater microalgae growth and Carbon dioxide Sequestration", Taichung, Taiwan, National Chung Hsing University, PhD thesis, 2013.

[10] W. J. Oswald, H. B. Gotaas, H. F. Ludwig, "Algae symbiosis in oxidation ponds II. Growth characteristics of Chlorella pyrenoidosa cultured in sewage", Sewage and Industrial Wastes, 1953, 25: 26-37.

[11] W. J. Oswald, "The coming industry of controlled photosynthesis", American Journal of Public Health ,1962, 52: 235–242.

[12] W. J. Oswald, "Microalgae and wastewater treatment", Borowitzka and Borowitzka ed., Cambridge University Press, UK. 1988.

[13] W. J. Oswald, "My sixty years in applied algology", Journal of Applied Phycology, 2003, 15: 99–106.

[14] R. Kothari, V. V. Pathak, V. Kumar, D. P. Singh, "Experimental study for growth potential of unicellular alga Chlorella pyrenoidosa on dairy waste water: an integrated approach for treatment and biofuel production", Bioresource Technology, 2012, 116, 466–470.

[15] S. Mandal, N. Mallick, "Waste Utilization and Biodiesel Production by the Green Microalga Scenedesmus obliquus" Applied and Environmental Microbiology, 2011, 77: 374–377

[16] A. C. Wilkie, W. W. Mulbry, 2002 "Recovery of dairy manure nutrients by benthic freshwater algae", Bioresource Technology, 84: 81–91.

[17] M. S. Kumar, Z. H. Miao, S. K. Wyatt, "Influence of nutrient loads, feeding frequency and inoculum source on growth of Chlorella vulgaris in digested piggery effluent culture medium" Bioresource Technology, 2010, 101: 6012–6018.

[18] N. Rashid, M. S. U. Rehman, S. Memon, Z. U. Rahman, K. Lee, J.-In Hana, "Current status, barriers and developments in biohydrogen production by microalgae", Renewable and Sustainable Energy Reviews, 2013, 22: 571–579.

[19] D. Das, T. N. Veziroğlu, "Hydrogen production by biological processes: a survey of literature", International Journal of Hydrogen Energy, 2001, 26: 13–28.

[20] B. K. Nayak, S. Roy, D. Das, "The potential of sustainable algal biofuel production using wastewater resources", International Journal of Hydrogen Energy, 2014, 39: 7553–7560.

[21] J. Komárek, V. Jankovská, "Review of the green algal genus Pediastrum: Implication for pollen-analytical research", Bibliotheca Phycologica, 2001, 108: 1-127.

[22] H. W. Bischoff, C. H. Bold, "Phycological studies. IV. Some soil algae from enchanted rock and related algal species", University of Texas Publication, 1963; 6318: 32–36.

[23] S. Kant, P. Gupta, "Algal Flora of Ladakh", Scientific Publishers, Jodhpur, India, 1998, p.341.

[24] APHA, AWWA, WPCF, "Standards Methods for the Examination of Water and Wastewater", 21st ed. APHA-AWWA-WPCF, Washington, DC, 2005.

[25] S. Kosourov, A. Tsygankov, M. Seibert, M. L. Ghirardi, "Sustained hydrogen photoproduction by Chlamydomonas reinhardtii: effects of culture parameters", Biotechnology and Bioengineering, 2002, 78: 731–740.

[26] M. Ni, D. Y. C. Leung, M.K.H. Leung, K. Sumathy, "An overview of hydrogen production from biomass", Fuel Processing Technology, 2006, 87: 461–472.

[27] A. Melis, T. Happe, "Hydrogen Production. Green Algae as a Source of Energy", Plant Physiology, 127: 740–748.

[28] G. Torzillo, A. Scoma, C. Faralon, L. Gianelli, "Advances in the biotechnology of hydrogen production with the microalga Chlamydomonas reinhardtii", Critical Reviews in Biotechnology, 2014, 1–12, DOI: 10.3109/07388551.2014.90073

# 20

# Environmental Pillars for Sustainable Management System in Ancient Olympia

**A. G. Stergiadou[1, *], V. Drosos[2], A. K. Douka[3]**

[1]Institute of Forest Engineering and Topography, Department of Agriculture, Forestry and Natural Environment, Aristotle University of Thessaloniki, Thessaloniki, Greece
[2]Department of Forestry and Management of Natural Recourses, Democritus University of Thrace, Orestiada, Greece
[3]Law Faculty, Aristotle University of Thessaloniki, Thessaloniki, Greece

**Email address:**

nanty@for.auth.gr (A. G. Stergiadou), vdrosos@fmenr.duth.gr (V. Drosos), sissi010591@hotmail.com (A. K. Douka)

**Abstract:** Sustainability and Environmental friendly management are meanings which were known since the beginning of ages; but were established by the UNEP in 1972 throughout the celebration of the World Environment Day (WED) every year on 5 June; in order to raise global awareness to take positive environmental action to protect nature and the planet Earth. The aim of this paper is to give a chronological series of changes at the unique place of Ancient Olympia, in order to show how nature was before and after the wild fire of Peloponnese at Southern Greece. As foresters and environmentalists we proposed some environmental friendly measures as a systematic treatment of the areas' sustainable management after a wild fire. The post-fire management of Kronius hill and a new plantation based on the existing species with the synergy of natural reforestation gave marvelous solutions after a decade. A series of technical works and protective measures against erosion are suggested in order to achieve the effective development of the area.

**Keywords:** Environmental Pillars, Sustainable Management, Post-Fire Measurements, Reforestation, Ancient Olympia

## 1. Introduction

Basic pillar for sustainable development is the protection of nature. The most recent reports of the EU Community Mechanism for Civil Protection point out a large increase in the number, severity and intensity of natural phenomena and man-made disasters resulting in the loss of human lives as properties and having catastrophic consequences on social and economic infrastructure, cultural heritage and on the environment [1]. The potential for severe soil erosion exists after a wildfire because as a fire burns it destroys plant material and the litter layer. Shrubs, forbs, grasses, trees and the litter layer break up the intensity of severe rainstorms. So fire can destroy the soil protection. There are several steps to take to reduce the amount of soil erosion [2].

In Greece during the summer periods' between 2007 – 2009 more than 71 people died and over 200.000ha of forest areas had been burned throughout the wild fire of Peloponnese. At Ancient Olympia (Figure 1) a part of historical forest burned down and due to immense efforts of the firefighters the stadium and the buildings of the museum

of Ancient Olympia remained intact.

*Figure 1. Archeological area of Ancient Olympia.*

Unfortunately the coniferous forest (Pinus halepensis and maki) which was surrounding the archeological area and the Kronious Hill was burned. The reforestation of the burned areas and a protection system against soil erosion are the emergency technical works that must be applied.

# 2. Erosion Control Techniques After Wildfire

The research on the fate of burnt forest ecosystems includes efforts to understand their post-fire evolution and the factors affecting it, methods for the protection and management of burnt areas aiming at their rehabilitation, the costs involved, and the way in which fire danger changes with time in the regenerating forest [3].

The main step after a wildfire like this at Peloponnese is to bring a team of specialists at forestry, wildlife, ecology, watershed, historic properties, forest engineering's, etc in order to apply effective soil stabilization techniques.

The US Government Technical Team based on BAER teams system proposed the following treatments for Ancient Olympia. A process was implemented for the Kladeos watershed case study and includes two phases. 1st Phase was the emergency stabilization, which involved a) the identification of potential values at risk from the effect of the fire including life, property, roads and cultural resources, b) the identification of how the fire has changed the watershed response based on the map of soil burn severity and specifically address changes to erosion, runoff and slope stability, c) the definition of the emergency of: threat, location, duration and extent, d) the treatment selection based on combinations of land, channel, road, protection and safety with the synergy of stabilizing factors, e) the monitoring within the next years and f) the implementation of recommendations. 2nd Phase was the long term rehabilitation and restoration [4].

The needs for an environmental friendly technique to mitigate the problem generated by soil instability and the incidence of erosion have been provoked the appearance in recent years of two different eco-technological concepts': ground bio-engineering and eco-engineering [5]. Slope stabilization after fire can be achieved by using different types of grass or vegetation and also environmentally friendly simple methods as: seeding, mulching or planting; to the most complex ones that integrate a variety of different engineering techniques using all types of materials (live cribs walls, vegetated gabions, etc) [6].

# 3. Transition to Restoration

The only safe route forward for restoration and reforestation after a wildfire is to abandon small-scale measurements and go for long term techniques in a timescale of the next decades. In the case study of Ancient Olympia as a forest engineering team we proposed techniques of stabilizing the archeological area by constructing a wall with existing materials of the area (stones, cement, etc) based on the soil stabilization techniques. The purpose of the slope stabilization measurements of Kladeos river which runs through the valley of archeological area is to minimize the sliding and to regenerate the hole area after the fire of 2007 (Fig. 3).

## 3.1. Eco-engineering Supportive Works

At the ancient walls of Olympia two problematic areas were located which are prone to collapse after some time due to deforestation. Therefore it is essential to give a solution to this problem, without however affecting of the aesthetic of the sanctuary area is needed (Fig. 2). Firstly, the supporting wall at the Sanctuary area has inclined from the vertical alignments and the upper part of it has fallen down. The thrust of the coniferous routes, the country road that passes over the archeological area and the deforestation after the fire made the possibility of the fallen wall more obvious (Fig. 3, 4).

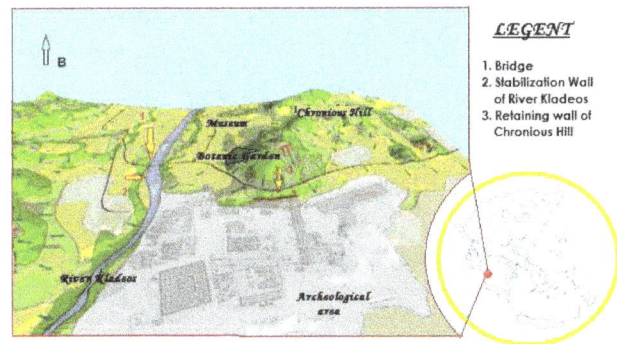

**Figure 2.** *Environmentally friendly Technical Works in Ancient Olympia after the fire in 2007.*

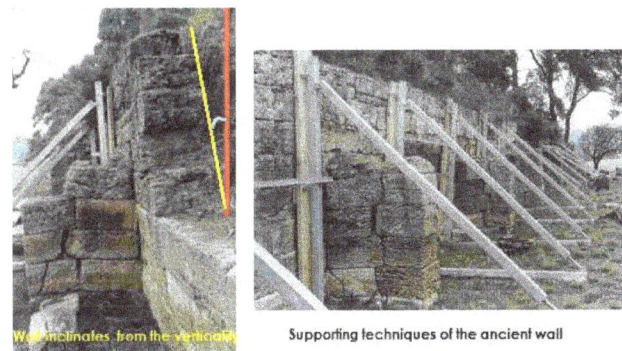

**Figure 3.** *Supporting techniques of the existing ancient wall.*

**Figure 4.** *A anti-aesthetic retaining stabilization system with metallic pillars at the existing part of the ancient wall of Olympia.*

The second measure that has been proposed to be taken was the slope stabilization of Kladeos River. The existing narrow bridge which is narrow and the riot vegetation in the river bed minimize the flow in case of overflow. So the ancient wall near the river bed needs maintenance, stabilization and raise height (Fig. 4, 5).

**Figure 5.** *The bridge at Kladeos and the vegetation over the ancient wall at the river aside.*

Kraus in 1997 has drawn how the ancient wall it was build in order to give us the opportunity to rebuild it or restore it (Fig. 6).

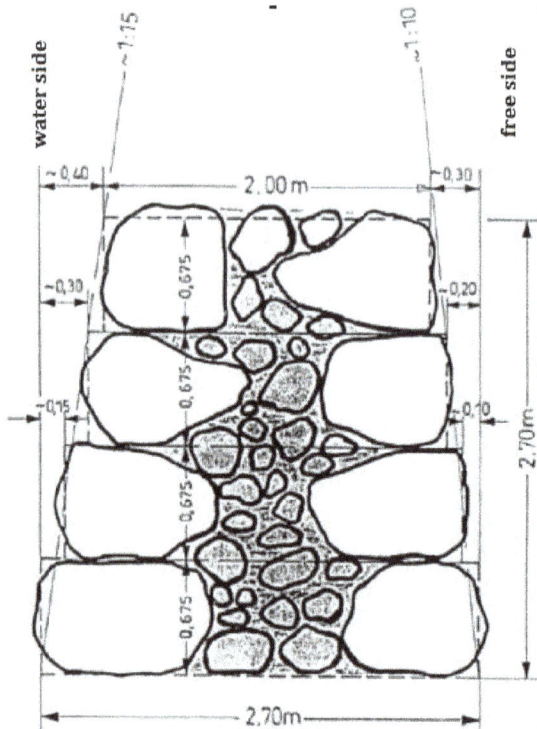

**Figure 6.** *Technical design of the ancient wall of Olympia at Kladeos riverside.*

The repair solution without affecting the sacred place proposes eco-friendly technical works. In the common areas weight walls and wire containers with suitable anchorage on steep terrain are proposed (Fig. 7, 8, 9, 10).

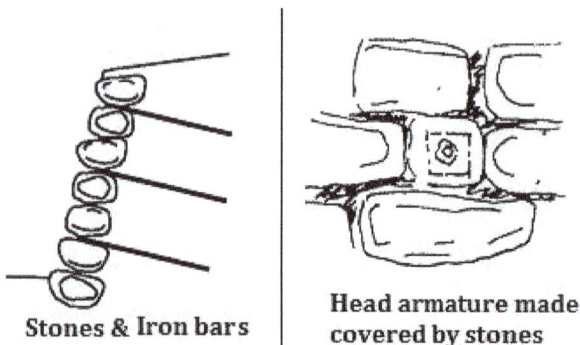

**Stones & Iron bars**

**Head armature made covered by stones**

**Figure 7.** *Stone walls with iron bars and armature made head.*

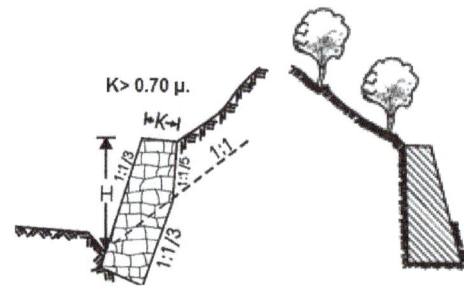

**Figure 8.** *Weight walls made of stone and concrete.*

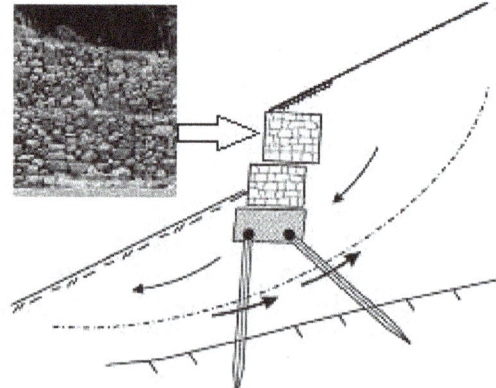

**Figure 9.** *Wire stonework underground anchorage.*

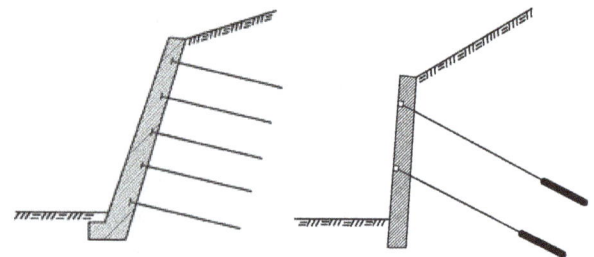

**Figure 10.** *Anchorage on the downhill.*

### 3.2. Restoration of Vegetation

As a general principle that nature knows best, restoration must be done with local species. However, since the natural process is slow, we decide with various manipulations to expedite the rhythms in order to have faster recovery.

The impact of fire on semi- mountainous forest species in the area are following:

*OAK:*

After the fire burned clumps of oaks have the ability of re-vegetation with suckers is a new path. In the case of partial destruction of the oak canvas it can reorganize the movement of juice, a few weeks after the fire. The fires in the oak forests can cause attenuation problems and attacks in neighboring blown clusters. Infestations of insects usually exhibit the greatest intensity in the first year after the fire. Fungi are installed in the first year but the attack culminates from the third year after the fire [7]. No negative effect is observed from the fire to oak atoms which are damaged by fire in an area of less than 20% of the circumference of the shank.

The deep grooves of the lower shell can be protected from

heat, while the formation of small square plates on the surface helps to transport fire there. However, the light grey color helps to absorb less heat than a species with darker crust color. The leaves are very favorable for the heat transfer area ratio / volume. The litter carpet in less favorable environments decomposes slowly providing a uniform layer of loose for the transportation of fire.

*CHESTNUTS:*

According to the records oak such as chestnut trees [8], grow rapidly after fire tillers, without human intervention. Chestnut trees are exposed to the danger of being displaced by better oak species, where chestnut trees and oak coexist. For this reason, reinforced vegetation of chestnut trees is needed. However, where the risk of an entry of fir trees in the ecosystem exist, the fire acts positively, because it reduces the completion between the tree species. The fir ecosystem fire and fir forests, act positively, since it reduces competition.

*PINUS HALEPENSIS:*

Pinus halepensis fits greatly after fires and easily regenerates after them. The halepensis pine flowering almost every year and cones ripen in the third spring after their fertilization. Maturation cones do not open, but they remain closed for 10 to 15 or more years so they are fertile seeds. Closed cones do not burn nor open during wild fire. However if someone visits a burnt forest of halepensis pine after the first two days (48 hours) of the fire then he will see a rain of seeds that fall to the ground.

These seeds germinate in autumn and after having fallen at least 2 mm of rain, which means that there is enough moisture in soil. The main role in the success of natural regeneration of halepensis pine plays the plant "Cistus incanus or zistrose tree", which occurs abundantly after fire. The "Cistus incanus"on the one hand protects young seedlings from direct sunlight and secondly, because the fungus that creates mycorrhizae in "Cistus incanus"also creates mycorrhizae in halepensis pine, rising up to 100 times the capacity of water uptake by the roots and therefore the possibility of survival of seedlings.

Since the trees were burned have an age greater than 15 years, the natural regeneration of halepensis pine is guaranteed no replanting is needed. As regards hardwoods, whether these evergreens such as holly, holms oak, etc. or are either deciduous tree such as the poplar, the oaks, etc., are updated very quickly after the fire. In Parnitha already trunk vegetation is giving the first promising message. Early autumn bulbs underground plants as cyclamens', the yolks, etc have also sprouted [9].

To achieve the vegetation recovery target and the configuration of the landscape, we suggest planting a numerous of trees and shrubs that participated in the composition of the ancient oak forest and from which most were deported or were oppressed by the halepensis pine that will dominate from now on.

After the fire of August 26 of 2007 that destroyed the natural environment, over time around the Archaeological Site and the New Museum was created a botanical garden by the forest service in order to restore the archaeological site and the surrounding wider region to the reconstitution of the Olympic landscape with immediate soil protection measures and restoration of vegetation, based on historical references, in conjunction with the pre-fire conditions and with particular emphasis on maintaining the geomorphology of Kronios hill.

The main plants that can be used for the reforestation of Ancient Olympia are the following:

- Oak (Quercus ithaburensis Decaisne Ssp. Macrolepis (Kotschy) Hedge)
- Fluffy oak (Quercus pubescens Willd.)
- Aria (Quercus ilexL.)
- Quercus (Quercus coccifera L.)
- Pine (Pinus pinea L.)
- White poplar (Populus alba L.)
- Cypress (Cupressus sempervirens L.)
- Apollo Daphne (Laurus nobilis L.)
- Wild olive (Olea europaea L.ssp. Oleaster Negodi)
- Schinos (Pistacia lentiscus L.)
- Ash (Fraxinus ornus L.)
- Oleander (Nerium oleander L.)
- Wicker (Vitex agnus - castus L.)
- Arbut (Arbutus unedo L.)
- Arbut (Arbutus andrachne L.)
- Myrtle (Myrtus communis L.)

In addition to the above plant species and the restoration of specific areas the redbud, the linden, maple and shrubs, the broom, different type of oak and populous can be used. In particular, for the needs of aesthetic improvement and increase of biodiversity of the area among other species (Pomegranate, fig) kinds of aromatic flora such as rosemary, lavender, the wormwood, marjoram, mint, sage and savory can be used.

### 3.3. Especially for the Kronio Hill of Ancient Olympia

The evolution of Kronios hill is shown in figure 11.

**Figure 11.** *Evolution of vegetation at Kronios hill (1:1900, 2:1910, 3:1930, 4:2004, 5: 2/9/2007, 6:10/23/2007, 7:07/12/2011, 8:08/11/2012, 9:07.27.2014).*

The result of this work was building a system of

specifications based on Mediterranean ecosystem needs. Despite the enduring damage, new shoots and evolution are shown. The ancient forest of Olympia successor is the halepensis pine to evergreen broadleaf subcarriers. The under storey today after the fire was restored, but not the arboreal vegetation except for a few cypresses. We hope to have the second stage of woodland and broadleaf evergreen and Aleppo pine for better protection.

The anticorrosion measures of log bars Kronio hill were almost successful (figure 12b), except for a small area which required an enhanced bracing (figure 12a).

*Figure 12. Unsuccessful, successful (a) and (b) retaining steps.*

## 4. Results

The restoration of the Archaeological and Greater Landscape of the Ancient Olympia has been declared one of the UNESCO World Heritage Monument, prerequisites research of certain tree species and implementation of technological projects aimed at a reconstitution and protection of the natural environment. The historical background of the operator space in a specific time period has in any case to be taken into account.

The entire project is largely characterized by originality; given the lack of national and international specific actions for Mediterranean forest scientific knowledge ecosystems concerning restoration of the natural environment and archaeological areas after a fire incident for drafting recovery proposal the project is remarkable and unique.

The burnt area, two months after the fire, was immediately filled with herbaceous species of flora and physical referred birth of halepensis pine which gave hope and an encouraging message that Olympia had begun to speed their growth. Also, as expected, the natural regeneration of evergreen - broadleaved species was quite satisfactory. Interventions to protect the soil against erosion and reduce flooding, in conjunction with the installation of vegetation and hydro-seeding, effectively protected all of the burnt area and improved the general aesthetic landscape.

The wire stone armature was an excellent idea for Kronios hill, where was given more attention to aesthetic and manufacture tics. Of course, there were found some localized dispatch success because of exaggeration which was to achieve the best possible result, but also because of the pressure of timetables which brought more pressure on supervisors.

The hydro-seeding was the technique of geo-textile and the application in very steep slopes was crucial. These techniques worked additionally soil for protection, improved considerably rein in semen, the hydraulic properties of the soil and thus the emergence conditions of natural regeneration.

Finally, with respect to the horticultural operations, survival today more than three years after the intervention can be count as a great choice. It was considered quite satisfactory if you think about adversity and problems that were on their irrigation, maintenance and the lack of staff support general. In conclusion, conifers first showed increase compared to broadleaves. Unlike the oaks which showed little growth compared to the other leaved while competition with the reeds is strong.

In order reduce any fu-Lodi risk of fire, which is a natural factor in the Mediterranean environment, planning and the execution of rehabilitation works adapted to local conditions and requirements of the area is taking into account the natural and human environment around the archaeological site.

Further maintenance and general management anticorrosion and horticultural works, as well as the semantic plantings and especially monitoring and impact assessment of projects, were considered necessary actions to be continued for at least four years after the fire, so as to prevent any change in the situation that has been achieved in the archaeological landscape of Olympia.

## Acknowledgements

This research is part of a Thales Program; which has been co-financed by the European Union (European Social Fund – ESF) and Greek national funds through the Operational Program "Education and Lifelong Learning" of the National Strategic Reference Framework (NSRF) - Research Funding Program: Thales - Investing in knowledge society through the European Social Fund.

## References

[1] European Union, (2007). Council decision of 8 November 2007 establishing a Community Civil Protection Mechanism, Official journal of the European Union, 2007/779/EC, Luxembourg, pp: 9-17.

[2] Tom De Gomez, (2011). Soil erosion control after wildfire, Extension Arizona Cooperative, AZ1293:12/11, The University of Arizona, College of Agriculture and Life Sciences, pp: 1-6.

[3] FRIA, (2014). Forest Fires, Hellenic ministry for rural development and food, http://www.fria.gr/EngPage/forest_fires.html

[4] USFS, (2007). Technical Collaboration between the united states and Greece in post – fire emergency response, Final report October 14-27, 2007. http://www.tee.gr

[5] Vicente Andreu, Hayfa Khuder, Slobodan B. Mickovski, Ioannis A. Spanos, Joanne E. Norris, Luuk Dorren, Bruce C. Nicoll, Alexis Achim, José Luís Rubio, Luc Jouneau, Frédéric Berger, (2008). Eco technological Solutions for unstable slopes: Ground Bio- and Eco-engineering Techniques and Strategies, Slope Stability and Erosion Control: Eco technological Solutions, pp. 211-275. http://link.springer.com/chapter/10.1007/978-1-4020-6676-4_7

[6]  Stergiadou A., Eskioglou P., (2008). Slope stabilization process of the forest roads, Proceedings of FORMEC'08 – KWF, pp.96, Germany.

[7]  Boyce W. And Handelman G. (1961). Vibrations of rotating beams with tip mass, Journal ZAMP, Vol. 12, Issue 5, pp.369-392, Springer, http://link.springer.com/article/10.1007/BF01600687

[8]  Dafis Sp., (2008). Vegetation rehabilitation monitoring program in the riparian forest of Nestos. EKBY, Vol. 5, pp. 1-25.

[9]  Kakouros, P. and S. Dafis, (2010). Guidelines for restoration of Pinus nigra forests affected by fires through a structured approach. Version 2. Greek Biotope-Wetland Centre. Thermi. 27 p.

# 21

# Evaluation from an educational perspective of the effects of waste batteries on the environment

**Dilek Çelikler, Zeynep Aksan**

Department of Elementary Science Education, Faculty of Education, Ondokuz Mayıs University, Samsun, Turkey

**Email address:**
dilekc@omu.edu.tr (D. Çelikler), zeynep.axan@gmail.com (Z. Aksan)

**Abstract:** The aim of study was to determine the locations where students studying Science Education (i.e. science teacher candidates) in Turkey dispose of waste batteries, and to identify their views regarding the effects of waste batteries on the environment. A total of 80 volunteer, third-year students from the department of science education participated in the study. Based on the study results, it was determined that the large majority of the students disposed of waste batteries in ordinary trash cans. It was also determined that the students lacked sufficient knowledge regarding the recycling of waste batteries and their effects on the environment. After waste batteries are disposed in ordinary trash cans, the chemicals they contain can mix with the soil and underground water in land fill areas, and cause pollution. To prevent waste batteries from having detrimental effects on the environment, they should be collected in containers specifically designed for waste batteries. In addition, students should be informed about practices relating to the proper disposal of waste batteries, and actively participate in them.

**Keywords:** Waste Batteries, Environment, Science Teaching Student

## 1. Introduction

Batteries represent one of the different types of solid wastes. Once they become solid wastes after being used, batteries are considered and classified as "hazardous wastes", since they can lead to significant problems if disposed improperly. By definition, hazardous wastes include all types and forms of wastes (solid, liquid, gas, sludge, etc.) which pose a hazard for human health and the environment. Hazardous wastes can be either of domestic or industrial origin, and display a wide range of different properties. Thus, hazardous wastes are solid wastes which represent a potential threat for public health and the environment (1).

Due to their toxicity, their prevalence, and their physical resistance to deterioration, waste batteries represent a significant threat for the environment and human health (2, 3). The toxicity of batteries is mainly due to their lead, mercury, and cadmium content. In addition to this, the other metals in batteries such as zinc, copper, manganese, lithium, and nickel can also pose a threat to the environment. Alkali and zinc-carbon batteries contain heavy metals such as mercury, zinc, and manganese; for this reason, it is necessary to recycle these types of batteries (4).

Considering that batteries, which are widely used in daily life, represent a form of hazardous waste, it is both necessary and important for batteries not to be disposed together with other types of waste. In this context, the Portable Battery Producers and Importers Association (*Taşınabilir Pil Üreticileri ve İthalatçıları Derneği*, TAP) in Turkey distributes materials, free of charge, to form battery collection points across the country, and also conducts efforts to ensure the widespread collection of waste batteries by using various types of containers. The association also ensures the participation and contribution of consumers to these efforts by informing them about the detrimental effects of waste batteries and their recycling. As science education students will be become science teachers in the future, it is very important for a sustainable future that these teachers are both aware and have knowledge regarding the collection of waste batteries and their effects on the environment. In this context, this study evaluated and reflected in detail the views of science education students regarding the collection of waste batteries and their effect on the environment. We believe that the findings of this study will contribute to and provide further depth to the literature on this subject.

## 2. Methodology

The study was conducted using the general screening model. The general screening model is a screening approach conducted on populations consisting of a large number of individuals in order to reach a general conclusion regarding the population. It is performed by screening the population as a whole, or a certain group or sample within the population (5).

The study was performed with the participation of 80 volunteer, third-year students receiving education at the Science Education Department of an Education Faculty in Turkish public university. To determine the students' views regarding the collection of waste batteries and their effects on the environment, the students were asked to answer two open-ended questions in writing. Examples of the answers provided by the students are shown below by keeping the students' name confidential and coding them as "$F_1$, $F_2$...$F_n$".

## 3. Results

The study results concerning the answers provided by the students to the questions on the collection and the environmental effects of waste batteries are provided in two sections.

### Section 1: Collection of Waste Batteries

The percentage distribution of the answers given by science teaching students to the question, "*Where do you dispose waste batteries?*" is provided in Graph 1.

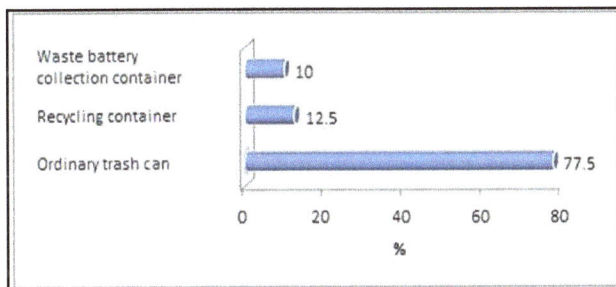

**Graph 1.** *Percentage distribution of the locations where students disposed of waste batteries*

As shown in Graph 1, most students explained that they disposed of waste batteries in ordinary trash cans. Some of the students expressed that they disposed of waste batteries in waste battery containers. Direct citations of the answers given by the students to this question are provided below:

"*I dispose of the batteries I use into the trash.*" ($F_{10}$)

"*I don't know much about recycling, so I throw waste batteries into trash cans.*" ($F_{43}$)

"*I don't think that I know enough about the collection of used batteries and the processes they have to go through. That's why I just throw the batteries I use into the trash.*" ($F_{23}$)

"*I just throw batteries into the trash; because it is more difficult to find waste battery collection containers and to throw used batteries in them.*" ($F_{74}$)

"*The only thing I pay attention to when it comes to recycling is the proper disposal of waste batteries.*" ($F_{56}$)

"*I throw used batteries into waste battery containers.*" ($F_{17}$)

"*I dispose of used batteries into a battery collection container at the grocery store. These containers should be more common.*" ($F_{27}$)

Table 1 provides the frequency distribution of the underlying reasons provided by the students concerning the locations where they disposed waste batteries.

**Table 1.** *The frequency distribution of the underlying reasons provided by the students concerning the locations where they disposed of waste batteries*

| Student's answers | Answer Frequency f |
|---|---|
| I throw waste batteries into ordinary trash cans, because I am not knowledgeable about their proper disposal. | 29 |
| I throw waste batteries into ordinary trash cans, because I think they are non-recyclable. | 18 |
| I throw waste batteries into ordinary trash cans, because I consider them as ordinary waste. | 16 |
| I throw waste batteries into ordinary trash cans, because I am not knowledgeable about recycling. | 11 |
| I throw waste batteries into ordinary trash cans, because there are no waste battery collection points near where I live. | 7 |
| I throw waste batteries into recycling containers, but have no knowledge about how they are recycled. | 10 |
| I throw waste batteries into was battery collection containers, but have no knowledge about what happens next. | 8 |

An evaluation of Table 1 indicates that the science education students threw waste batteries into trash cans mainly because they lacked knowledge regarding waste batteries and the disposal and recycling of waste batteries. In addition, it was noted that some of the students threw waste batteries into trash cans because of the lack of waste battery collection containers near the places they lived. Furthermore, it was observed that students who threw waste batteries into recycling and waste battery containers lacked any knowledge regarding the processes which the collected waste batteries underwent. These results demonstrated that the students lacked adequate knowledge regarding waste batteries. Direct citations of the answers given by the students to this question are provided below:

*"I don't know much about waste batteries, and I don't know what else can be done with them other than throwing them into the trash." (F₆₃)*

*"I throw them into trash cans, because I think that they would then be taken for recycling." (F₇₁)*

*"I dispose of waste batteries in battery collection containers. I have no idea about what the processes that are performed afterward." (F₃₇)*

*"They are generally thrown into the trash along with other domestic waste. However, owing to a project initiated by some schools, waste battery collection containers are now being placed within the school premises, and teachers are informing students that they should dispose any waste batteries at home into these containers." (F₁₂)*

### Section 2: The Effects of Waste Batteries on the Environment

The frequency distribution of the answers given by the science teaching students to the question, *"What effect do waste batteries thrown into the trash have on the environment?"* is provided in Table 2.

**Table 2.** *The views of students regarding the effects of waste batteries on the environment*

| Student's answers | Answer Frequency |
| --- | --- |
|  | f |
| Lack sufficient knowledge regarding the detrimental effects of waste batteries on the environment. | 51 |
| Throwing waste batteries into trash cans will lead to ground pollution. | 8 |
| The heavy metals in batteries pose a threat for the environment. | 6 |
| Throwing waste batteries into trash cans will lead to water pollution. | 5 |
| Waste batteries are harmful for aquatic creatures. | 5 |
| The heavy metals in batteries are harmful for the environment and human health. | 5 |
| Throwing waste batteries into trash cans will lead to environmental pollution. | 4 |
| Throwing waste batteries into the ground/soil will reduce its fertility. | 2 |
| In nature, the elimination of batteries takes many years. | 2 |
| Throwing waste batteries into trash cans will lead to air pollution. | 1 |

An evaluation of Table 2 indicates that most of the science education students were not sufficiently knowledgeable about the effects that waste batteries have on the environment when thrown into trash cans. On the other hand, some of the students were aware of the hazards associated with waste batteries, stating that throwing waste batteries into trash cans would lead to ground and water pollution due to the heavy metals they contain; that waste batteries represent a threat for the environment; that waste batteries are harmful for the environment and human health; and that waste batteries would reduce the fertility of soils/grounds. Direct citations of the answers given by the students to this question are provided below:

*"If they are not recycled, batteries will lead to environmental pollution. They will especially cause ground pollution." (F₈).*

*"When batteries are thrown into the trash, their chemicals will mix with the soil and waters. The heavy metals in batteries will thus pass into the food chain." (F₈₀)*

*"If waste batteries find their way into the ground, the metals they contain may pass into the soil and reduce its fertility." (F₆₇)*

*"Throwing waste batteries into the trash causes air, water, and ground pollution. It reduces the fertility of the soil, and harms plant life." (F₄₃).*

*"Because they contain elements such as mercury and lead, waste batteries will harm the environment; the metals they*

*contain will cause ground pollution." (F₃₉)*

*"Because they contain metals, batteries have detrimental effects on the ground, water, plants, animals, and humans." (F₇₁)*

*"When they are thrown into the environment, waste batteries take many years to be eliminated, significantly harming the environment in the process." (F₂₄)*

## 4. Discussion and Conclusions

It was determined that the large majority of the students participating in the study disposed waste batteries by throwing them into ordinary trash cans. However, as waste batteries represent hazardous wastes, they should not normally be thrown into trash cans along with domestic waste. The students threw waste batteries into trash cans mainly due to their lack of knowledge regarding the recycling and disposal of waste batteries. In addition, students who disposed of waste batteries into recycling containers erroneously assumed that waste batteries are collected together with solid wastes. This situation illustrated that the students were generally unaware that waste batteries are collected using separate waste battery collection containers.

This study demonstrated that most of the students were not knowledgeable about the detrimental effects of waste batteries on the environment. On the other hand, students who were able to describe the detrimental effects of waste batteries especially mentioned that they can cause air, water and ground pollution; reduce soil fertility; lead to pollution and environmental hazards; and endanger the health and life of living creatures. However, only a few of the students were able to provide explanations concerning how waste batteries led to such detrimental effects. When batteries are disposed into the environment in an uncontrolled manner, the heavy metals they contain present the risk of causing ground and water pollution. In addition, these chemicals are toxic substances for both the environment and human health. Throwing waste batteries into trash cans thus leads to environmental pollution as well as detrimental effects for human health. For this reason, it is essential for used/flat batteries to be disposed into battery collection containers and to be processed in recycling facilities.

When waste batteries are thrown into bodies of water or buried into the ground, the external casing of the battery will eventually erode or become pierced, causing the heavy metals and chemicals the battery contains to mix with the surrounding water or ground. For this reason, it is necessary to collect waste batteries separately and to dispose of them in waste battery collection containers. Disposing of waste batteries in such a way, and then recycling them as necessary, not only reduces the risk of having the various chemicals inside batteries mix with the underground waters and ground in landfill areas, but also allows the efficient use of natural resources through the recycling of reusable materials within batteries (6, 7).

The first step for ensuring a sustainable environment is raising individuals who are conscious, sensitive, and aware of environmental issues. Raising such individuals can only be achieved through the efforts of our teachers, to whom we entrust the education of future generations. Considering that a sustainable future will be made possible by the activities of our teachers, it is necessary for science education students – who will become the teachers of the future – to be knowledgeable regarding the collection of waste batteries and their effects on the environment. For this reason, student-centered methods and techniques, as well as effective teaching materials, should be used in the environment-related classes which science teaching students attend during their education. In addition, these students (as well as the general public) should be informed about waste batteries through the visual and written media, and measures should be taken to ensure the active participation of students into the collection of waste batteries into waste battery collection containers to prevent them from having detrimental effects on the environment.

# References

[1]   W.A. Suk, Hazardous waste: assessing, detecting, and remediation. In: Ed. Wallace R.B. Public health and preventive medicine, 15th edition USA: Mc Graw Hill, 2008, 901-908.

[2]   A.M. Bernandes, D.C.R. Espinosa and J.A.S. Tenorio, Recycling of batteries: A review of current processes and Technologies, Journal of Power Sources, 2004,130, 291-298.

[3]   S. Kierkegaard, EU Battery Directive, Charging up the batteries: Squeezing more capacity and power into the new EU Battery Directive, Computer Law & Security Report, 2007, 23, 357-364.

[4]   M. Bartolozzi, The recovery of metals from spent alkaline–manganese batteries: A review of patent literature, Resources, Conservation and Recycling, 1990, 4, 233–240.

[5]   N. Karasar, Bilimsel araştırma yöntemleri. (22th ed). Ankara: Nobel, 2011.

[6]   Taşınabilir Pil Üreticileri ve İthalatçilari Derneği (TAP), Atık pillerin toplanması ve bertarafı, Genel Eğitim Sunumu, 2014.

[7]   URL-1. Taşınabilir Pil Üreticileri ve İthalatçıları Derneği (TAP). Taşınabilir pillerin kullanımında dikkat edilmesi gereken hususlar. (http://tap.org.tr/tasinabilir_pillerin_kullaniminda_dikkat_edil mesi_gereken_hususlar-185.html)

# 22

# Evaluation through the use of drawings of the knowledge of science teacher candidates in Turkey regarding the recycling of waste batteries

**Zeynep Aksan, Gonca Harman, Dilek Çelikler**

Department of Elementary Science Education, Faculty of Education, Ondokuz Mayıs University, Samsun, Turkey

**Email address:**
zeynep.axan@gmail.com (Z. Aksan), gonca.harman@omu.edu.tr (G. Harman), dilekc@omu.edu.tr (D. Çelikler)

**Abstract:** The aim of this study was to determine through the use of drawings the knowledge of science teacher candidates regarding the recycling of waste batteries. The study was conducted with 47 third-year science teacher candidates attending the faculty of education of a public university in Turkey. In this study, the science teacher candidates were asked to demonstrate their knowledge regarding the recycling of waste batteries through the use of drawings and written descriptions. The drawings and written descriptions collected during the study were divided into different groups based on the 5 different levels previously used by Bartoszeck, et al. (1), Uzunkavak (2,3), and the answers provided by the students were evaluated using descriptive analysis. Based on the study results, the large majority of the teacher candidates' drawings and written descriptions were determined as being between level 2 and 4. The study results indicated that most of the science teacher candidates had either limited or erroneous information regarding the recycling on waste batteries.

**Keywords:** Recycling, Waste Batteries, Drawing, Science Teacher Candidates

## 1. Introduction

In the present-day world, the combined effects of industrialization, rising world population and the reckless use and destruction of natural environments have altered the ecological balance, leading to numerous environmental problems that threaten the lives of countless organisms on earth. One of the most important environmental issues in today's world is the disposal of solid wastes. The term "solid wastes" refer to materials generated by domestic, commercial and industrial activities which are disposed by consumers following use, but which, due to their adverse effects on human and environmental health and other public concerns, also need to be regularly removed urban settings (4, 5).

Batteries represent one of the different types of solid wastes. Once they become solid wastes after being used, batteries are considered and classified as "hazardous wastes", since they can lead to significant problems if disposed improperly. By definition, hazardous wastes include all types and forms of wastes (solid, liquid, gas, sludge, etc.) which pose a hazard for human health and the environment. Hazardous wastes can be either of domestic or industrial origin, and display a wide range of different properties. Thus, hazardous wastes are solid wastes which represent a potential threat for public health and the environment (6).

Recycling is a process that mainly involves the reduction of the amount of waste generated, and which can be summarized with the concept of "reduce, reuse, recycle." One definition of recycling is "the sorting, collection and grouping of recyclable wastes, followed by their conversion into other products or energy through physical and chemical methods" (7). As a concept, recycling also encompasses the "reuse and utilization" of wastes. The recycling and reuse of many materials we use in our daily lives is performed differently based on considerations such as the environmental problems they cause, the related economic factors, and the cautious use of existing natural resources. Some of these everyday materials consist a large variety of different substances, and may require complex processes for proper recycling. Batteries contain both recyclable metals such as nickel, and toxic metals such as cadmium; consequently, the recycling processes of batteries need to consider both of these types of metals (8).

The recycling process of waste batteries begins with the disposal of flat batteries by consumers into battery collection containers located at battery collection points. The waste batteries are taken from the battery collection points by authorized persons, after which they are sorted and stored as necessary. Following this, the waste batteries undergo various process in order to extract the valuable and reusable substances within them, resulting in the formation of new products (9). The recycling process of waste batteries is illustrated in Figure 1.

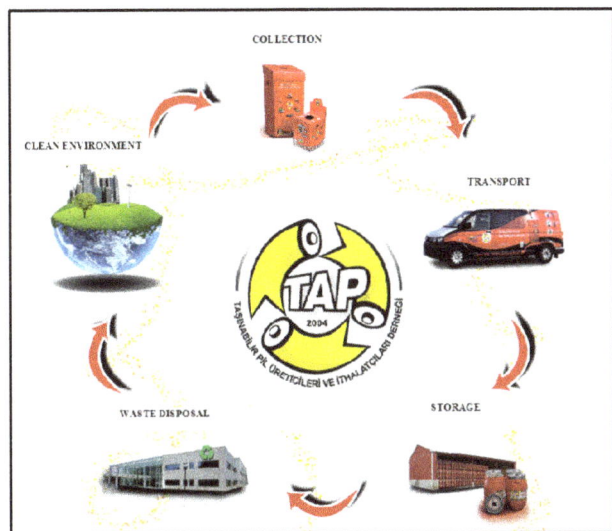

**Figure 1.** *The recycling process of waste batteries (9)*

Batteries contain numerous substances we commonly use in our daily lives for a variety of different purposes. The fact that batteries are widely used in everyday life further magnifies the problems associated with the hazardous waste they generate. For this reason, it is necessary to be cautious regarding the use of batteries, to avoid utilizing them when not necessary, and to dispose of them in a manner that results in the least harm for the environment or human health. Such measures regarding battery disposal are essential for a cleaner and healthier environment. It is important to bear in mind that, in addition to individual efforts, public regulations and responsibilities also play an important part in the proper disposal and recycling of batteries (10).

The drawing method is often used to illustrate the knowledge, misconceptions and conceptual changes of individuals regarding a particular subject (11) Compared to other methods used for illustrating changes in thought, the drawing method is more efficient in that it takes less time to complete, effectively reflects numerous different types of information, and is readily understood and performed by individuals (12). In addition, drawings allow students to reflect and express, independently of words, thoughts and beliefs which might otherwise remain concealed (13). It also enables students who otherwise do not like to answer questions to have a more enjoyable time, and to thus answer the relevant questions more willingly and rapidly (14). In this context, this study aimed to identify through the use of drawings the knowledge of science teacher candidates regarding battery recycling processes, and to thereby classify the level of knowledge of these teacher candidates.

## 2. Methodology

The study was performed at the faculty of education of a public university in Turkey, with the participation of 47 third-year science teacher candidates. During the study, the science teacher candidates were asked to describe the recycling process of batteries through drawings and written descriptions. The names of the science teacher candidates were kept confidential and coded using a "$F_n$" format. Several examples of the responses provided by the science teacher candidates are shown in this manuscript.

The drawings and written descriptions obtained during the study were separated into different groups based on the 5 different levels previously determined and used Bartoszeck et al. (1) and Uzunkavak (2,3). The answers to the questions were evaluated using descriptive analysis. The levels which were employed for evaluating the study data are shown in Table 1.

**Table 1.** *The levels and descriptions used for assessing the study participants' knowledge and drawings*

| Levels | Statements |
|--------|-----------|
| Level 1 | No theoretical knowledge/drawing |
| Level 2 | Wrong theoretical knowledge/drawing |
| Level 3 | Partially correct theoretical knowledge/drawing |
| Level 4 | Incomplete theoretical knowledge/drawing |
| Level 5 | Entirely correct and complete theoretical knowledge/drawing |

## 3. Results

The science teacher candidates were asked to make drawings regarding the recycling process of waste batteries, and to then describe these drawings in writing. The data obtained from the answers provided by the science teacher candidates are shown in Table 2.

**Table 2.** *The theoretical knowledge and drawing levels of the science teacher candidates regarding the recycling of batteries, provided in percentage (%) and frequency (f)*

| Levels | Theoretical Knowledge | | Drawing | |
|--------|------|------|------|------|
| | % | f | % | f |
| Level 1 | 12.8 | 6 | 8.5 | 4 |
| Level 2 | 25.5 | 12 | 38.3 | 18 |
| Level 3 | 17.0 | 8 | 8.5 | 4 |
| Level 4 | 44.7 | 20 | 44.7 | 21 |
| Level 5 | 0 | 0 | 0 | 0 |

According to the data on Table 2, 12.8% of the science teacher candidates provided no written theoretical information, while 25.5% provided incorrect theoretical information, 17.0% provided partially correct theoretical information, and 44.7% provided correct yet insufficient theoretical information. Evaluation of the teacher candidates' drawings revealed that 12.8% provided no drawings, while 8.3% provided incorrect drawings, 8.5% provided partially correct drawings, 44.7% provided correct yet insufficient drawings. The study results thus showed that none of the science teacher candidates were able to provide fully accurate and complete theoretical knowledge and drawings regarding the recycling of waste batteries.

The frequency distribution of the levels of the science teacher candidates' drawings and knowledge regarding the recycling process of waste batteries is provided in Graph 1.

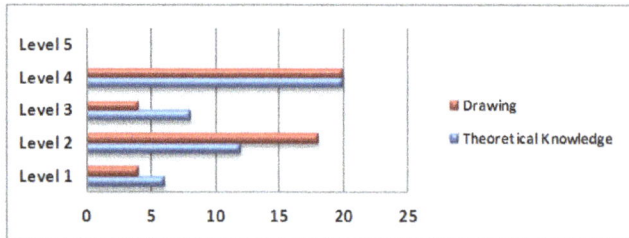

Graph 1. *The frequency distribution of the levels of the science teacher candidates' drawings and theoretical knowledge regarding the recycling process of waste batteries*

Evaluation of the teacher candidates' answers and drawings regarding the recycling process of waste batteries indicated that some of them provided incorrect (Level 2) answers and drawings concerning this process. Several examples of the Level 2 answers provided by the teacher candidates were as follows:

"*Batteries can be recycled by recharging them.*" $(F_{24})$

"*During recycling, the chemicals inside batteries are removed and replaced.*" $(F_4)$

"*When batteries are throw into trash cans, they will tend to expand and burst.*" $(F_{23})$

"*It is wrong to burn batteries. This is because when burned, the smoke released by the battery due to its composition will pollute the air.*" $(F_{36})$

"*Batteries can be recycled by restoring the activity of the substances between the (+) and (-) end.*" $(F_{18})$

Examples of Level 2 drawings performed by the teacher candidates are provided in Figure 2:

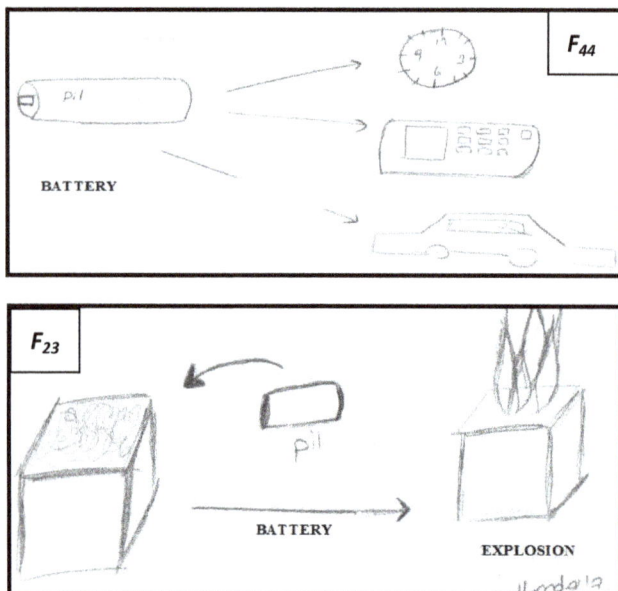

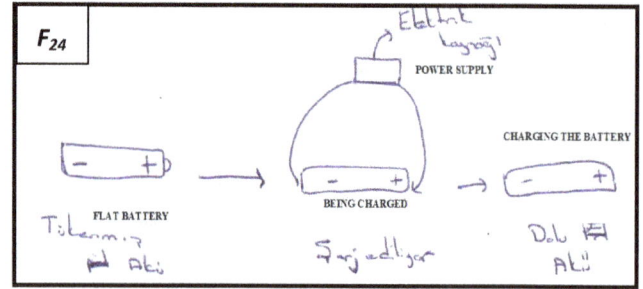

Figure 2. *Irrelevant wrong drawing which are example level 2*

Evaluation of the teacher candidates' answers and drawings regarding the recycling process of waste batteries indicated that some of these candidates provided partially correct (Level 3) answers and drawings concerning this process. Several examples of the Level 3 answers provided by the teacher candidates were as follows:

"*There are waste battery containers. We are supposed to dispose used batteries into waste battery containers. Batteries can be recycled for reuse. During recycling, batteries will be recharged once again, thus making them ready for reuse.*" $(F_{12})$

"*We should not throw the batteries we use into regular trash cans. Doing so may lead to pollution. Instead, we should recycle batteries in order to recharge them.*" $(F_{22})$

Examples of Level 3 drawings performed by the teacher candidates are provided in Figure 3:

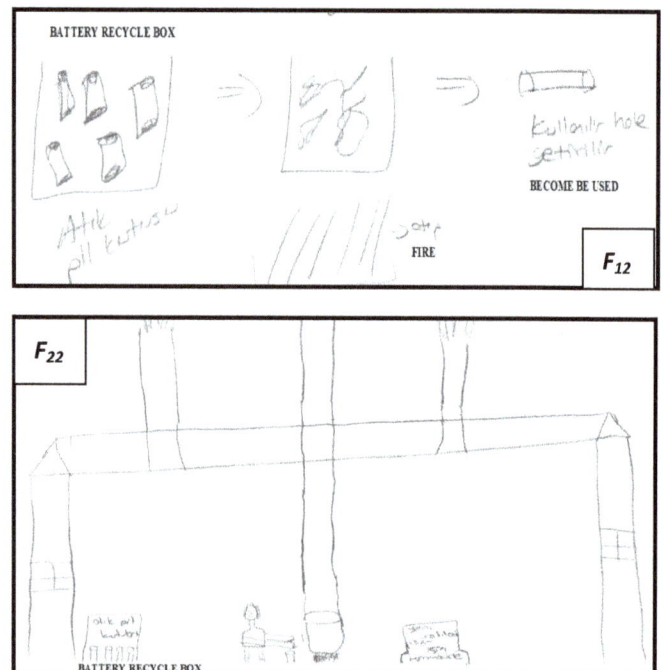

Figure 3. *Partly accurate drawing which are example level 3*

Evaluation of the teacher candidates' answers and drawings regarding the recycling process of waste batteries indicated that many of these candidates had correct yet insufficient (Level 4) knowledge concerning this process. Several examples of the Level 4 answers provided by the teacher candidates were as follows:

"*We should dispose flat batteries in recycling containers. Batteries* *collected in recycling containers will be processed in specialized facilities and converted into new batteries." ($F_{35}$)*

*"We should dispose batteries in recycling containers. Organizations responsible for recycling will then collect these batteries, take them to factories, and perform the necessary recycling processes." ($F_7$)*

*"Batteries are first collected inside waste battery containers. Companies then collect these waste batteries and store them." ($F_{19}$)*

*"Waste batteries are collected inside waste battery containers. These are then taken to recycling facilities to obtain new products." ($F_{46}$)*

Examples of Level 4 drawings performed by the teacher candidates are provided in Figure 4:

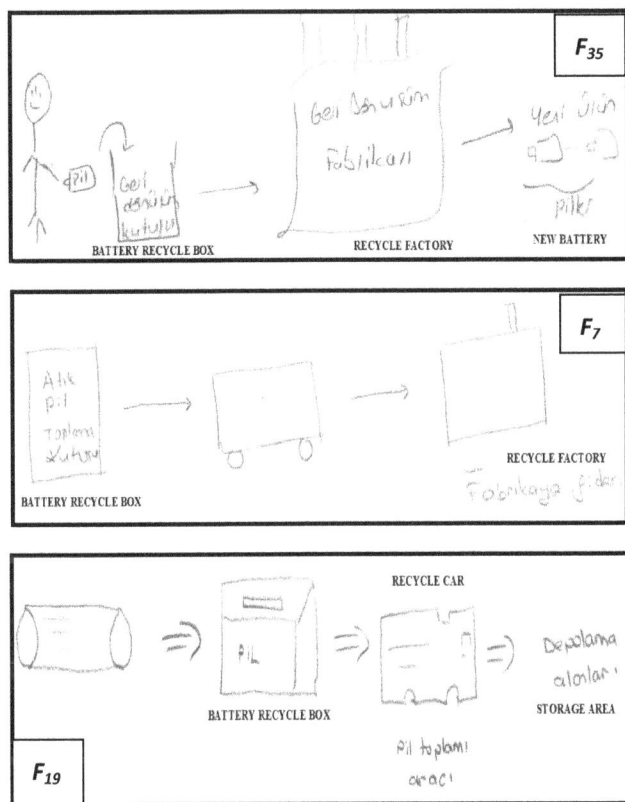

***Figure 4.*** *Incomplete theoretical drawing which are example level 4*

## 4. Conclusions and Recommendations

Based on the study results, 25.5% of the science teacher candidates provided incorrect information regarding the recycling process of waste batteries, and were accordingly classified as Level 2; while 44.7% provided correct yet insufficient information, and were accordingly classified as Level 4. In addition, 12.8% of the science teacher candidates provided no drawings on the subject, while 44.7% provided Level 4 drawings – in other words, illustrations which were correct yet insufficient. During the study, none of the teacher candidates were able to provide theoretical information and drawings which were correct and fully complete. These results indicated that the teacher candidates generally have incorrect and incomplete information regarding the recycling process of

waste batteries, and that their drawings were generally not sufficient.

The study results showed the effectiveness of the drawing method in reflecting knowledge on a particular subject without the limitations of words. Other studies which used the drawing method to evaluate students' knowledge on a given subject included Bartoszeck et al.'s study (1) regarding the human organs, Kara et al.'s study (15) on the concept of light, Kara's study (16) and Uzunkavak's study (2) on Newton's Laws, Köse's study (17) regarding photosynthesis and respiration in plants, Uzunkavak's study (3) on the concept of light, Çelikler and Topal's study (18) on the carbon dioxide and water cycles, Çelikler and Kara's study (19) on the period table, Çelikler and Aksan's study (20) regarding the greenhouse effect. All of these studies demonstrated that the students' knowledge on the subject could be identified more easily and effectively with the drawing method. In the current study, it was observed that although the teacher candidates had difficulties in expressing their knowledge regarding the recycling process of waste batteries, they were able to reflect their thoughts more easily through drawings.

Considering that batteries are commonly used in many of the objects and devices we routinely employ in our daily lives, and that waste batteries represent a hazardous form of waste; it is both necessary and important to raise the awareness of individuals regarding waste batteries and the recycling of batteries. For this reason, the recycling of batteries should be taught to students starting from elementary school and all the way up to higher education programs, though the use of student-centered methods and techniques. Education on this subject should be provided by organizing suitable educational activities, by creating suitable learning environment, and by ensuring a good and comprehensive classes regarding the environment. Such educational approaches would allow individuals to develop the necessary awareness towards the environment, thus enabling them to become individuals who display a healthy concern for the future, and who understand that the scope of environmental problems are not limited by time or space.

## References

[1]  A.B. Bartoszeck, D.Z. Machado and M. Amann-Gainotti, Representations of internal body image: A study of preadolescents and adolescent students in Araucaria, Paraná, Brazil. Ciências & Cognição, 2008, 13(2), 139-159.

[2]  M. Uzunkavak, Öğrencilerin newton kanunları bilgilerinin yazı ve çizim metoduyla karşılaştırılması. SDU International Journal of Technologic Sciences, 2009, 1(1), 29-40.

[3]  M. Uzunkavak, Öğrencilerin iş kavramında pozitiflik-negatiflik ayrımı becerilerinin yazı ve çizim metoduyla ortaya çıkarılması. SDU International Journal of Technologic Sciences, 2009, 1(2), 10-20.

[4]  K.C. Clayton and J.M. Huie, Solid wastes management the ragional approach. Ballinger Publisher Company, Cambridge, 1973.

[5]    H. Palabıyık and D. Altunbaş, "Kentsel katı atıklar ve yönetimi", Çevre sorunlarına çağdaş yaklaşımlar: Ekolojik, ekonomik, politik ve yönetsel perspektifler. C. Marin and U. Yıldırım (Ed.), Beta, İstanbul, 2004, pp. 103-124.

[6]    W.A. Suk, Hazardous waste: assessing, detecting, and remediation. Robert B. Wallace (Ed.). Public health and preventive medicine. 15th edition USA: Mc Graw Hill, 2008, pp. 901-908.

[7]    Ç. Güler, Geri dönüşüm. Ç. Güler (Ed.), Çevre sağlığı, çevre ve ekoloji bağlantılarıyla. Ankara: Yazıt Yayıncılık, 2012, 561-566.

[8]    C.A. Nogueira and F. Margarido, Chemical and physical characterization of electrode materials of spent sealed ni–cd batteries. Waste Manag, 2007, 27, 1570-1579.

[9]    Taşınabilir Pil Üreticileri ve İthalatçıları Derneği (TAP). Atık pillerin toplanması ve bertarafı. Genel Eğitim Sunumu, 2014.

[10]   C.I. Yavuz, S. Acar Vaizoğlu and Ç. Güler, Hayatımızdaki piller. Sürekli Tıp Eğitimi Dergisi (STED), 2013, 21(6), 19-25.

[11]   R.T. White and R.F. Gunstone, Probing understanding. London: The Falmer Pres., 1992.

[12]   B. Atasoy, Fen ve teknoloji öğretimi. Ankara: Asil, 2004.

[13]   A. Ayas, Kavram öğrenimi, "Fen ve teknoloji öğretimi". S. Çepni (Ed.), Ankara: Pegema Yayıncılık, 2006.

[14]   G.V. Thomas and A.M.J. Silk, An introduction to the psychology of children's drawings. Hemel Hempstead: Harvester Wheat Sheaf, 1990.

[15]   İ. Kara, E. D. Avcı and Y. Çekbaş, Investigation of the scinence teacher candidates' knowledge level about the concept of light (Fen bilgisi öğretmen adaylarının ışık kavramı ile ilgili bilgi düzeylerinin araştırılması). Mehmet Akif Ersoy Üniversitesi Eğitim Fakültesi Dergisi, 2008. Retrieved from http://efd.mehmetakif.edu.tr/arsiv/aralik2008/aralik2008/46-5 7.pdf

[16]   İ. Kara, Revelation of general knowledge and misconceptions about Newton's laws of motion by drawing method. World Applied Sciences Journal, 2007, 2(S), 770-778.

[17]   S. Köse, Diagnosing student misconceptions: Using drawings as a research method. World Applied Sciences Journal, 2008, 3(2), 283-293.

[18]   D. Çelikler and N. Topal, Determination of the knowledge of pre-service elementary science teachers about the cycle of carbondioxide and water by drawing (İlköğretim fen bilgisi öğretmen adaylarının karbondioksit ve su döngüsü konusundaki bilgilerinin çizim ile saptanması). Journal of Educational and Instructional Studies in the World, 2011, 1(1), 72-79.

[19]   D. Çelikler and F. Kara, To determinate of the knowledge of pre-service elementary science teachers about the periodic table by drawing (İlköğretim fen bilgisi öğretmen adaylarının periyodik çizelge konusundaki bilgilerinin çizim yoluyla saptanması). Journal of Research in Education and Teaching, 2012, 1(3), 70-76.

[20]   D. Çelikler and Z. Aksan, Determination of knowledge and misconceptions of pre-service elementary science teachers about the greenhouse effect by drawing. Procedia-Social and Behavioral Sciences, 2014, 136.

# 23

# Biological purification processes for biogas using algae cultures

**Rameshprabu Ramaraj, Natthawud Dussadee**

School of Renewable Energy, Maejo University, Sansai, Chiang Mai-50290, Thailand

**Email address:**

rrameshprabu@gmail.com , rameshprabu@mju.ac.th (Ramaraj R.), natthawu@yahoo.com, natthawu@mju.ac.th (Dussadee N.)

**Abstract:** Bioenergy is a type of renewable energy made from biological sources including algae, trees, or waste from agriculture, wood processing, food materials, and municipalities. Currently, the uses of renewable fuels (bioethanol, biodiesel, biogas and hydrogen) are increased in the transport sector worldwide. From an environmental and resource-efficiency perspective biogas has several advantages in comparison to other biofuels. The main components of biogas are methane ($CH_4$) and carbon dioxide ($CO_2$), but usually biogas also contains hydrogen sulphide ($H_2S$) and other sulphur compounds, water, other trace gas compounds and other impurities. Purification and upgrading of the gas is necessary because purified biogas provides reductions in green house gas emissions as well as several other environmental benefits when used as a vehicle fuel. Reducing $CO_2$ and $H_2S$ content will significantly improve the quality of biogas. Various technologies have been developed and available for biogas impurity removal; these include absorption by chemical solvents, physical absorption, cryogenic separation, membrane separation and biological or chemical methods. Since physiochemical methods of removal are expensive and environmentally hazardous, and biological processes are environmentally friendly and feasible. Furthermore, algae are abundant and omnipresent. Biogas purification using algae involved the use of algae's photosynthetic ability in the removal of the impurities present in biogas. This review is aimed at presenting the algal characteristics, scientific approach, gather and clearly explain the main methods used to clean and purify biogas, increasing the calorific value of biogas and making this gas with characteristics closest as possible to natural gas through algae biological purification processes.

**Keywords:** Algae, Biogas, Biological Purification, Renewable Energy

## 1. Introduction

Bioenergy should play an essential part in reaching targets to replace petroleum-based transportation fuels with a viable alternative, and in reducing long-term $CO_2$ emissions, if environmental and economic sustainability are considered carefully. The world continues to increase its energy use, brought about by an expanding population and a desire for a greater standard of living. This energy use coupled with the realization of the impact of $CO_2$ on the climate, has led us to reanalyze the potential of plant-based biofuels [1]. The term biofuel is referred to as liquid or gaseous fuels for the transport sector that are predominantly produced from biomass. A variety of fuels can be produced from biomass resources including liquid fuels, such as ethanol, methanol, biodiesel, Fischer-Tropsch diesel, and gaseous fuels, such as biohydrogen and biogas.

The process of biogas production from algal biomass is an alternative technology that has larger potential energy output compared to green diesel, biodiesel, bioethanol, and hydrogen production processes. Moreover, anaerobic digestion can be integrated into other conversion processes. The organic fraction of almost any form of biomass (from plants, algae and other microorganisms), including sewage sludge, animal wastes and industrial effluents, can be broken down through anaerobic digestion (AD) into $CH_4$ and $CO_2$ mixture called as "biogas". The first methane digester plant was built at Bombay, India in 1859 [2, 3]. AD approaches steadily growing role in the renewable energy mix in many countries. AD is the process by which organic materials are biologically treated in the absence of oxygen by naturally occurring bacteria to produce 'biogas' which is a mixture of $CH_4$ (40-70%) and $CO_2$ (30-60%) with traces of other gases such as hydrogen, hydrogen sulphide and ammonia [4]; the biogas process also produces potentially useful by-products in the form of a liquid or solid 'digestate' [5].

Normally, biogas is comprised of $CH_4$, $CO_2$, and other trace gas compounds gases such as water vapour, $H_2S$, halogenated hydrocarbons, siloxanes, ammonia, nitrogen, and oxygen [4]. Biogas is a valuable fuel which is produced in digesters filled with the feedstock like dung or sewage. All types of biomass can be used as substrates for biogas production as long as they contain carbohydrates, proteins, fats, cellulose, and hemicelluloses as main components. The composition of biogas and the methane yield depends on the feedstock type, the digestion system, and the retention time. In general, the use of plant biomass for energy generation today is problematic because of the competition with food or feed production. This is because most of the plants used for energy generation today (crop plants, sugar cane, sugar beets, canola, etc.) have to be grown on arable land. Low demand alternatives like switchgrass are only beginning to emerge. Algae have got a number of potential advantages compared to higher plants because of faster growth rates and the possibility of cultivation on non-arable land areas or in lakes or the ocean, therefore attenuating food and feed competition [6,7]. Of the potential sources of biogas the most efficient producers of biomass are the photosynthetic algae (micro and macroalgae).

Photosynthetic pigments, including chlorophyll, have an important role since it provides the oxygen and the source of energy for all living things. Plant and algae growth is affected by the photosynthesis speed which depends on the availability of $CO_2$. Biological $CO_2$ fixation by algae is another such form; i.e. sunlight being used to reduce $CO_2$ to carbon. Capturing $CO_2$ from flue gases is the precautionary principle which needs preventive action, at both national and international levels to minimize this potential action [8]. A promising approach therefore seems to be the use of fast-growing algae species for anaerobic fermentation to produce biogas, which then can substitute natural gas resources.

To utilize biogas as a transport fuel, $CO_2$ and $H_2S$ must be removed from the concentration to leave biomethane. Biogas purification is the process where any impurities are removed such as sulphides and ammonia. Biogas upgrading on the other hand is the process which removes $CO_2$ and the end product is bio-methane. The bio-methane which has been upgraded is suitable for injection into the national gas grid or vehicle fuel [4]. Biogas needs cleaning for two main reasons; the first is to improve the calorific value of the product gas and the second is to reduce the chance of damaging downstream equipment which is due to the formation of harmful compounds [9]. Thus, biogas has a wide availability and renewable nature due to the organic materials and microorganisms required for biogas synthesis. Biogas purification methods can be divided into two generic categories:

1 Those involving physicochemical phenomena (reactive or non-reactive absorption; reactive or non-reactive adsorption).

Those involving biological processes (contaminant consumption by living organisms and conversion to less harmful forms). Biological processes are widely employed for $CO_2$ and $H_2S$ removal, especially in biogas applications.

For $CO_2$ capture from biogas, physical and chemical absorption methods are generally applied with fewer complications; however, these methods are needed to post treat the waste materials for regeneration of cycling utilization. The biological methods of $CO_2$ capture from biogas are potentially useful [10]. Biological processes are widely employed for $H_2S$ removal, especially in biogas applications [11]. Furthermore, biogas is an environment friendly, clean, cheap and versatile fuel. Consequently, the purpose of the current paper is to present an integrated review of the biogas production methodologies and purification process, algal characteristics, approaches and clearly explain the main methods used to clean and purify biogas, increasing the calorific value of biogas and making this gas with characteristics closest as possible to natural gas through algae biological purification processes

## 2. Growth Characteristics of Algae and Importance

Algae are the most important primary producer in aquatic ecosystem [12]. Many species of algae are present such as; green, red and brown algae which belong to the group of Chlorophyta, Rhodophyta and Phaeophyta, respectively. Algal growth is found in a wide range of habitats, like fresh water, marine water, in deep oceans, in rocky shores, the plank-tonic and benthic algae can become important constituents of soil flora and can exist even in such extreme conditions as in snow, sands/desert or in hot springs, open and closed ponds, photo bioreactors, sewage and wastewater, desert as well as $CO_2$ emitting industries etc [13]. Generally they are found in damp places or water bodies and are common in terrestrial as well as aquatic environments. Algae, a broad category encompassing eukaryotic microalgae, cyanobacteria and macroalgae, can be cultivated to produce biomass for a wide range of applications [14].

Algae are a very diverse group of predominantly aquatic photosynthetic organisms of tremendous ecological importance, because they were the beginning of the food chain for other animals. Algae played an important role in self-purification of contaminated natural waters and offered an alternative for advance nutrition removal in water or wastewater [15, 16]. The idea to incorporate microalgae as an agent of bioremediation was firstly proposed by Oswald and Gotaas in 1957 [17]; the biomass recovered was converted to methane, which was a major source of energy [18]. Hence, algae provided the basis of the aquatic food chain and they were fundamental to keep $CO_2$ of carbon cycle via photosynthesis as a substantial role in biogeochemical cycles [12]. Most algae were photoautotrophic, converting solar energy into chemical forms through photosynthesis.

The mechanisms of algal photosynthesis were very similar to photosynthesis in higher plants and their products are molecularly equivalent to conventional agricultural crops [19]. The main advantages of culturing algae as a source of biomass were as follows: (1) high photosynthetic yields (up to a maximum of 5-6% conversion of light c.f. 1-2% for the majority of terrestrial plants); (2) the ability to grow in fresh, salt and wastewater; (3) high oil content; (4) the ability to produce non-toxic and biodegradable biofuels; (5) many species of algae can be induced to produce particularly high concentrations of chosen compounds–proteins, carbohydrates, lipids and pigments - that are of commercial value; (6) the ability to be used in conjunction with wastewater treatment [13,17–20]. Since algae was a key primary producer global-wide, algae biomass was essential biological natural resources which played an important role in nutrient, food, fertilizer, pharmaceutics and biofuel.

In addition, algae application is widely accepted in practice as one of the best strategies in bioengineering. There are several reasons for this approach: (1) the best growth rate among the plants, (2) low impacts on world's food supply, (3) specificity for $CO_2$ sequestration without gas separation to save over 70% of total cost, (4) excellent treatment for combustion gas exhausted with NOx and SOx, (5) high value of algae biomass including of feed, food, nutrition, pharmaceutical chemicals, fertilizer, aquaculture, biofuel, etc [13, 20]. Algae an important application for the cultivation of algae is the production of biomass for energy purposes. Due to the energy crisis, renewable energy becomes a popular issue in this world today and there are several alternatives such as bioenergy, solar, wind, tide, geothermal, etc. For bioenergy, algae are the third generation biofuel [20]. For the reasons of the best energy conversion efficiency of sunlight [15] and the highest growth rate [18], algae have the best potential among all the energy crops. Because of the fast growth, many high valuable products are generated, e.g. food, biofuel, etc [Figure 1].

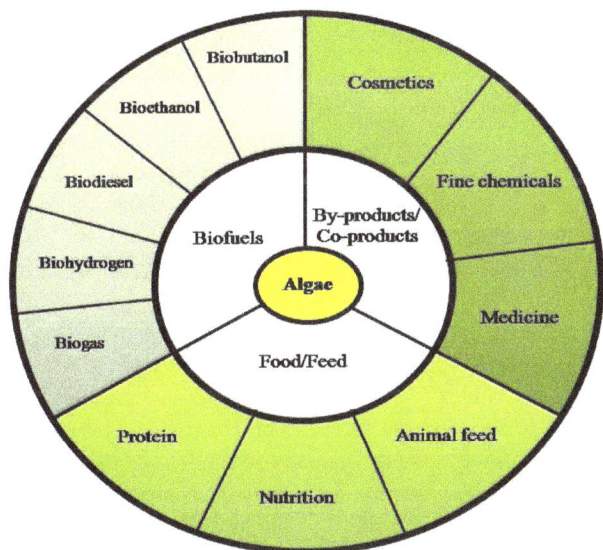

*Figure 1. Potential products from algae*

Algae produce biomass, which can be converted into energy or an energy carrier through a number of energy conversion processes. They include thermochemical conversion (gasification, direct combustion and pyrolysis), biochemical conversion (anaerobic fermentation, anaerobic digestion and

photobiological hydrogen production) and esterification of fatty acids to produce biodiesel [13,18,20]. A lot of studies was indicated the importance of algae in carbon dioxide fixation [12–16,18,20]. Driver et al. [21] stated that algae are an attractive feedstock for the production of liquid and gaseous biofuels that do not need to directly compete with food production. Figure 2 illustrated the detailed information, process including algal stain selection, water type, cultivation methods, growth mode and harvesting methods. Furthermore, the various scenarios for biofuel development from algae are represented. Many options are available with regard to algae type and strain choice, including both eukaryotic algae and prokaryotic cyanobacteria, the source of water for cultivation, cultivation method and mode of growth, the method of algae harvesting and the biofuel conversion process. The understanding of biological phenomena, algal genetics, carbon storage metabolism, photosynthesis and algal physiology, have the potential for significant advances in algal biofuel feasibility [21]. This is being driven by advances in genomic technologies to provide the potential for genetic and metabolic engineering, plus the development of high-throughput techniques for the screening of natural strains for suitable biofuel characteristics.

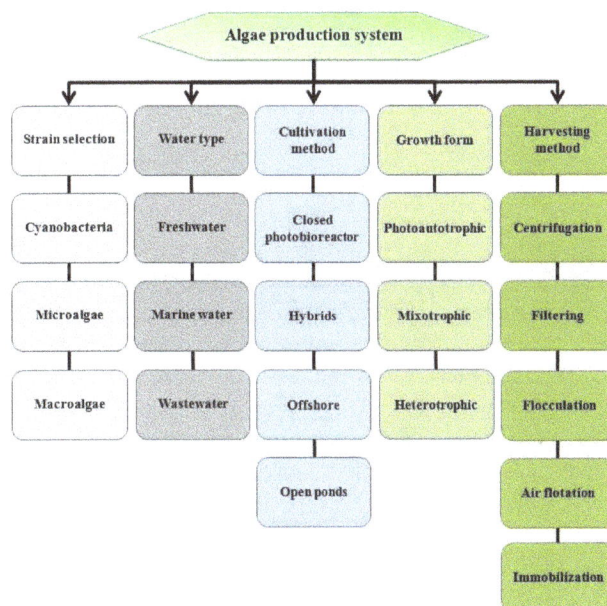

*Figure 2. Algae production system*

## 3. Algae Biogas Production Process and Technology

Anaerobic digestion (AD) is a common process for the treatment of a variety of organic materials and biogas production. Macroalgal and microalgal biomass can be AD to produce methane. Recently, microalgae have also become a topic of interest in the production of biogas through anaerobic fermentation [22].The AD of algae is a prospective environmentally feasible option for creating a renewable source of energy for industrial and domestic needs. Algal AD on is a key unit process that integrates efficiency and beneficially into the production of algal derived biofuels. Both, macro- and microalgae are suitable renewable substrates for the anaerobic digestion process. The process of biogas production from algal biomass is an alternative technology that has larger potential energy output compared to green diesel, biodiesel, bioethanol, and hydrogen production processes [4]. Moreover, anaerobic digestion can be integrated into other conversion processes and, as a result, improve their sustainability and energy balance. Opposite to biohydrogen, bioethanol or biodiesel that only uses determined macromolecules (carbohydrates and lipids), biogas is produced by biological means under anaerobic conditions that converts all algae macromolecules into methane [5, 8].

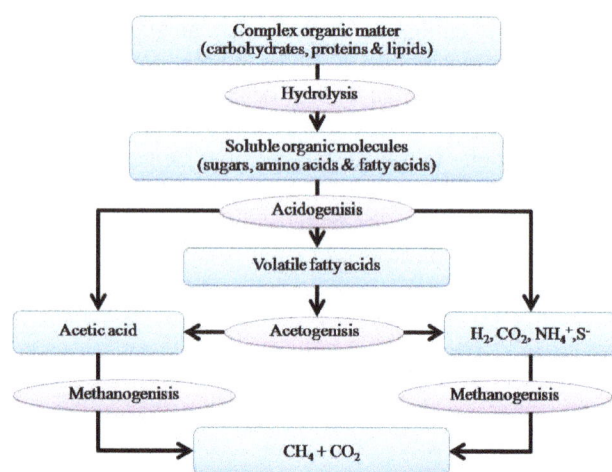

*Figure 3. Stages of Anaerobic Digestion (methane fermentation process)*

AD is an application of biological methanogenesis which is an anaerobic process responsible for degradation of much of the carbonaceous matter in natural environments where organic accumulation results in depletion of oxygen for aerobic metabolism. Since AD is a process by which almost any organic waste can be biologically converted in the absence of oxygen. This process, which is carried out by a consortium of several different microorganisms, is found in numerous environments, including sediments, flooded soils, animal intestines, and landfills. Accordingly, this is a complex process, which requires specific environmental conditions and different bacterial populations. Mixed bacterial populations degrade organic compounds, thus producing, as end-product; a valuable high energy mixture of gases (mainly $CH_4$ and $CO_2$) termed biogas [9]. Methane fermentation is a complex process, which can be divided up into four phases: hydrolysis, acidogenesis, acetogenesis/dehydrogenation, and methanation (Figure 3). These four stages are involved in the breakdown of organic matter on the path to methane production; stages include hydrolysis, fermentation (or acidogenesis), acetogenesis and eventual methanogensis (1). Hydrolysis involves the conversion of complex molecules and compounds–carbohydrates, lipids and proteins – found in organic matter into simple sugars, long chain fatty acids and amino acids, respectively. Acidogenesis in turn converts these

into volatile fatty acids, acetic acid, $CO_2$ and $H_2$. Acetogenesis converts the volatile fatty acids into more acetic acid, carbon dioxide and hydrogen gas. Methanogens have the ability to produce methane by using the carbon dioxide and hydrogen gas or the acetic acid produced from both the acetogenic or acidogenic phases [10,11].

### 3.1. Anaerobic Digestion of Macroalgae Biomass

Macroalgae is one such source of aquatic biomass and potentially represents a significant source of renewable energy. The average photosynthetic efficiency of aquatic biomass is 6–8%, which is much higher than that of terrestrial biomass (1.8–2.2%). Macroalgae are fast growing marine and freshwater plants that can grow to considerable size (up to 60 m in length). Annual primary production rates (grams C m$^{-2}$ yr$^{-1}$) are higher for the major marine macroalgae than for most terrestrial biomass [23]. Macroalgae can be subdivided into the blue algae (Cyanophyta), green algae (Chlorophyta), brown algae (Phaeophyta) and the red algae (Rhodophyta). Either Freshwater macroalgae or marine macroalgae (kelp or seaweed) could be used for solar energy conversion and biofuel production [23]. Macroalgae received a large amount of attention as a biofuel feedstock due to its prolific growth in natural habitat of freshwater system, eutrophic coastal water fouling beaches and coastal waterways.

Macroalgae can be converted to biogas by process of AD to biogas (~ 60% $CH_4$) [24]. Research conducted in the 1980's on macroalgae (giant brown kelp (*Macrocystis*)) [25] still provides a bench mark for biogas yields for a number of macroalgal species, but since this time there have been developments in AD technology and an enormous increase in its use. In comparison to terrestrial biomass crops, macroalgae contain little cellulose and no lignin and therefore undergo a more complete hydrolysis. AD has been used to dispose and process this material for the production of biogas; the AD of macroalgae biomass could meet two currently important needs, the mitigation of the eutrophication effects and the production of renewable energy. Because of the abundance of seaweed/ freshwater macroalgae biomass its conversion can be highly desirable and convenient, mostly for countries with long coastlines or eutrophic environments [26].

Investigations on the use of macroalgae of the brown algae division in processes of methane fermentation were conducted by Vergara-Fernàndez [27]. He was examining the possibility of applying to this end the biomass of *Macrocystis pyrifera* and *Durvillea antarctica* macroalgae and a substrate based on the mixture of these species. His study proved that for all substrates tested the yield of biogas production was comparable and reached 180.4±1.5 dm$^3$/kg d.m.d. Singh and Gu [28] and Parmar et al. [29] were also analyzing the yield of biogas production with the use of microphytobenthos plants as an organic substrate. They achieved the highest technological effects during fermentation of *Laminaria digitata* brown algae belonging to the order *Laminariales*. In that case, methane production was high and reached 500 dm$^3$ $CH_4$/kg o.d.m. The use of *Macrocystis sp.* enabled achieving

390–410 dm$^3$ CH$_4$/kg o.d.m., whereas upon the use of *Gracilaria sp.* and *Laminaria sp.* methane production accounted for 280–400 dm$^3$ CH$_4$/kg o.d.m. and 260–280 dm$^3$ CH$_4$/kg o.d.m., respectively [30].

The feasibility of biogas production from macroalgae collected from the Orbetello lagoon. Maroalgae biomass collected from the same lagoon was used for biogas production in batch reactors. He demonstrated that it is possible to produce CH$_4$ directly from macroalgae, preserving the spontaneous epiphytic microorganisms, as microbial starter of the digestion process. Moreover, it is possible to foster CH$_4$ yield by using anoxic sediments collected from the same lagoon as a further microbial inoculum. In fact, the addition of sediment improved the degradation activity, accelerating the removal of volatile fatty acids (VFA) from the medium and their conversion into methane, reducing the digestion time and increasing CH$_4$ yield [31]. The promising results obtained despite the harsh conditions (high salts, sulphur and heavy metals concentration) have been favoured, in our opinion, thanks to a pre-existing adaptation and mutual interactions within the native microorganisms. The bacterial pool was highly adapted both to biotic and abiotic factors, that is to macroalgal tissue composition and to the salts and toxic components present in water and sediments. Furthermore, this approach solely based on the exploitation of the intrinsic degradation potential of the reference ecosystem, proved to be suitable for a selective and non-intensive anaerobic digestion of macroalgae. In the review by Dębowski et al. [30] presented the effectiveness of biogas production with the use of macroalgae as a substrate in methane fermentation processes (Table 1). Huesemann et al. [32] stated that AD of macroalgae was technically feasible at scale and it has been suggested that it could be a cost-competitive with anaerobic digestion of terrestrial biomass and municipal solid waste.

*Table 1. Effectiveness of biogas production with the use of macroalgae as a substrate in methane fermentation processes.*

| Macroalgae taxon | Quantity of biogas/methane |
|---|---|
| *Durvillea antarctica* | 179.3±80.2 dm$^3$ CH$_4$/kg d.m. d |
| *Gracilaria* sp. | 280–400 dm$^3$/kg o.d.m. |
| *Laminaria* sp. | 260–280 dm$^3$/kg o.d.m. |
| *Laminaria digitata* | 500 dm$^3$/kg o.d.m. |
| *Macrocystis* | 390–410 dm$^3$/kg o.d.m. |
| *Macrocystis* sp. | 189.9 dm$^3$ CH$_4$/kg o.d.m. |
| *Macrocystis pyrifera* | 181.4±52.3 dm$^3$ CH$_4$/kg d.m. d |
| *M. pyrifera+Durvillea antarctica* | 164.2±54.9 dm$^3$ CH$_4$/kg d.m. d |
| *Pilayella+Ectocarpus+Enteromarpha* | 40.0–54.0 dm$^3$/kg |
| | 29.2–39.4 dm$^3$ CH$_4$/kg |
| *Ulva* sp. | 200 dm$^3$/kg o.d.m. |
| *Ulva lactuca* | 157–271 dm$^3$ CH$_4$/kg o.d.m. |

### 3.2. Anaerobic Digestion of Microalgae Biomass

Microalgae are highly productive and are able to produce large quantities of biomass more efficiently [13,14,16]. Generally, the composition of microalgae is $CO_{0.48}H_{1.83}N_{0.11}P_{0.01}$ [13], and microalgae have been found to have several constituents, mainly including lipids (7–23%), carbohydrates (5–23%), and proteins (6–52%). The chemical compositions of microalgae are mainly dependent on the

species and culture conditions. Microalgae AD is a key unit process that integrates efficiency and beneficially into the production of microalgae derived biofuels. The first authors to report on the anaerobic digestion of microalgae biomass were Golueke et al. [33]. They investigated the anaerobic digestion of *Chlorella vulgaris* and *Scenedesmus*, microalgae species grown as part of a wastewater treatment process.

The technical feasibility data on the anaerobic digestion of algal biomass have been reported for many species of algae. Among the microscopic algae, the following cultures have been successfully used for the production of methane: the mixed culture of *Scenedesmus* sp. and *Chlorella* sp., the mixed culture of *Scenedesmus* sp., *Chlorella* sp., *Euglena* sp., *Oscillatoria* sp., and *Synechocystir* sp., the culture of *Scenedesmus* sp. alone, and together with either *Spirulina* sp., *Euglena* sp., *Micractinium* sp., *Melosira* sp., or Oscillatoria SP. The production of biogas through AD offers significant advantages over other forms of bioenergy production. Since AD consists of organic carbon degradation into organic acids and biogas. Biogas mainly consists of methane (around 65%), which is carbon most reduced state, and carbon dioxide (around 35%), which is its most oxidized state. Other gases (normally less than 1%), such as nitrogen, nitrogen oxides, hydrogen, ammonia and hydrogen sulphide are also formed [34, 35].

*Table 2. Effectiveness of biogas production with the use of microalgae as a substrate in methane fermentation processes.*

| Macroalgae taxon | Quantity of biogas/methane |
|---|---|
| *Arthrospira platensis* | 481±13.8 dm$^3$/kg o.d.m. |
| *Chlamydomonas reinhardtii* | 587±8.8 dm$^3$/kg o.d.m. |
| *Chlorella kessleri* | 335±7.8 dm$^3$/kg o.d.m. |
| *Chlorella vulgaris* | 150 dm$^3$ CH$_4$/kg o.d.m. |
| | 240 dm$^3$ CH$_4$/kg o.d.m. |
| *Dunaliella salina* | 505±24.8 dm$^3$/kg o.d.m. |
| *Euglena gracilis* | 485±3.0 dm$^3$/kg o.d.m. |
| *Phaeodactylum tricornutum* | 350±3.0 dm$^3$ CH$_4$/kg o.d.m. |
| *Scenedesmus obliquus* | 210±3.0 dm$^3$ CH$_4$/kg o.d.m. |
| *S. obliquus* | 287±10.1 dm$^3$/kg o.d.m. |
| *Scenedesmus* sp.+*Chlorella* sp. | 986 dm$^3$/kg o.d.m. |
| | 180±8 dm$^3$/dm$^3$ d |
| *Scenedesmus sp*+*Chlorella* sp. | 573±28 cm$^3$/dm$^3$ d |
| | 818±96 cm$^3$/dm$^3$ d |
| *Spirulina maxima* | 240 dm$^3$ CH$_4$/kg o.d.m. |
| *Spirulina platensis* | 280±0.8 dm$^3$ CH$_4$/kg o.d.m. |

Sialve et al. [35] stated that an organic matter composition can be converted stoichiometrically into methane for calculating the theoretical methane yield. Thus, lipids (1.014 L/g VS), followed by proteins (0.851 L/g VS) and carbohydrates (0.415 L/g VS) have the highest theoretical methane yield. Indeed, inducing a particular macromolecule accumulation in microalgae cells has proven to successfully increase the methane yield. Research conducted with carbohydrate-enriched cyanobacteria *Arthrospira platensis* by phosphorus limitation attained a methane yield of 0.203 L/g COD when biomass had 60% of carbohydrates in respect to 0.123 L/g COD when the carbohydrate content was 20% [36]. In the review by Dębowski et al. [30] presented the effectiveness of biogas production with the use of macroalgae as a substrate in methane fermentation processes (Table 2).

The biogas yield of plants is generally limited by the greater or lesser proportion of lignocellulose, which is difficult to recycle. Efficiency of biogas production is related to the species-dependent, efficiency of cell degradation and presence or absence of molecules. However, the use of microalgae with a low lignocellulose content, for example *Chlorella vulgaris*, *Phaeodactylum tricornutum* and *Spirulina platensis*, permits an almost complete utilization of the organic substance. Golueke et al. [33] demonstrated the ability of microalgae to pass through an anaerobic digester intact and remain undigested. The authors noted that microalgal cells are known to be able to effectively resist bacterial attack and found intact microalgae cells in digestate leaving a digester after a 30-day hydraulic retention time. The composition of the biogas and the yield could be varied depending on the cell contents, the cell wall components and the stability of the cell wall. In particular the protein content of the cell plays a decisive role. Depending on the type of algae, the biogas yield was between 280 and 400 L/kg total volatile solids. Generally the variability is related to two main aspects: (i) the macromolecular composition, and (ii) the cell wall characteristics of each microalgae species. The difference in anaerobic biodegradability due to the macromolecular composition lies on the methane potential of different organic compounds in microalgae cells. Consequently, pretreatment techniques have been used to solubilize particulate biomass and improve the anaerobic digestion rate and extent.

# 4. Pretreatment Methods for Increased Biogas Production from Algae

Algae anaerobic biodegradability is limited by their complex cell wall structure. Thus, pretreatment techniques are being investigated to improve algal methane yield. Various pretreatment technologies have been developed in recent years. These pretreatment technologies aim to make AD faster, potentially increase biogas yield, and make use of new and/or locally available substrates, and prevent processing problems such as high electricity requirements for mixing or the formation of floatinglayers. Pretreatment methods can be divided into four categories: thermal, mechanical, chemical and biological processes (Figure 4).

Pretreatment methods have been studied in order to disintegrate microalgae cells, solubilise the organic content, and increase the anaerobic digestion rate and extent. Thermal pretreatments have been the most widely investigated already in continuous reactors and leading to net energy production [36, 37]. Mechanical pretreatments have mostly been investigated in batch assays using algae cultures [38]. Thermal pretreatments have been the most widely studied already in continuous reactors and leading to net energy production [39]. Mechanical pretreatments were less dependent on algae species, but required a higher energy input if compared with chemical, thermal and biological methods [38]. Chemical pretreatments have been proved successful, particularly when

combined with heat [39]. Enzymatic pretreatment seem to improve microalgae hydrolysis [40], which is promising due to its low energy input.

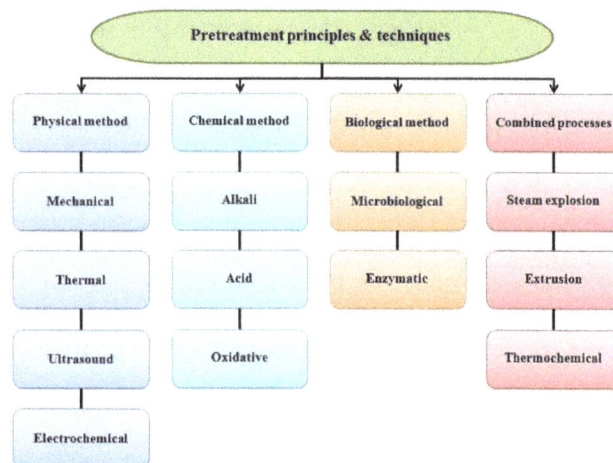

*Figure 4. Pretreatments for improving algae biogas production*

## 4.1. Pretreatment Methods for Increased Biogas Production from Macroalgae

Pretreatment of the algae is thus needed to aid both mechanical transport (pumping) as well as microbiological AD. Biogas can be derived via anaerobic fermentation of any organic matter, including the cellulose and hemicellulose within macroalgae, although the biomass must be subjected to pretreatment processes in order to liberate the sugars needed for fermentation. The effect of the pretreatment technologies, thermal treatment, thermochemical treatment, mechanical treatment, wet oxidation, hydrothermal pretreatment, steam explosion, plasma-assisted pretreatment and ball milling. One option is mechanical pretreatment of the algae; however a method which can handle the long fibrous material in macroalgae species is needed. Another method, which is relatively untested but promising, is enzymatic pretreatment which during recent years has been tested on many substrates to investigate effect on biogas potential [41].

The mechanical pretreatment effectively broke up the structure of all macroalgae into homogenous slurry. Mechanical pretreatment could increase the soluble COD-concentration of the tested algae by 1.5 to 3 times compared to raw algae. Enzymatic treatment increased it by 1.3 to 1.7 times. The best results were achieved by combining mechanical and enzymatic treatment where the concentration could was increased 3.5 times compared to raw algae [42]. A mechanical pretreatment phase is usually the first step not only for methane [43]. Nielsen and Heiske [44] was discussed the effect on methane yield of *U. lactuca* by various pretreatments including mechanical maceration and autoclavation. Sodium hydroxide soaking at room temperature prior to AD led to a 18% increase in methane potential in macroalgae as (*Palmaria palmata*), possess a high methane potential ($308 \pm 9$ mL $g_{VS}^{-1}$) [45]. Nielsen and Heiske [44] studied four macroalgae species-harvested in Denmark-for their suitability of bioconversion to methane. In batch experiments (53 °C) methane yields varied from 132 ml g volatile solids(-1) (VS) for *Gracillaria vermiculophylla*, 152 ml gVS(-1) for *Ulva lactuca*, 166 ml g VS(-1) for *Chaetomorpha linum* and 340 ml g VS(-1) for *Saccharina latissima* following 34 days of incubation. With an organic content of 21.1% (1.5-2.8 times higher than the other algae) *S. latissima* seems very suitable for anaerobic digestion. However, the methane yields of U. lactuca, G. vermiculophylla and C. linum could be increased with 68%, 11% and 17%, respectively, by pretreatment with maceration. Nielsen and Heiske [44] data of methane potentials in different macroalgae with pretreatments were presented in Table 3.

*Table 3. Methane potentials of different macroalgae with pretreatments.*

| Macroalgae taxon | Pretreatment | Methane yield (ml g VS$^{-1}$) | Methane production (ml g algae$^{-1}$) |
|---|---|---|---|
| *Batch screening of methane potentials of different macroalgae*[a] | | | |
| Chaetomorpha linum | Washed, chopped | 166 ± 43.5 | 11.4 ±2.98 |
| Chaetomorpha linum | Washed, macerated | 195 ± 8.7 | 13.4 ±1.46 |
| Saccharina latissima | Washed, chopped | 340 ± 48.0 | 68.2 ± 9.63 |
| Saccharina latissima | Washed, macerated | 333 ± 64.1 | 66.8 ± 12.87 |
| Gracilaria vermiculophylla | Washed, chopped | 132 ± 60.0 | 17.3 ±4.88 |
| Gracilaria vermiculophylla | Washed, macerated | 147± 56.3 | 19.3 ± 7.39 |
| Ulva lactuca | Washed, chopped | 152 ± 18.7 | 9.9 ± 1.21 |
| Ulva lactuca | Washed, macerated | 255 ±47.7 | 16.5 ± 3.08 |
| *Pretreatments of U. lactuca*[b] | | | |
| Ulva lactuca | Unwashed, chopped | 174± 23.3 | 12.8 ± 3.33 |
| Ulva lactuca | Unwashed, macerated | 271 ± 16.2 | 17.6 ±1.12 |
| Ulva lactuca | Washed, chopped | 171 ±22.3 | 12.2 ± 1.06 |
| Ulva lactuca | Washed, macerated | 200 ±11.0 | 14.3 ± 1.53 |
| Ulva lactuca | Washed, 110 C/20 min | 157 ± 13.4 | 11.3 ±0.96 |
| Ulva lactuca | Washed, 130 C/20 min | 187 ± 23.2 | 13.4 ± 1.72 |
| Ulva lactuca | Dried, grounded | 176 ±17.3 | 95.6 ± 9.42 |

Note: [a]34 days of incubation; [b]42 days of incubation (source: Nielsen and Heiske, 2011)

### 4.2. Pretreatment Methods for Increased Biogas Production from Microalgae

The digestibility of microalgal biomass varies significantly even between closely related species [46]. CH$_4$ yields from microalgae vary due to variation in cellular protein, carbohydrate and lipid content, cell wall structure, and process parameters such as the bioreactor type and the digestion temperature. Regarding the cell wall characteristics, it is mostly composed of organic compounds with low biodegradability and/or bioavailability, such as cellulose and hemicellulose. This tough cell wall hinders the methane production, since organic matter retained in the cytoplasm is not easily accessible to anaerobic bacteria [47]. AD is carried out by heterogeneous microbial populations involving multiple biological and substrate interactions. Anaerobic biodegradation can be divided into four main phases: hydrolysis, acidogenesis, acetogenesis and methanogenesis (before mentioned). AD (sometimes also called methanogenic fermentation) is widely applied in digestion of manure, sewage sludge and organic fraction of municipal solid wastes in industrial and agrarian societies. Anaerobic digestion of microalgal biomass has been studied from many freshwater and marine microalgae in various combinations. Rigid eukaryotic cell walls of microalgae can limit the anaerobic digestion of the biomass [33,47]. Pretreatment techniques were pointed out as a necessary step for microalgae cell disruption and biogas production by Chen and Oswald [47]. The effectiveness of pretreatment methods on biogas production depends on the characteristics of microalgae, i.e., the toughness and structure of the cell wall, and the macromolecular composition of cells. For instance, *Scenedesmus* sp. has one of the most resistant cell walls, since it is composed by multilayers of cellulose and hemicellulose on the inside, and sporopollenin and politerpene on the outside [48].

Microalgae complex cell wall structure confers a resistance to biological attack. In fact, species without cell wall (e.g. *Dunaliella* sp. and *Pavlova_cf* sp.) or containing a glycoprotein cell wall (e.g. Chlamydomonas sp., *Euglena* sp. and *Tetraselmis* sp.) showed higher methane yields than those with a more complex cell wall, containing recalcitrant compounds (e.g. *Scenedesmus* sp. and *Chlorella* sp.) [49]. Rates and yields of CH$_4$ formation from microalgal biomass often increase with digestion temperature. For example, [33] reported 5–10% increase in digestibility of microalgal biomass, when the digestion temperature was increased from 35 to 50 °C. Chen and Oswald [47] increased the CH$_4$ yield by 33% by heat pretreating microalgal biomass at 100 °C for 8 h. In both examples, however, the amount of energy consumed in the heating and pretreatment was higher than the corresponding energy gain from increased CH$_4$ production [50].

Retention times required to obtain high CH$_4$ yields from untreated microalgal biomass are relatively long, 20–30 days [51,52]. AD of microalgal biomass has been investigated in batch and fed-batch systems as well as in continuously stirred tank reactors [50]. Zamalloa et al. [52] suggested that anaerobic sludge blanket reactors, anaerobic filter reactors and anaerobic membrane bioreactors should be tested due to their high volumetric conversion rates. In the review by Passos et al. [30] presented the effectiveness of biogas production with the main pros and cons of microalgae pretreatment methods (Table 4). As can be seen, thermal pretreatment seems effective at increasing biogas production, while energy demand is low compared to mechanical ones. Nevertheless, biomass thickening or dewatering is crucial. Scalability may be a handicap for microwave pretreatment. Regarding thermo-chemical pretreatment, studies have shown positive results on microalgae biodegradability increase; however further studies should evaluate the risk of contamination in continuous bench and pilot-scale reactors.

*Table 4. Comparison of pretreatment methods for increasing microalgae anaerobic biodegradability.*

| Pretreatment | Control parameters | Biomass solubilization | Methane yield increase | Pros | Cons |
|---|---|---|---|---|---|
| Thermal (<100 °C) | Temperature; exposure time | √√√ | √√ | Low energy demand; Scalability | High exposure time |
| Hydrothermal (>100 °C) | Temperature; exposure time | √√√ | √√ | Scalability | High heat demand; thickened or dewatered biomass; risk of formation of refractory compounds |
| Thermal with steam explosion (>100 °C) | Temperature; exposure time; pressure | √√√ | √√√ | Scalability | High electricity demand; scalability; biomass dewatering |
| Microwave | Power; exposure time | √√ | √√ | – | |
| Ultrasound | Power; exposure time | √√ | √√ | Scalability | High electricity demand; biomass dewatering |
| Chemical | Chemical dose; exposure time | √ | √ | Low energy demand | Chemical contamination; risk of formation of inhibitors; Cost |
| Thermo-chemical | Chemical dose; exposure time; temperature | √√√ | √√ | Low energy demand | Chemical contamination; risk of formation of inhibitors; Cost |
| Enzymatic | Enzyme dose; exposure time; pH, temperature | √ | √ | Low energy demand | Cost, sterile conditions |

# 5. Algae Biogas Impurity Removal and Upgrade Technology

Biogas produced in AD plants or landfill sites is primarily composed of $CH_4$ and $CO_2$ with smaller amounts of $H_2S$, $NH_3$ and $N_2$. Trace amounts of $H_2$, VOCs and $O_2$ may be also present in biogas and landfill gas. Usually, the gas is saturated with water vapor and may contain dust particles. Additionally, organic silicon compounds are usually present in particular with reference to landfill gas, however their presence was highlighted also in AD biogas. The heating value of biogas is determined mainly by the methane content of the gas [53].

The main impurities are $CO_2$ which lowers the calorific value of the gas and sulfuric acid ($H_2S$) which could cause several problem on the plants and for human health, in fact on the plants it causes corrosion (compressors, gas storage tank and engines), while it's toxic after its inhalation. Although $CO_2$ is a major problem in the biogas as its removal is useful to adjust the calorific value and the relative density, and the removal of $H_2S$ can be of crucial point to the technological and economic feasibility of upgrading process of the gas [54]. Biogas production is growing and there is an increasing demand for upgraded biogas, to be used as vehicle fuel or injected to the natural gas grid. To enable the efficient use of biogas in these applications the gas must be upgraded. Since separation of $CO_2$ and $N_2$ from $CH_4$ is significantly important in natural gas upgrading, and capture/removal of $CO_2$, $CH_4$ from air ($N_2$) is essential to greenhouse gas emission control. Removal of $CO_2$ is done in order to reach the required Wobbe index of gas. As methane has a 23-fold stronger greenhouse gas effect than $CO_2$, it is important to keep methane losses low, for both economic and environmental reasons.

In general, in the standards requirements on Wobbe index values and limits on the concentration of certain components such as sulfur, oxygen, dust and the water dew point, as well as a minimum methane volumetric concentration of 96% are defined. There are several different commercial methods for reducing the $CO_2$ content of biogas. Two common methods of removing carbon dioxide from biogas are absorption (water scrubbing, organic solvent scrubbing) and adsorption (pressure swing adsorption, PSA). Less frequently used are membrane separation, cryogenic separation and process internal upgrading, which are a relatively new method, currently under development. The upgraded biogas is often named biomethane. Various technologies can be applied for removal of contaminants.

When $CO_2$ and other impurities are removed during the upgrading process, the methane concentration increases and thus the resulting biomethane can be utilized as an alternative to natural gas. Starr et al. [55] articulated on the carbon capture technologies that upgrade biogas by removing its $CO_2$ content. There are quite a few different technologies on the market today. The main unit operations used are absorption, adsorption, membrane separation and cryogenic separation; further information about these unit operations and their associated technologies shown in Table 5. A common factor of all of these techniques is that the removed $CO_2$ is normally released back into the atmosphere. In some cases, if its quality is high enough, it can be used for industrial purposes such as increasing the $CO_2$ concentration for photosynthesis in greenhouses or for carbonation in food production.

Strevett et al. [56] investigated the mechanism and kinetics of chemo-autotrophic biogas upgrading. In this experiment, different methanogens using only $CO_2$ as a carbon source and $H_2$ as an energy source were examined. The selection between mesophilic and thermophilic operation temperatures is typically based on whether the completion of reaction or the rate of reaction is of primary concern. Thermophilic methanogens exhibit rapid methanogenesis, while mesophilic bacteria give more complete conversion of the available $CO_2$ [56]. They selected *Methanobacterium thermoautotrophicum.* The organism works optimally at temperatures of 65–70 °C and has a specific requirement for $H_2S$, so both unwanted components are removed. A synthetic biogas of 50–60% $CH_4$, 30–40% $CO_2$ and 1–2% $H_2S$ was mixed with $H_2$ to a final mole fraction of $H_2$: $CO_2$ equaling 0.79:0.21. The gas mixture was fed to the hollow fibers packed with organisms. This biological system can effectively remove $CO_2$ and $H_2S$, while approximately doubling the original $CH_4$ mass. Alternative physicochemical treatment methods only remove the contaminating gas components, without changing $CH_4$ mass. Furthermore, physicochemical treatment generates additional waste and unwanted end products. The purified biogas contains about 96% $CH_4$ and 4% $CO_2$, while $H_2$ and $H_2S$ were not detected [56].

*Table 5. Current biogas upgrading technologies (adopted from Starr et al. [55]).*

| Unit operation | Technology | Acronym | Description of process |
|---|---|---|---|
| Absorption | High pressure water scrubbing | HPWS | Water absorbs $CO_2$ under high pressure conditions. Regenerated by depressurizing |
| | Chemical scrubbing | AS | Amine solution absorbs $CO_2$. The amine solution is regenerated by heating |
| | Organic physical scrubbing | OPS | Polyethylene glycol absorbs CO2. It is regenerated by heating or depressurizing |
| Adsorption | Pressure swing adsorption | PSA | Highly pressurized gas is passed through a medium such as activated carbon. Once the pressure is reduced the $CO_2$ is released from the carbon, regenerating it |
| Membrane | Membrane separation | MS | Pressurized biogas is passed through a membrane which is selective for $CO_2$ |
| Cryogenic | Cryogenic separation | Cry | Biogas is cooled until the $CO_2$ changes to a liquid or solid phase while the methane remains a gas. This allows for easy separation |

*Table 6. Comparison table of $CO_2$ fixation in the uptake rate and consumption efficiency.*

| Culture | Species | $CO_2$ source | biomass | Uptake rate (mg/L/day) | consumption efficiency (w/w) | Reference |
|---|---|---|---|---|---|---|
| Pure | *Chlorophyta* sp. | 10% $CO_2$ | 8.2 g/m²/d | 4 (10% $CO_2$) | 9% | Hase et al. [57] |
| | *Chlorella* sp. | | 13.2 g/m²/d | 6 (10% $CO_2$) | | |
| | *Chlorella* sp. | 1, 5, 10% $CO_2$ | 2.25 g/L | 83 (10% $CO_2$) | 4% | Ramanan et al. [58] |
| | *Spirulina platensis* | | 2.91g/L | 70 (10% $CO_2$) | 2% a | |
| Mixed | - | Diffusion from air | 22 t as C | 19 (50cm) | 44% | Green et al. [59] |
| | - | Injected $CO_2$ 11.5 kg/day as C | 10.2 kg/day as C | 187 | 148% | Weissman and Tillett [60] |
| | Dominant species [a] | Diffusion from air | 0.126 g/L (TSS[c]) | 162 | 123% | Tsai [20] |
| | Dominant species [b] | Diffusion from air | 0.136 g/L (TSS[c]) | 175 | 131% | Ramaraj [13] |

Note: [a] the genera *Chlorella, Oscillatoria, Oedogonium, Anabaena, Microspora* and *Lyngbya*
[b] the genera *Anabaena, Chlorella, Oedogonium* and *Oscillatoria*; [c]total suspended solids (as a biomass)

# 6. Biogas Purification Using Algae Biological Biogas Purification Methods and Techniques

Microalgae are a group of unicellular or simple multicellular photosynthetic microorganisms that can fix $CO_2$ efficiently from different sources [12–16], including the atmosphere, industrial exhaust gases, and soluble carbonate salts. Furthermore, combination of $CO_2$ fixation, biofuel production, and wastewater treatment may provide a very promising alternative to current $CO_2$ mitigation strategies. Presence of chlorophyll and other pigments help in carrying out photosynthesis. The true roots, stems or leaves are absent. Mostly they are photoautotrophic and carry on photosynthesis, some of these are chemo heterotrophic and obtain energy from chemical reactions as well as nutrients from preformed organic matter. Beside the plants, since algae had high potential $CO_2$ fixation in the current knowledge.

Microalgae can fix $CO_2$ using solar energy with efficiency ten times greater than terrestrial plants [13, 16]. The issue of greenhouse gas attracts an enormous attention worldwide recently. When atmospheric $CO_2$ concentration increased, it would gradually disturb the balance of global climate to cause unusual and astounding phenomena on earth. Therefore, we require the rapid development of bio-carbon-fixation technology to eliminate the adverse effects of $CO_2$, to transfer atmospheric $CO_2$ through the carbon cycle and to promote carbon balancing ecologically. Currently, many innovative alternatives of physical, chemical and biological technologies of $CO_2$ mitigation are rapidly developed.

At present, algae application of $CO_2$ sequestration has developed as a popular topic and the current interests are including: species, power plant flue gas utilization, reactor design, growth condition, growth kinetics and modeling. The most studies in the literature concerned the maximum $CO_2$ uptake rate by the artificial photo-bioreactors [12, 13, 20]. Among those techs, bio-eco-technology is the most natural and ecological way to accomplish the designed targets by the utilization of "self-designed" bio-functions of nature [12, 13, 15, 16]. The different sources and approaches of algal $CO_2$ uptake rate and consumption efficiency was presented in Table 6. Accordingly, algae production has a great potential for $CO_2$ bio-fixation process and deserves a close look.

Biogas purification/scrubbing using algae involved the use of algae's photosynthetic ability in the removal of the impurities (mainly $CO_2$ and $H_2S$) present in biogas, leaving a purified biogas containing almost pure methane, which could be used for energy generation. Biological purification technology is worth examining because has double impact. The method about removing $CO_2$ from biogas by microalgal culturing using the biogas effluent as nutrient medium and effectively upgrade biogas also simultaneously reduce the biogas effluent nutrient [61]. Using biogas as a source of carbon dioxide has two main advantages: the biomass production costs are reduced and the produced biomass does not contain harmful compounds, which can occur in flue gases. Hendroko et al. [62] verified xhibit that microalgae (*Scenedesmus* sp.) in laboratory experiments using biogas slurry as growing medium and biogas are given periodically generate 21% of $CO_2$ compared with 24% of controls. They summarized: digestion slurry with seed cake JatroMas cultivar as raw material is able to increase growth of microalgae *Scenedesmus* sp. higher than standard media; microalgae *Scenedesmus* sp. is able to capture $CO_2$ gas in bio-methane; with integration of slurry and bio-methane intake, there is tendency *Scenedesmus* sp. growth is more increasing; Mutualism symbiosis among slurry, bio-methane and microalgae *Scenedesmus* sp. will give impact to increasing of $CH_4$ content in bio-methane. In other word, microalgae can be work as purification biologic from bio-methane [62].

There are several authors [10, 62, 63] reported that *Arthrospira* sp, *Chololera vulgaris* SAG 211-11b, *Chlorella* sp. MM-2, *Chlorella* sp. MB-9, *Chlorella vulgaris* ARC1, *Chlamydomonas* sp. dan *Scenedesmus* sp. was a positive synergy with biogas. The productivity of the system with Zarrouk media and biogas almost 5 times higher than that for the same media without biogas when piggery waste was used, the utilization of biogas brings a productivity gain of about 2–5 times higher [63].

Kao et al [64] demonstrates that the microalga *Chlorella* sp. MB-9 was a potential strain which was able to utilize $CO_2$ for growth when aerated with desulfurized biogas ($H_2S < 50$ ppm) produced from the anaerobic digestion of swine wastewater. The demonstrated system can be continuously used to upgrade biogas by utilizing a double set of photobioreactor systems and a gas cycle-switching operation. Furthermore, they demonstrated that the efficiency of $CO_2$ capture from biogas could be maintained at 50% on average, and the $CH_4$ concentration in the effluent load could be maintained at 80% on average, i.e., upgrading was accomplished by increasing the $CH_4$ concentration in the biogas produced from the anaerobic digestion of swine wastewater by 10%.

Some literatures mentioned about the cultivation microalgae using biogas as $CO_2$ provider. Kao et al. [64] used biogas that contained $20\pm2\%$ $CO_2$ for *Chlorella* sp. culture with variation of light intensity which was at cloudy and at sunny day. Kao et al. [10] used biogas that contained $20\pm1\%$ $CO_2$ for *Chlorella* sp. culture with variation flow rate of biogas which was 0.05; 0.1; 0.2; 0.3 vvm. Douškova et al. [65] investigated the potential of biogas as $CO_2$ provider for *Chlorella vulgaris*; and optimization of biogas production from distillery stillage is described. The growth kinetics of microalgae *Chlorella* sp. consuming biogas or mixture of air and $CO_2$ in the concentration range of 2–20% (v/v)

(simulating a flue gas from biogas incineration) in laboratory-scale photo-bioreactors. It was proven that the raw biogas (even without the removal of $H_2S$) could be used as a source of $CO_2$ for growth of microalgae. The growth rate of microalgae consuming biogas was the same as the growth rate of the culture grown on a mixture of air and food-grade $CO_2$. Several species of algae can metabolize $H_2S$ [66]. Using a biological system to remove $H_2S$ has similar benefits to using one to remove $CO_2$: lower upkeep costs, more environmentally sustainable and non-hazardous waste.

Furthermore, Tongprawhan et al. [67] used oleaginous microalgae to capture $CO_2$ from biogas for improving methane content and simultaneously producing lipid. They screened several microalgae for identify their ability to grow and produce lipid using $CO_2$ in biogas. Finally, they reported a marine *Chlorella* sp. was the most suitable strain for capturing $CO_2$ and producing lipid using biogas (50% v/v $CO_2$ in methane) as well as using 50% v/v $CO_2$ in air. Sumardiono et al. [68] established to evaluate the design of the photobioreactor system for purifying biogas through the culturing of microalgae. This system represented a simple promising way for the current forthcoming technologies of biogas purification. It helps to decrease the concentration of $CO_2$ in biogas concomitantly producing microalgae biomass. The microalgae *Nannochloropsis* is able to use $CO_2$ from biogas produced from the anaerobic digestion of tannery sludge. The results show that cultivation of microalgae under the biogas to scrub out $CO_2$ and promote enrichment of methane in the biogas in this work and obtained scrubbing of 27% from 30%.

The biocapture of $CO_2$ by microalgae can be applied to improve the quality of biogas by reducing the $CO_2$ content as this would lead to an increase in the methane content [69]. The microalgae Chlorella sp.was analysed in terms of conditioning biogas. As a result the biogas components $CO_2$ and $H_2S$ could be reduced up to 97.07% and100%, respectively. Also an increase of microalgae cell count could be documented, which provides interesting alternatives for the production of algae ingredients. Consequently, the algae biological purification is an alternative to other biogas purification methods.

# 7. Conclusion

Biogas is a promising and valuable renewable energy source. Biogas can be utilized in several ways; either raw or upgraded. As a minimum, biogas has to be cooled, drained and dried immediately after production, and almost always it has to be cleaned for the content of $CO_2$, $H_2S$ and other impurities. Using the photosynthesis of algae to remove the $CO_2$ from biogas is an alternative method that solves the problems of the common non-biological methods. Algae are self-sustaining with the addition of minimal nutrients and light. Algae were used as a biological method to remove $CO_2$ through photosynthesis. Algae has several advantages over conventional chemical $CO_2$ removal methods because algae is inexpensive to obtain, requires only light and minimal nutrients in addition to the $CO_2$ for growth, and the waste can be harvested for biofuels. Several species of algae can metabolize $H_2S$. The $H_2S$ content in biogas, at levels higher than 300–500 ppm, damages the energy conversion technique. Today biological cleaning reduces the content of hydrogen sulphide to a level below 100 ppm. Using a biological system to remove $H_2S$ has similar benefits to using one to remove $CO_2$: lower upkeep costs, more environmentally sustainable and non-hazardous waste. Maintaining a pure culture would increase the efficiency of the algae in processing $CO_2$. Using biological metabolism to purify biogas is a promising means of biofuel production. The incorporation of algae in photobioreactors to purify biogas has several advantages over conventional chemical methods of $CO_2$ removal. Obtaining algae is relatively inexpensive because culturing algae requires minimal nutrients for their growth. Growth of the algae requires a light source as well, which does not necessarily have to be expensive if illumination is provided by natural sunlight, which is not limited in supply.

# References

[1] C. S. Jones, S. P. Mayfield, "Algae biofuels: versatility for the future of bioenergy", Current Opinion in Biotechnology, 2012, 23: 346–351.

[2] P. J. Meynell, "Methane: planning a digester", New York: Schocken Books, 1976.

[3] A. Demirbas, "Biofuels sources, biofuel policy, biofuel economy and global biofuel projections", Energy Conversion and Management, 2008, 49: 2106–2116.

[4] N. Dussadee, K. Reansuwan, R. Ramaraj, "Potential development of compressed bio-methane gas production from pig farms and elephant grass silage for transportation in Thailand", Bioresource Technology, 2014, 155: 438–441.

[5] G. Lastella, C. Testa, G. Cornacchia, M. Notornicola, F. Voltasio, V. K. Sharma, "Anaerbic digestion of semi-solid organic waste: biogas production and its purification", Energy Conversion and Management, 2002, 43: 63–75.

[6] B. E. Rittmann, "Opportunities for renewable bioenergy using microorganisms", Biotechnology and Bioengineering, 2008, 100, 203–212.

[7] E. Stephens, I. L. Ross, Z. King, J. H. Mussgnug, O. Kruse, C. Posten, M. A. Borowitzka, B. Hankamer, "An economic and technical evaluation of microalgal biofuels", Nature Biotechnology, 2010, 28, 126–128.

[8] J. H. Mussgnug, V. Klassen, A. Schlüter, O. Kruse, "Microalgae as substrates for fermentative biogas production in a combined biorefinery concept", Journal of Biotechnology, 2010, 150:51–56.

[9] E. Ryckebosch, M. Drouillon, H. Vervaeren, "Techniques for Transformation of Biogas to Biomethane", Journal of Biomass and Bioenergy, 2011, 35: 1633–1645.

[10] C. Y. Kao, S. Y. Chiu, T. T. Huang, L. Dai, G. H. Wang, C. P. Tseng, C. H. Chen, C. S. Lin, "A mutant strain of microalga Chlorella sp. for the carbon dioxide capture from biogas", Biomass and Bioenergy, 2012, 36: 132–140.

[11] N. Abatzoglou, S. A Boivin, "A review of biogas purification processes", Biofuels, Bioproducts and Biorefining, 2009, 3: 42–71.

[12] R. Ramaraj, D. D-W. Tsai, P. H. Chen, "Freshwater microalgae niche of air carbon dioxide mitigation", Ecological Engineering, 2014; 68: 47–52.

[13] R. Ramaraj, Freshwater microalgae growth and Carbon dioxide Sequestration, Taichung, Taiwan, National Chung Hsing University, PhD thesis, 2013.

[14] R. Ramaraj, D. D-W. Tsai, P. H. Chen, "Algae Growth in Natural Water Resources", Journal of Soil and Water Conservation, 2010, 42: 439–450.

[15] R. Ramaraj, D. D-W. Tsai, P. H. Chen, "Chlorophyll is not accurate measurement for algal biomass", Chiang Mai Journal of Science, 2013, 40: 547–555.

[16] R. Ramaraj, D. D-W. Tsai, P. H. Chen, "An exploration of the relationships between microalgae biomass growth and related environmental variables", Journal of Photochemistry and Photobiology B: Biology, 2014, 135: 44–47.

[17] C. Vílchez, I. Garbayo, M. V. Lobato, J. M. Vega, "Microalgae-mediated chemicals production and waste removal. Enzyme and Microbial Technology, 1997, 20: 562–572.

[18] W. J. Oswald, "My sixty years in applied algology", Journal of Applied Phycology, 2003, 15: 99–106.

[19] L. E. Graham, L. W. Wilcox, "Algae", Prentice Hall Inc. Upper Saddle River, New Jersey, 2000.

[20] D. D-W. Tsai, Watershed Reactor Analysis, $CO_2$ Eco-function and Threshold Management Study, Taichung, Taiwan, National Chung Hsing University, PhD thesis, 2012.

[21] T. Driver, A. Bajhaiya, J. K. Pittman, "Potential of Bioenergy Production from Microalgae", Current Sustainable/Renewable Energy Reports, 2014, 1: 94-103.

[22] L. Yang, Ge. Xumeng, C. Wan, F. Yu, Y. Li, "Progress and perspectives in converting biogas to transportation fuels", Renewable & Sustainable Energy, 2014, 40: 1133–1152.

[23] K. W. Gellenbeck, D. J. Chapman, "Seaweed uses: the outlook for mariculture", Endeavour, 1983, 7: 31–37.

[24] A. B. Ross, J. M. Jones, M. L. Kubacki, "Since macroalgae are again receiving attention as a substrate for anaerobic digestion", Bioresource Technology, 2008, 99: 6494–6504.

[25] D. P. Chynoweth, D. L. Klass, S. Ghosh, "Anaerobic digestion of kelp", In Biomass conversion processes for energy and fuels, Edited by Sofer SS, Zaborsky OR. New York: Plenum Press; 1981: 315–318.

[26] A. D. Hughes, M. S. Kelly, K.D. Black, M. S. Stanley, "Biogas from Macroalgae: is it time to revisit the idea?", Biotechnology for Biofuels, 2012, 5: 86.

[27] A. Vergara-Fernàndez, G. Vargas, N. Alarcon, A. Antonio, "Evaluation of marine algae as a source of biogas in a two-stage anaerobic reactor system", Biomass & Bioenergy, 2008, 32: 338–344.

[28] J. Singh, S. Gu, "Commercialization potential of microalgae for biofuels production", Renewable & Sustainable Energy Reviews, 2010, 14: 2596–2610.

[29] A. Parmar, N. K Singh, A. Pandey, E. Gnansounou, D. Madamwar, "Cyanobacteria and microalgae: a positive prospect for biofuels", Bioresource Technology, 2011, 102: 10163–10172.

[30] M. Dębowski, M. Zieliński, A. Grala, M. Dudek, "Algae biomass as an alternative substrate in biogas production technologies—Review", Renewable and Sustainable Energy Reviews, 2013, 27: 596–604.

[31] G. Migliore, C. Alisi, A. R. Sprocati, E. Massi, R. Ciccoli, M. Lenzi, A. Wang, C. Cremisini, "Anaerobic digestion of macroalgal biomass and sediments sourced from the Orbetello lagoon, Italy", Biomass and Bioenergy, 42: 69–77.

[32] M. Huesemann, G. Roesjadi, J. Benemann, F. B. Metting, "Biofuels from Microalgae and Seaweeds. In Biomass to Biofuels", Blackwell Publishing Ltd.: Oxford, UK, 2010, 165–184.

[33] C. G. Golueke, W. J. Oswald, H. B. Gotaas, "Anaerobic digestion of algae", Applied Microbiology, 1957, 5: 47–55.

[34] I. Angelidaki, B. K. Ahring, "Thermophilic anaerobic digestion of livestock waste: the effect of ammonia", Applied Microbiology and Biotechnology, 1993, 38: 560–564.

[35] B. Sialve, N. Bernet, O. Bernard, "Anaerobic digestion of microalgae as a necessary step to make microalgal biodiesel sustainable", Biotechnology Advances, 2009, 27: 409–416.

[36] G. Markou, I. Angelidaki, E. Nerantzis, D. Georgakakis, "Bioethanol Production by Carbohydrate-Enriched Biomass of Arthrospira (Spirulina) platensis", Energies, 2013, 6: 3937–3950.

[37] S. Schwede, Z.-U. Rehman, M. Gerber, C. Theiss, R. Span, "Effects of thermal pretreatment on anaerobic digestion of Nannocloropsis salina biomass", Bioresource Technology, 2013, 143: 505–511.

[38] S. Cho, S. Park, J. Seon, J. Yu, T. Lee, "Evaluation of thermal, ultrasonic and alkali pretreatments on mixed-microalgal biomass to enhance anaerobic methane production", Bioresource Technology, 2013, 143: 330–336.

[39] F. Passos, M. Solé, J. Garcia, I. Ferrer, "Biogas production from microalgae grown in wastewater: effect of microwave pretreatment", Applied Energy, 2013, 108:168–175.

[40] A. Mahdy, L. Mendez, S. Blanco, M. Ballesteros, C. González-Fernández, "Protease cell wall degradation of Chlorella vulgaris: effect on methane production", Bioresource Technology, 2014, 171: 421–427.

[41] K. Ziemiński, I. Romanowska, M. Kowalska, "Enzymatic pretreatment of lignocellulosic wastes to improve biogas production", Waste Management, 2012, 32: 1131–1137.

[42] H. Li, H. Kjerstadius, E. Tjernström, Å. Davidsson, "Evaluation of pretreatment methods for increased biogas production from macro algae (Utvärdering av förbehandlingsmetoder för ökad biogasproduktion från makroalger)", SGC Rapport 2013, 278: 136 [http://www.sgc.se/ckfinder/userfiles/files/SGC278.pdf ]

[43] S. Tedesco, T. M. Barroso, A. G. Olabi, "Optimization of mechanical pre-treatment of Laminariaceae spp. biomass-derived biogas", Renewable and Sustainable Energy Reviews, 2013, 27: 596–604.

[44] G. Jard, C. Dumas, J. P. Delgenes, H. Marfaing, B. Sialve, J. P. Steyer, H. Carrère, "Effect of thermochemical pretreatment on the solubilization and anaerobic biodegradability of the red macroalga Palmaria palmate", Biochemical Engineering Journal, 2013, 79: 253–258.

[45] H. B Nielsen, S. Heiske, "Anaerobic digestion of macroalgae: methane potentials, pre-treatment, inhibition and co-digestion", Water Science Technology, 2011, 64: 1723–1729.

[46] A. M. Lakaniemi, O. H. Tuovinen, J. A. Puhakka, "Anaerobic conversion of microalgal biomass to sustainable energy carriers--a review", Bioresource Technology, 2013, 135: 222–231.

[47] P. H. Chen, W. J. Oswald, "Thermochemical treatment for algal fermentation", Environment International, 1998, 24: 889–897.

[48] C. González-Fernández, B. Sialve, N. Bernet, J.P. Steyer, "Impact of microalgae characteristics on their conversion to biofuel. Part II: Focus on biomethane production", Biofuels, Bioproducts & Biorefining, 2011, 6: 205–218.

[49] P. Bohutskyi, M. J. Betenbaugh, E. J. Bouwer, "The effects of alternative pretreatment strategies on anaerobic digestion and methane production from different algal strains", Bioresource Technology, 2014, 155: 366–372.

[50] B. Sialve, N. Bernet, O. Bernard, "Anaerobic digestion of microalgae as a necessary step to make microalgal biodiesel sustainable", Biotechnology Advance, 2009: 27, 409–416.

[51] M. Ras, L. Lardon, B. Sialve, N. Bernet, J. P. Steyer, "Experimental study on a coupled process of production and anaerobic digestion of Chlorella vulgaris", Bioresource Technology, 2011, 102: 200–206.

[52] C. Zamalloa, E. Vulsteke, J. Albrecht, W. Verstraete, "The techno-economic potential of renewable energy through the anaerobic digestion of microalgae", Bioresource Technology, 2011, 102: 1149–1158.

[53] L. Lombardi, E. Carnevale, "Economic evaluations of an innovative biogas upgrading method with $CO_2$ storage", Energy, 2013, 62: 88–94.

[54] P. Iovane, F. Nanna, Y. Ding, B. Bikson, A. Molino, "Experimental test with polymeric membrane for the biogas purification from $CO_2$ and $H_2S$", Fuel, 2014, 135: 352–358.

[55] K. Starr, X. Gabarrell, G. Villalba, L. Talens, L. Lombardi, "Life cycle assessment of biogas upgrading technologies", Waste Management, 2012, 32: 991–999.

[56] K. A. Strevett, R. F. Vieth, D. Grasso, "Chemo-autotrophic biogas purification for methane enrichment: mechanism and and Kinetics", The Chemical Engineering Journal and the Biochemical Engineering Journal, 1995, 58: 71–79.

[57] R. Hase, H. Oikawa, C. Sasao, M. Morita, Y. Watanabe, "Photosynthetic production of microalgal biomass in a raceway system under greenhouse conditions in Sendai City", Journal of Bioscience and Bioengineering, 2000. 89:157–163.

[58] R. Ramanan, K. Kannan, A. Deshkar, R. Yadav, T. Chakrabarti, "Enhanced algal CO2 sequestration through calcite deposition by Chlorella sp. and Spirulina platensis in a mini-raceway pond", Bioresource Technology, 2010, 101: 2616–2622.

[59] F. B. Green, L. Bernstone, T. J. Lundquist, J. Muir, R. B. Tresan, W. J. Oswald, "Methane fermentation, submerged gas collection, and the fate of carbon in advanced integrated wastewater pond systems", Water Science Technology, 1995, 31: 55–65.

[60] J. C. Weissman, D. M. Tillett, "Aquatic Species Project Report, NREL/MP-232-4174", In: Brown LM, Sprague S (eds) National Renewable Energy Laboratory, 1992, 41–58.

[61] C. Yan, Z. Zheng, "Performance of photoperiod and light intensity on biogas upgrade and biogas effluent nutrient reduction by the microalgae *Chlorella* sp.", Bioresource Technology, 2013, 139: 292–299.

[62] R. Hendroko, M. Kawaroe, Salafudin, G. Saefurahman, N. E. Fitrianto, D. W. Sari, Y. Sakri, "Biorefinery preliminary studies: integration of slurry and $CO_2$ as biomethane digester waste for microalgae Scenedesmus sp. growth", International seminar on chemical engineering Soehadi Reksowardojo, Bandung, October 5–7, 2011.

[63] L. Travieso, E. P. Sanchez, F. Benitez, J. L. Conde, "Arthrospira sp. intensive cultures for food and biogas purification", Biotechnology Letters; 1993; 15:1091–1094.

[64] C. Y. Kao, S. Y. Chiu, T. T. Huang, L. Dai, L. K. Hsu, C. S. Lin, "Ability of a mutant strain of the microalga Chlorella sp. to capture carbon dioxide for biogas upgrading", Applied Energy, 2012, 93: 176–183.

[65] I. Doušková, F. Kaštanek, Y. Maleterova, P. Kaštanek, J. Doucha, V. Zachleder, "Utilization of distillery stillage for energy generation and concurrent production of valuable microalgal biomass in the sequence: Biogas-cogeneration-microalgae-products", Energy Conversion and Management, 2010, 51: 606–611.

[66] H. Biebl, N. Pfennig, "Growth of sulfate-reducing bacteria with sulfur as electron acceptor", Archives of Microbiology, 1977, 112: 115–117.

[67] W. Tongprawhan, S. Srinuanpan, B. Cheirsilp, "Biocapture of $CO_2$ from biogas by oleaginous microalgae for improving methane content and simultaneously producing lipid", Bioresource Technology, 2014, 170: 90–99.

[68] S. Sumardiono, I. S. Budiyono, S. B. Sasongko, "Utilization of Biogas as Carbon Dioxide Provider for Spirulina platensis Culture", Current Research Journal of Biological Sciences, 2014, 6: 53–59.

[69] G. Mann, M. Schlegel, R. Schumann, A. Sakalauskas, "Biogas-conditioning with microalgae", Agronomy Research, 2009, 7: 33–38.

# 24

# Microalgae biomass as an alternative substrate in biogas production

**Rameshprabu Ramaraj[1, *], Natthawud Dussadee[1], Niwooti Whangchai[2], Yuwalee Unpaprom[3]**

[1]School of Renewable Energy, Maejo University, Sansai, Chiang Mai-50290, Thailand
[2]Faculty of Fisheries Technology and Aquatic Resources, Maejo University, Sansai, Chiang Mai 50290, Thailand
[3]Program in Biotechnology, Faculty of Science, Maejo University, Sansai, Chiang Mai-50290, Thailand

**Email address:**

rrameshprabu@gmail.com, rameshprabu@mju.ac.th (Ramaraj R.), natthawu@yahoo.com (N. Dussadee), natthawu@mju.ac.th (Dussadee N.)

**Abstract:** The running down of fossil energy sources makes the production of bioenergy an expected need worldwide. Therefore, energy crops have gained increasing attention in recent years as a source for the production of bioenergy because they do not compete with food crops. Microalgae have numerous advantages such as fast growth rates and not competing with food production. Because of the fast growth, many high valuable products are generated, e.g. food, biofuel, etc. Due to the energy crisis, renewable energy becomes a popular issue in this world today and there are several alternatives such as bioenergy, solar, wind, tide, geothermal, etc. For bioenergy, algae are the third generation biofuel crop. There is an increased demand for biogas in the society and one way to meet this is to use cultivated microalgae as fermentation substrate. In the present study, we maintained algae growth process and biomass production in autotrophic condition continuously for over 2 month's period. Growth system (photobioreactor) was setup under room temperature and continuous illumination light through fluorescent lamps; light intensity was average as 48.31 [$\mu mol^{-1} m^{-2}$ per $\mu A$]. In reactor, dominant microalgae species were including *Anabaena* sp., *Chlorella* sp., *Oscillatoria* sp., *Oedogonium* sp. and *Scenedesmus* sp. The content of total solids (TS) and volatile solids (VS) in the algae biomass was measured; the results were average as 12500 $g/m^3$ and 6320 $g/m^3$, respectively. Furthermore, microalgal biomass is a potentially valuable fermentation substrate, and produce over 60% of methane gas.

**Keywords:** Microalgae, Biomass, Bio-Reactor, Fermentation Feedstock, Biogas

## 1. Introduction

Algae are the most important primary producer in aquatic ecosystem [1]. They are a diverse group of photosynthetic organisms with a range of unicellular to multicellular forms that are found in the ocean, freshwater bodies, on rock, soils and vegetation [2]. Algae can be broadly divided into macroalgae, which include multicellular seaweeds, and microalgae, which are small unicellular algae, found in a wide variety of environments and comprising of many evolutionarily distinct organisms. It provides food and oxygen for many species in the aquatic Environment and it's vitally crucial to keep carbon dioxide ($CO_2$) of carbon cycle via photosynthesis to balance the $CO_2$ concentration in atmosphere. Through photosynthesis they fix light energy and reduce simple inorganic molecules into complex organic molecules supporting the whole community of living organisms occupying higher trophic levels in the ecosystem [3].

Microalgae are ideal organisms for biological monitoring [4]. Microalgal density, abundance, and diversity are ideal indicators of the health of aquatic ecosystems and water quality. Hence, algal biomass measurement is important in many biological and ecological studies and in algal industry [5]. Therefore, algal community plays critical roles as the primary producer and as a major biotic component in the nutrient/energy cycle in aquatic ecosystems [3]. Microalgae have the ability to fix carbon dioxide, nutrients and store the solar energy into their cells via photosynthesis which makes them interesting as an alternative energy source; some studies had also indicated the importance of algae in carbon dioxide fixation [1–3].

Microalgae growth rate is the highest compared with the other plants. Due to the energy crisis, renewable energy becomes a popular issue in this world today and there are several alternatives such as bioenergy, solar, wind, tide, geothermal, etc. Since it is an excellent biomass producer, the biomass is broadly extracted to obtain various biochemical used as medicine, nutrition, food etc. For energy crisis, algae provide an excellent biomass as a renewable energy source, so called "bioenergy", and turn algae as the most efficient bio-component (carbohydrates, proteins, lipids and other mineral source) and bio-oil maker [2]. It is one of the most important bio-technological species currently.

For bioenergy, algae are the third generation biofuel [6]. For the reasons of the best energy conversion efficiency of sunlight and the highest growth rate, algae have the best potential among all the energy crops [2, 5]. Regarding biofuel production, microalgae can provide different types of biofuels, including: biodiesel (from algal fatty acids); ethanol (produced by fermentation of starch); hydrogen (produced biologically); and methane (produced by anaerobic digestion of algal biomass). Some authors are more assertive, and suggest that the production of methane via anaerobic digestion (AD) is the most feasible and cost-effective route to an energy product [7,8]. Feasibility analysis of biodiesel production from algae underlined the importance of including AD technology into the overall fuel production process, either by utilizing defatted algae or whole algal biomass for conversion to methane. Economic analysis conducted with respect to the cost of lipid extraction and conversion to biofuels suggests that for algae with lipid content below 40% direct methane production is the most economically feasible approach [9]. This is supported by Harun et al. [10] who demonstrated that more energy could be generated from the production of methane from microalgae (14.04 MJ/kg), rather than biodiesel (6.6 MJ/kg) or ethanol (1.79 MJ/kg) where their unit "kg" is assumed to be "kg of dry weight algae". Furthermore, up to 65% of the chemical energy stored in the algal biomass can be potentially recovered through AD to methane [11].

Recent studies are increasing our knowledge about anaerobic digestion of microalgae. Theoretical calculations [9] as well as bottle and digester experiments [12] have shown the great potential of anaerobically digesting microalgae for methane production which can be further converted into a clean and renewable biofuel [8]. Accordingly, the biofuel feedstock source to be considered is algae. This paper highlights some of the current details of biofuel generation, focusing mainly on microalgae, which may allow biofuel production from these organisms to be economically and sustainably viable in the future. Consequently, the primary objective of this study was to investigate the potential of the freshwater mixed culture microalgae biomass substrate to estimation and production of biogas.

# 2. Materials and Methods

The methodology is illustrated in Figure 1. In the first stage of reactor setup, we selected the standard bioreactor, "CSTR (continuously stirred tank reactor)". Two bench scale CSTRs were set up in the Energy Research Center (ERC), Maejo University, Sansai, Chiang Mai, Thailand. All the reactors were operated by batch feed conditions and other operational factors were listed in Table 1.

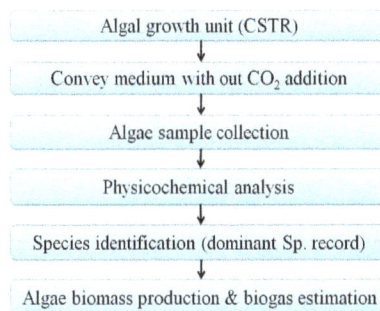

**Figure 1.** *Flow chart of methodology.*

**Table 1.** *Operational parameters*

| Operational parameter | Photo-bioreactor |
|---|---|
| Scale | Laboratory |
| Reactor design | 1L flask |
| Water volume | 1L |
| Feeding | Batch |
| Filter size | 0.45 μm |
| Mixing speed | Magnetic mixer |
| Light source | Fluorescent tube panels |
| Temperature | Room temperature |
| Algae species | Mixed microalgal culture |
| Operation period | 60 days |

## 2.1. Algae Culture, Medium Preparation and Species Identification

Algae samples were collected by plankton net (20-μm pore size) from solar freshwater fish pond at School of Renewable Energy, Maejo University, Sansai, Chiang Mai, Thailand. Algae medium was prepared using method of Ershad-Langroudi et al. [13] to make the specific medium of Conway. One liter solutions of chemical compounds demonstrated in Table 2 were prepared and then mixed and sterilized. The mixture of chemical compounds was added to filter by 0.2 μ filter and distributed to 1000 ml Erlen Meyer flasks to culture the algae. All experiments were carried out in triplicate. The microalgae were identified microscopically using light microscope with standard manual for algae [14,15].

**Table 2.** *Chemical compounds utilized in Conway medium preparation Concentration*

| Chemical compound | Concentration |
|---|---|
| $KNO_3$ | 116 g |
| NaEDT | 45 g |
| $H_3BO_3$ | 33.6 g |
| $MnCl_2.4H_2O$ | 0.36 g |
| $ZnCl_2$ | 2.1 g |
| $CoCl_2.6H_2O$ | 2 g |
| $(NH_4)6MoO_7.4H_2O$ | 0.9 g |
| $CuSO_4.H_2O$ | 2 g |
| Vitamin B1 | 200 mg |
| Vitamin B12 | 100 mg |
| $NaSiO_3$ | 20 g |
| $Na_2H_2PO_4.2H_2O$ | 20 g |
| $FeCl_3.6H_2O$ | 1.3 g |

## 2.2. Analytical Methods

All the physicochemical indexes including pH, chemical oxygen demand (COD); total suspended solids (TSS), total solids (TS), volatile solids (VS), were continuously monitored according to the standard method [16]. Light intensity measured by light meter (LI-COR light meter (LI-250)) and temperature was measured by laboratory thermometer. Biogas estimation method was adopted from von Sperling & Chernicharo [16] and Pavlostathis & Giraldo-Gomez [17].

# 3. Results and Discussion

## 3.1. Algae Growth and Measurement

Algae were the polyphyletic, simple microscopic or macroscopic, unicellular to multicellular, motile or immotile organisms which grew in abundance in any water body such as lakes, ponds, rivers, streams, marine etc (before mentioned). Algae varied in size and shape, from microalgae of less than 1μm to macroalgae over 30 m in length. They grew in any aquatic environment and used light and $CO_2$ to create biomass. Ecologically, algae were the most widespread of the photosynthetic plants, constituting the bulk of carbon assimilation through microscopic cells [1,2,4,5].

There were several main groups of freshwater algae, which differed primarily in pigment composition, biochemical constituents and life cycle. The most important algae in term of known species are classified in four groups: green algae (Chlorophyceae), blue-green algae (Cyanophyceae), brown algae (Chrysophyceae) and red algae (Rhodophyceae), and the brief explanations were shown in Table 3.

Table 3. The known species of algae species in freshwater.

| Main groups | Class | Known species | Size | Chlorophyll | Carotenes | Storage material |
|---|---|---|---|---|---|---|
| Green algae | Chlorophyceae | 5000 | 3-10 μm | Chl-a, Chl-b and Chl-c1 | α, β, γ | Starch and triglycerides |
| Blue-green algae | Cyanophyceae | 2000 | 0.5-60 μm | Chl-a and Chl-b | β | Starch and triglycerides |
| Brown algae | Phaeophyceae | 1000 | up to 30 m | Chl-a and Chl-c1&c2 | α, β | triglycerides and carbohydrates |
| Red algae | Rhodophyceae | 164 | up to 10 cm | Chl-a and Chl-d | β, ε | Starch |

The biomass of microalgae contains many various chemical compounds, which were significant in different aspects such as pharmaceutics, human food and energy. Carbohydrates, proteins, lipids, nucleic acids and pigments were the basic and major components of algae; beside those, Acylglycerides, Glycolipids, phospholipids, fatty acids, methyl esters, polysaccharides and all had important roles. Table 4 gave the cell content of these major fractions with their elemental composition and energetic properties [2].

Table 4. Elemental composition of algal biochemical components.

| Biochemical component | Characteristic elemental composition | Calculated calorific value/kJ $g^{-1}$ | Range of typical cell content (%) |
|---|---|---|---|
| Lipids | $C_1H_{1.83}O_{0.17}N_{0.0031}P_{0.006}S_{0.0014}$ | 36.3 | 15–60 |
| Acylglycerides | $C_1H_{1.83}O_{0.096}$ | 40.2 | — |
| Glycolipids | $C_1H_{1.79}O_{0.24}S_{0.0035}$ | 33.4 | — |
| Phospholipids | $C_1H_{1.88}O_{0.173}N_{0.012}P_{0.024}$ | 35.3 | — |
| Fatty acids | $C_1H_{1.91}O_{0.12}$ | 39.6 | — |
| Methyl esters | $C_1H_{1.92}O_{0.05}$ | 43.0 | — |
| Protein | $C_1H_{1.56}O_{0.3}N_{0.26}S_{0.006}$ | 23.9 | 20–60 |
| Nucleic acids | $C_1H_{1.23}O_{0.74}N_{0.40}P_{0.11}$ | 14.8 | 3–5 |
| Polysaccharides | $C_1H_{1.67}O_{0.83}$ | 17.3 | 10–50 |

Biomass was a critical measurement in the microalgal harvesting process for applications. A number of methods had been developed to estimate and quantify, which were useful in different cases. Different methods were available such as dry weight: Total suspended solids (TSS), volatile suspended solids (VSS) and fixed suspended solids (FSS); wet weight method; chlorophyll (Chl) method: Chl-a, Chl-b and Chl-a+b), epifluorescence microscopy, bioluminescence, photometric, turbidity, packed cell volume and cell count etc [5]. In general, algal biomass measurement indexes were classify into two groups, (1) direct index such as TSS and VSS and (2) indirect index such as chlorophyll, so-called proxy index. They all were the popular indexes [2,5]. In this study, we used gravimetric analysis (by TSS) measurement was called as direct biomass method. Algae biomass was obtained by filtering a sample through 0.45 μm Whatman filter paper followed by drying in oven for 1h at 103°C [15]. The dry weight method is the most widely applied biomass estimation. The dry weight measurement usually gave a much more consistent result than the wet weight and was usually used as a standard method [2,5,6,15]. It was an important parameter for estimating biomass concentration, productivity and percentages of cell components. In this study, for aeration with atmospheric air there is a significant yield in both the growth rate and final biomass concentration (3.04–3.6 g $L^{-1}$); the average biomass production was 3.6 g $L^{-1}$ with the lowest agitation and aeration conditions used whole study period.

## 3.2. Microalgal Culture Conditions and System

The batch system of algal cultures of 1000 ml, in duplicate reactor, was grown for 2 months, in continuously-stirred tank reactors (CSTR) under room temperature. All the reactors (1000 ml Erlen Meyer flasks) were continuously illuminated with the fluorescent lamps day and night for the photoautotrophic growth.

The light intensity and temperature were monitored through the study and the average was 48.31 [$\mu mol^{-1}m^{-2}$ per $\mu A$] and 27.5 °C. In reactor, dominant algae species were including *Anabaena* sp., *Chlorella* sp., *Oscillatoria* sp., *Oedogonium* sp. and *Scenedesmus* sp. Dominant algal species microscopic structure shown in Figure 3.

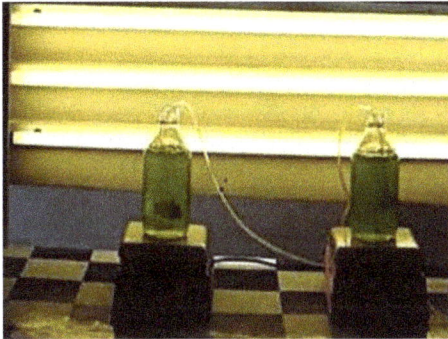

*Figure 2. Photobioreactor*

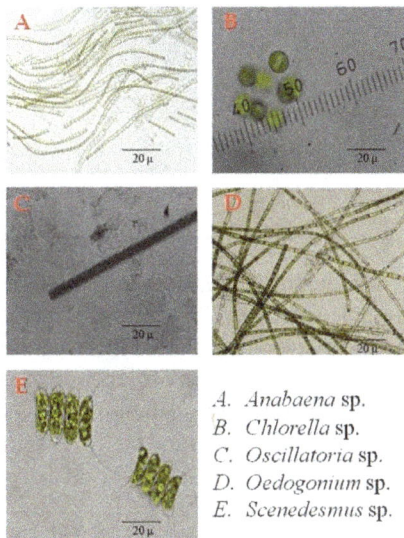

A. *Anabaena* sp.
B. *Chlorella* sp.
C. *Oscillatoria* sp.
D. *Oedogonium* sp.
E. *Scenedesmus* sp.

*Figure 3. Light microscopic pictures of dominant microalgae spices (identified before the fermentation stage).*

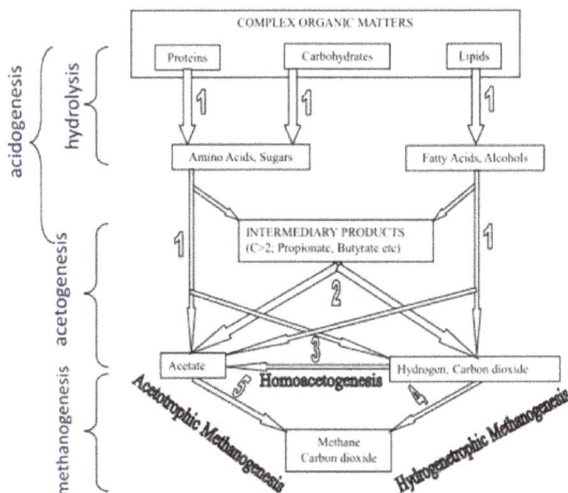

*Figure 4. Organics conversion of microalgae biomass and anaerobic system process.*

The many studies in the literature concerned the maximum $CO_2$ uptake rate by the artificial photobioreactors used with artificial medium [16–21] and natural water medium [1–6] such as continuously stirred tank reactor (CSTR) which was the standard reactor [2,22]; and this utilized air $CO_2$ to grow microalgae with Conway medium. In the literature, many studies used Conway Medium to cultivate marine microalgae (23, 24). Higher cell density was achieved by genus *Dunaliella*, *Chlorella* and *Isochrysis* in Conway Medium. There is very few studies applied Conway Medium in freshwater microalgae cultivation [25,26]. However, this study results showed better results (3.04–3.6 g $L^{-1}$ of algae biomass) using Conway medium with mixed culture of freshwater microalgae cultivation.

### 3.3. Biochemical Methane Potential (BMP) of Microalgae

Biomass can be considered as solar energy collected and stored by plants termed as "energy crops", such as algae. Algae are an efficient tool to trap solar energy into biomass for later conversion into biogas. Biomethanation is an anaerobic microbiological process by which biomass can be microbiologitally converted into methane. Hence through methane fermentation the chemical energy fixed from solar energy by algae may be converted into the readily available chemical energy in methane gas [27].

The biochemical methane potential (BMP) assay constitutes a useful tool to determine both the ultimate biodegradability and the methane conversion yield of organic substrates [28]. The BMP evaluates the ultimate amount of methane produced by any given waste or biomass under anaerobic conditions. The information provided by the BMP value is important when evaluating potential substrates and for optimizing the design and functioning of an anaerobic digester. Apparently, the BMP of microalgae depends mainly on its composition, which itself depends on the growth conditions and and is specific species [29].

Algal biomass contains considerable amount of biodegradable components such as carbohydrates, lipids and proteins. This makes it a favorable substrate for anaerobic microbial flora and can be converted into methane rich biogas [9]. In spite of the fact that microalgae have high potential for biogas production, there are some studies on anaerobic digestion of microalgal biomass utilizing *Chlamydomonas reinhardtii*, *Scenedesmus obliquus*, *Chlorella vulgaris*, *Dunaliella tertiolecta*, *S. obliquus* and *Phaeodactylum triconutum* biomass. The concentration of substrate in the BMP assay also impacts on the final biodegradability and methane productivity [9].

When the C, H, O and N composition of a wastewater or substrate is known, the stoichiometric relationship reported by Buswell and Boruff [30] and Angelidaki and Sanders [31], and can be used to estimate the theoretical gas composition on a percentage molar basis. However, it must be kept in mind that this theoretical approach does not take into account needs for cell maintenance and anabolism.

In this equation, the organic matter is stoichiometrically converted to methane, carbon dioxide and ammonia. The specific methane yield expressed in liters of $CH_4$ per gram of volatile solids (VS) can thus be calculated as:

$$C_aH_bO_cN_d + \left(\frac{4a-b-2c+3d}{4}\right)H_2O \rightarrow \left(\frac{4a+b-2c-3d}{4}\right)CH_4 + \left(\frac{4a-b+2c+3d}{4}\right)CO_2 + dNH_3$$

Equation (1)

$$B_o \frac{4a+b-2c-3d}{12a+b+16c+14d} * V_m$$

Equation (2)

where $V_m$ is the normal molar volume of methane.

The ratio $r_G$ of methane to carbon dioxide can therefore be computed from

$$n = \frac{-b+2c+3d}{a},$$

the average carbon oxidation state in the substrate [32] as follows:

$$r_G = \frac{4}{4}$$

Equation (3)

The biogas composition however also depends on the amount of $CO_2$ which is dissolved in the liquid phase through the carbonate system, and is therefore strongly related to pH.

The ammonium production yield in the digester can be evaluated using Eq. (1):

$$Y_{N-NH_3}\ (mg\ g\ VS^{-1}) = \frac{d*17*1000}{12a+b+16c+14d}$$

Equation (4)

Eq. (1) is a theoretical approach that allows estimation of the maximum potential yields. Using Eq. (1), it is possible to compute a theoretical specific methane yield associated to a theoretical ammonia release (Table 5).

**Table 5.** *Theoretical methane potential and theoretical ammonia release during the anaerobic digestion of the total biomass [adopted from 9,33]*

| Species | Proteins (%) | Lipids (%) | Carbohydrates (%) | $CH_4$ (L $CH_4$ g $VS^{-1}$) | $N-NH_3$ (mg g $VS^{-1}$) |
|---|---|---|---|---|---|
| *Euglena gracilis* | 39–61 | 14–20 | 14–18 | 0.53–0.8 | 54.3–84.9 |
| *Chlamydomonas reinhardtii* | 48 | 21 | 17 | 0.69 | 44.7 |
| *Chlorella pyrenoidosa* | 57 | 2 | 26 | 0.8 | 53.1 |
| *Chlorella vulgaris* | 51–58 | 14–22 | 12–17 | 0.63–0.79 | 47.5–54.0 |
| *Dunaliella salina* | 57 | 6 | 32 | 0.68 | 53.1 |
| *Spirulina maxima* | 60–71 | 6–7 | 13–16 | 0.63–0.74 | 55.9–66.1 |
| *Spirulina platensis* | 46–63 | 4–9 | 8–14 | 0.47–0.69 | 42.8–58.7 |
| *Scenedesmus obliquus* | 50–56 | 12–14 | 10–17 | 0.59–0.69 | 46.6–42.2 |

Gross composition (Table 2) of several microalgae species adopted from Becker, [33]. As expected, the species that can reach higher lipid content (e.g. *C. vulgaris*) have a higher methane yield.

COD is commonly used in the water and wastewater industry to measure the organic strength of liquid effluents. It is a chemical procedure using strong acid oxidation. Organics conversion of microalgae biomass and anaerobic system process shown in Figure The strength is expressed in 'oxygen equivalents' i.e. the mg $O_2$ required to oxidise the C to $CO_2$. However, the COD concept could be estimate the methane yield [16, 17]. One mole of methane requires 2 moles of oxygen to oxidise it to $CO_2$ and water, so each gram of methane produced corresponds to the removal of 4 grams of COD.

$$CH_4 + 2O_2 \rightarrow CO_2 + H_2O$$

$$16 \qquad 64$$

or

1kg COD is equivalent to 250g of methane.

1kg COD $\Rightarrow$ 250g of $CH_4$

250g of $CH_4$ is equivalent to 250/16 moles of gas = 15.62 moles

1 mole of gas at NTP = 22.4 liters

Therefore 15.62 x 22.4 = 349.8 liters = 0.35 m$^3$.

In our study, the content of total solids (TS) and volatile solids (VS) in the algae biomass was measured; the results were average as 12500 g/m$^3$ and 6320 g/m$^3$, respectively. The average pH was 8.2 and average COD 2190 (ml/L). Methane

formation takes place within a relatively narrow pH interval, from about 6.5 to 8.5 with an optimum interval between 7.0 and 8.0. The process is severely inhibited if the pH decreases below 6.0 or rises above 8.5. The pH value increases by ammonia accumulation during degradation of proteins, while the accumulation of VFA decreases the pH value. The accumulation of VFA will often not always result in a pH drop, due to the buffer capacity of the substrate [34]. According to the COD estimation, our study shows the mixed culture microalgal biomass is a potentially valuable fermentation substrate, and produce 60.3% of methane gas.

## 4. Conclusions

Production of biofuels is undoubtedly one of the best solutions for declining the crude oil reserves and global warming due to excessive greenhouse gasses emissions. As fossil fuel prices increase and environmental concerns gain prominence, the development of alternative fuels from biomass has become more important. Biogas is considered a renewable energy carrier. As demonstrated here, microalgal biogas is technically feasible. Microalgae have several advantages over terrestrial plants such as higher photosynthetic efficiencies, lower need for cultivation area, higher growth rates, more continuous biomass production, no direct competition with food production, and possibility to use artificial medium, natural water medium (freshwater/marine water) and wastewater for biomass production. The algae biomass thus produced will constitute an additional source of organic substrate in the installation for biogas production.

# References

[1] R. Ramaraj, D. D-W. Tsai, P. H. Chen, "An exploration of the relationships between microalgae biomass growth and related environmental variables", Journal of Photochemistry and Photobiology B: Biology, 2014, 135: 44–47.

[2] R. Ramaraj, Freshwater microalgae growth and Carbon dioxide Sequestration, Taichung, Taiwan, National Chung Hsing University, PhD thesis, 2013.

[3] R. Ramaraj, D. D-W. Tsai, P. H. Chen, "Freshwater microalgae niche of air carbon dioxide mitigation", Ecological Engineering, 2014; 68: 47–52.

[4] R. Ramaraj, D. D-W. Tsai, P. H. Chen, "Algae Growth in Natural Water Resources", Journal of Soil and Water Conservation, 2010, 42: 439–450.

[5] R. Ramaraj, D. D-W. Tsai, P. H. Chen, "Chlorophyll is not accurate measurement for algal biomass", Chiang Mai Journal of Science, 2013, 40: 547–555.

[6] D. D-W. Tsai, Watershed Reactor Analysis, $CO_2$ Eco-function and Threshold Management Study, Taichung, Taiwan, National Chung Hsing University, PhD thesis, 2012.

[7] K. C. Park, C. Whitney, J. C. McNichol, K. E. Dickinson, S. MacQuarrie, B. P. Skrupski, J. Zou, K. E. Wilson, S. J. B. O'Leary, P. J. McGinn, "Mixotrophic and photoautotrophic cultivation of 14 microalgae isolates from Saskatchewan, Canada: potential applications for wastewater remediation for biofuel production", Journal of Applied Phycology, 2012, 24: 339–348.

[8] J.-C. Frigon, F. Matteau-Lebrun, R. Hamani Abdou, P. J. McGinn, S. J. B. O'Leary, S. R. Guiot, "Screening microalgae strains for their productivity in methane following anaerobic digestion", Applied Energy, 2013, 108: 100–107.

[9] B. Sialve, N. Bernet, O. Bernard, "Anaerobic digestion of microalgae as a necessary step to make microalgal biodiesel sustainable", Biotechnology Advance, 2009, 27: 409–416.

[10] R. Harun, M. Davidson, M. Doyle, R. Gopiraj, M. Danquah, G. Forde, "Technoeconomic analysis of an integrated microalgae photobioreactor, biodiesel and biogas production facility", Biomass Bioenergy, 2011, 35: 741–747.

[11] P. J. McGinn, K. E. Dickinson, K. C. Park, C. G. Whitney, S. P. MacQuarrie, F. J. Black, F. Jean-Claude, S. R. Guiot, S. J. B. O'Leary, "Assessment of the bioenergy and bioremediation potentials of the microalga Scenedesmus sp. AMDD cultivated in municipal wastewater effluent in batch and continuous mode", Algal Research, 2012, 1: 155–65.

[12] J. H. Mussgnug, V. Klassen, A. Schlüter, O. Kruse, "Microalgae as substrates for fermentative biogas production in a combined biorefinery concept", Journal of Biotechnology, 2010, 150: 51–56.

[13] H. Ershad-Langroudi, M. Kamali, B. Falahatkar, "The independent effects of ferrous and phosphorus on growth and development of Tetraselmis suecica; an in vitro study", Caspian Journal of Environmental Sciences, 2010, 8: 109–114.

[14] S. Kant, P. Gupta, "Algal Flora of Ladakh". Scientific Publishers, Jodhpur, India, 1998.

[15] APHA, AWWA, WPCF, "Standards Methods for the Examination of Water and Wastewater", 21st ed. APHA-AWWA-WPCF, Washington, DC, 2005.

[16] M. von Sperling, S. C. Oliveira, "Comparative performance evaluation of full-scale anaerobic and aerobic wastewater treatment processes in Brazil", Water Science Technology, 2009, 59: 15–22.

[17] S. G. Pavlostathis, E. Giraldogomez, "Kinetics of anaerobic treatment: a critical review", Water Science Technology, 1991, 24: 35–59.

[18] D. Bilanovic, A. Andargatchew, T. Kroeger, G. Shelef, "Freshwater and marine microalgae sequestering of $CO_2$ at different C and N concentrations – response surface methodology analysis", Energy Conversion and Management, 2009, 50: 262–267.

[19] A. Kumar, S. Ergas, X. Yuan, A. Sahu, Q. Zhang, J. Dewulf, F. X Malcata, H. van Langenhove, "Enhanced $CO_2$ fixation and biofuel production via microalgae: recent developments and future directions", Trends in Biotechnology, 2010, 28: 371–380.

[20] A. Toledo-Cervantes, M. Morales, E. Novelo, S. Revah, "Carbon dioxide fixation and lipid storage by Scenedesmus obtusiusculus", Bioresource Technology, 2013, 130: 652–658.

[21] D. J. Farrelly, L. Brennan, C. D. Everard, K. P. McDonnell, "Carbon dioxide utilisation of Dunaliella tertiolecta for carbon bio-mitigation in a semicontinuous photobioreactor", Applied Microbiology Biotechnology, 2014, 98: 3157–3164.

[22] D. D-W. Tsai, R. Ramaraj, P. H. Chen, "Growth condition study of algae function in ecosystem for $CO_2$ bio-fixation", Journal of Photochemistry and Photobiology B: Biology, 2012, 107: 24–34.

[23] F. Lananan, A. Jusoh, N. Ali, S. S. Lam, A. Endut, "Effect of Conway Medium and f/2 Medium on the growth of six genera of South China Sea marine microalgae", Bioresource Technology, 2013, 141:75-82.

[24] C. Tantanasarit, A. J. Englande, S. Babel, "Nitrogen, phosphorus and silicon uptake kinetics by marine diatom Chaetoceros calcitrans under high nutrient concentrations", Journal of Experimental Marine Biology and Ecology, 2013, 446: 67–75.

[25] R. W. Vocke, K. L. Sears, J. J. O'Toole, R. B. Wildman, "Growth responses of selected freshwater algae to trace elements and scrubber ash slurry generated by coal-fired power plants", Water Research, 1980, 14: 141–150.

[26] I. Orhan, P. Wisespongpand, T. Atici, B. Şener, "Toxicity propensities of some marine and fresh-water algae as their chemical defense", Journal of Faculty of Pharmacy of Ankara, 2003, 32: 19–29.

[27] P. H. Chen, W. J. Oswald, "Thermochemical treatment for algal fermentation", Environment International, 1998, 24: 889–897.

[28] I. Angelidaki, M. Alves, D. Bolzonella, L. Borzacconi, J. Campors, A. Guwy, S. Kalyuzhnyi, P. Jenicek, J. Van Lier, "Defining the biomethane potential (BMP) of solid organic wastes and energy crops: a proposed protocol for batch assays", Water Science Technology, 2009, 59: 927–934.

[29] M. E. Alzate, R. Muñoz, F. Rogalla, F. Fdz-Polanco, S. I. Pérez-Elvira, "Biochemical methane potential of microalgae biomass after lipid extraction", Chemical Engineering Journal, 2014, 243: 405–410.

[30] A. M. Buswell, C. S. Boruff, "The relationship between chemical composition of organic matter and the quality and quantity of gas produced during sludge digestion", Sewage Works Journal, 1932, 4: 454–460.

[31] I. Angelidaki, W. Sanders, "Assessment of the anaerobic biodegradability of macropollutants", Reviews in Environmental Science and Biotechnology, 2004, 3: 117–129.

[32] R. F. Harris, S. S. Adams, "Determination of the carbon-bound electron composition of microbial cells and metabolites by dichromate oxidation", Applied and Environmental Microbiology, 1979, 37: 237–243.

[33] E. W. Becker, "Microalgae in human and animal nutrition", A. Richmond (Ed.), Handbook of microalgal culture, Blackwell Publishing, Oxford, 2004.

[34] P. Weiland, "Biogas production: current state and perspectives", Applied Microbiology and Biotechnology, 2010, 85: 849–860.

# 25

# An educational approach to the recycling and disposal of waste batteries

**Dilek Çelikler, Filiz Kara**

Department of Elementary Science Education, Faculty of Education, Ondokuz Mayıs University, Samsun, Turkey

**Email address:**
dilekc@omu.edu.tr (D. Çelikler), filiz.kara@omu.edu.tr (F. Kara)

**Abstract:** The aim of this study was to determine the opinions of students studying Science Teaching in Turkey regarding the recycling and disposal of waste batteries. A total of 80 volunteer, third-year students from the department of science education participated in the study. Based on the study results, it was determined that the large majority of the students had insufficient knowledge regarding the recycling and disposal of waste batteries. To change the students' views that waste batteries can be converted into batteries once again by being processing in factories, or that waste batteries can be disposed of through melting, it is necessary to increase the students' knowledge and awareness on this subject.

**Keywords:** Waste Batteries, Recycling, Disposal, Science Teaching Student

## 1. Introduction

As an indispensable part of daily life in the modern world, technology not only facilitates the lives of humans, but also alters the conditions and environment in which they live. In parallel to the developments in technology, a large variety portable devices are nowadays used in daily life, either personally or at home and workplaces. Many of the most commonly used and encountered devices in daily life – such as laptop computers, telephones, watches, cameras and TV remote controls – employ portable cells and batteries as energy source. To prevent any harmful effects on living creatures and the environment, used batteries must be collected in specific containers rather than being thrown into ordinary trash cans.

As a concept, recycling also encompasses the "reuse and utilization" of wastes. The recycling and reuse of many materials we use in our daily lives is performed differently based on considerations such as the environmental problems they cause, the related economic factors, and the cautious use of existing natural resources. Some of these everyday materials consist a large variety of different substances, and may require complex processes for proper recycling. Batteries contain both recyclable metals such as nickel, and toxic metals such as cadmium; consequently, the recycling processes of batteries need to consider both of these types of metals (1).

Due to their toxicity, their prevalence, and their physical resistance to deterioration, waste batteries represent a significant threat for the environment and human health (2, 3). The toxicity of batteries is mainly due to their lead, mercury, and cadmium content. In addition to this, the other metals in batteries such as zinc, copper, manganese, lithium, and nickel can also pose a threat to the environment. Alkali and zinc-carbon batteries contain heavy metals such as mercury, zinc, and manganese; for this reason, it is necessary to recycle these types of batteries (4).

In case they are stored inappropriately, the toxic compounds in waste batteries may gradually seep into water sources and the ground (4). When waste batteries are thrown into bodies of water or buried into the ground, the external casing of the battery will eventually erode or become pierced, causing the heavy metals and chemicals the battery contains to mix with the surrounding water or ground. For this reason, it is necessary to collect waste batteries separately and to dispose of them in waste battery collection containers. Disposing of waste batteries in such a way, and then recycling them as necessary, not only reduces the risk of having the various chemicals inside batteries mix with the underground waters and ground in landfill areas, but also allows the efficient use of natural resources through the recycling of reusable materials within batteries (5, 6).

Waste batteries can be distinguished from one another depending on whether there are or are not rechargeable. For rechargeable batteries (NiCd, NiMh and Li-ion), there are currently no facilities in Turkey that can recycle the valuable metals (such as nickel and cobalt) in these batteries; for this reason, such batteries are sent abroad for recycling under the supervision of the TAP Association. As the recycling of non-rechargeable batteries is not economical, they are disposed in a controlled manner by the TAP Association in solid waste storage areas constructed underground or on the surface, with no other type of waste being kept in these areas (5).

For a sustainable future, it is important for science teaching students – who will be become teachers in the future – to be aware and knowledgeable regarding the collection, recycling and disposal of waste batteries. In this context, this study evaluated and reflected in detail the views of science teaching students regarding the recycling and disposal of waste batteries. We believe that the findings of this study will contribute and provide further depth to the literature on this subject.

# 2. Methods

The study was conducted using the general screening model. The general screening model is a screening approach conducted on populations consisting of a large number of individuals in order to reach a general conclusion regarding the population. It is performed by screening the population as a whole, or a certain group or sample within the population (7).

The study was performed with the participation of 80 volunteer, third-year students receiving education at the Science Education Department of an Education Faculty in Turkish public university. To determine the students' views regarding the recycling and disposal of waste batteries the students were asked to answer two open-ended questions in writing. Examples of the answers provided by the students are shown below by keeping the students' name confidential and coding them as "$F_1, F_2...F_n$".

# 3. Results

The study results concerning the answers given by the students on the recycling and disposal of waste batteries are provided in two sections.

### Section 1: The Recycling of Waste Batteries

The percentage distribution of the answers given by science teaching students to the question "How are waste batteries recycled?" is provided in Table 1.

*Table 1. The students' views regarding the recycling of waste batteries*

| Student's answers | Answer Frequency f |
|---|---|
| Waste batteries can be reprocessed in factories to be converted into batteries once again | 54 |
| Batteries can be recycled | 11 |
| Waste batteries are separated into their components in factories to obtain metals | 8 |
| The metals inside waste batteries is separated and obtained by melting them with heat | 5 |
| A battery is recycled when it is recharged. | 2 |

An evaluation of Table 1 indicates that the students were generally aware of the recyclability of waste batteries. However, many students described that waste batteries can be reprocessed in factories to be converted into batteries once again, which indicated either a lack of knowledge concerning this subject, or the fact that they confused the recycling of batteries with the recycling of packaging wastes such as plastic, glass, metal and paper. It was noted that a relatively low number of students mentioned that the metals within waste batteries could be separated and recycled. In addition, some of the students confused the concepts of recycling and recharging, and erroneously described that rechargeable batteries could be recycled by recharging them. Direct citations of the answers given by the students to this question are provided below:

*"When batteries are used and spent, we throw them into battery collection containers for recycling. Waste batteries are thus collected in these containers, and then sent to factories, where they are reprocessed and converted into batteries once again" ($F_{17}$)*

*"I am certain that batteries are being recycled. However, where I live, I have not seen any recycling container for batteries" ($F_{10}$)*

*"I know that batteries can be recycled, since I have seen a number of battery recycling containers until now. However, I have no idea how waste batteries are recycled" ($F_{57}$)*

*"Many of the batteries that we use today can be recharged. Recycling batteries probably involves recharging them" ($F_{69}$)*

*"Batteries can be recycled. As batteries are composed of metals, these metals can be separated by heating them under suitable conditions" ($F_{71}$)*

*"The recycling of waste batteries is more common in developing countries. This practice is not so common in Turkey" ($F_{48}$)*

### Section 2: Disposal of Waste Batteries

The percentage distribution of the answers given by science teaching students to the question "How are waste batteries disposed?" is provided in Table 2.

*Table 2. The students' views regarding the disposal of waste batteries*

| Student's answers | Answer Frequency f |
|---|---|
| I do not know how waste batteries are disposed | 44 |
| Waste batteries are disposed of through melting | 32 |
| Waste batteries are disposed of by being buried into the ground | 4 |

An evaluation of Table 2 indicates that the large majority of the students did not know how waste batteries were disposed; moreover, some of students erroneously described that waste batteries were disposed of through melting, which illustrated that they lacked knowledge regarding this subject. In addition, a limited number of students described that waste batteries were disposed of by being buried into the ground. This indicated these students' failure to consider and appreciate the fact that burying waste batteries would cause the heavy metals they contain to seep into ground and harm the environment. Direct citations of the answers given by the students to this question are provided below:

*"Because waste batteries do not break down in nature, I think that they are buried into the ground"* (F$_{31}$)

*"Waste batteries are melted into a liquid at high temperatures. But I don't know how they are disposed after they are melted"* (F$_{50}$)

*"I don't know how waste batteries are processed after they are collected"* (F$_{75}$)

*"I think that batteries are disposed and eliminated by melting them at high temperature"* (F$_{61}$)

## 4. Conclusions and Recommendations

Based on the study results, it was determined that the large majority of the students had insufficient knowledge regarding the recycling and disposal of waste batteries. To change the students' views that waste batteries can be converted into batteries once again by being processing in factories, or that waste batteries can be disposed of through melting, it is necessary to increase the students' knowledge and awareness on this subject. The first step of a healthy recycling system involves the sorting of waste materials at their source prior to collection. Waste batteries are not only hazardous wastes, but they are also relatively valuable items due to the reusable metals they contain. For this reason, waste batteries need to be collected in waste battery collection containers (without being thrown into ordinary trash cans or to the environment), and then sorted according to battery types; after these processes, waste batteries will undergo recycling to extract and reuse the valuable metals they contain. To increase the knowledge and awareness of students regarding the proper recycling and disposal of waste batteries, emphasis must be placed on environment-related education starting from elementary school classes, preferably by including environment classes into the curriculum. In addition to this; TAP, the authorized organization for the collection, sorting and storage of waste batteries in Turkey, places emphasis on educational activities with the aim of increasing the collection of waste batteries. In many provinces of Turkey, TAP performs seminars and presentations at elementary and secondary schools, while organizing conferences for various private and public organizations. In addition, the TAP also cooperates with the academic staff of universities, and conducts educational activities for universities students.

The first step for ensuring a sustainable environment is raising individuals who are conscious, sensitive, and aware of environmental issues. Raising such individuals can only be achieved through the efforts of our teachers, to whom we entrust the education of future generations. Considering that a sustainable future can be achieved largely through the efforts of future teachers, it is necessary for science teaching students – who will become teachers in the future – to be both aware and knowledgeable regarding the recycling and disposal of waste batteries. For this reason, to allow science teaching students to grasp the importance of the subject of waste batteries; student-centered methods and techniques, as well as effective teaching materials, should be used in the environment-related classes which these students attend during their education. In addition, these students (as well as the general public) should be informed regarding waste batteries through the visual and written media, and measures should be taken to ensure the active participation of students into the recycling and disposal of waste batteries before these wastes have any detrimental effects on the environment.

## References

[1] C.A. Nogueira and F. Margarido, Chemical and physical characterization of electrode materials of spent sealed ni–cd batteries, Waste Manag., 2007, 27, 1570–1579.

[2] A.M. Bernandes, D.C.R. Espinosa and J.A.S. Tenorio, Recycling of batteries: A review of current processes and Technologies, Journal of Power Sources, 2004, 130, 291-298.

[3] S. Kierkegaard, EU Battery Directive, Charging up the batteries: Squeezing more capacity and power into the new EU Battery Directive, Computer Law & Security Report, 2007, 23, 357-364.

[4] M. Bartolozzi, The recovery of metals from spent alkaline–manganese batteries: A review of patent literature, Resources, Conservation and Recycling, 1990, 4, 233–240.

[5] Taşinabilir Pil Üreticileri ve İthalatçilari Derneği (TAP), Atık pillerin toplanması ve bertarafı, Genel Eğitim Sunumu, 2014.

[6] URL-1. Taşınabilir Pil Üreticileri ve İthalatçıları Derneği (TAP). Taşınabilir pillerin kullanımında dikkat edilmesi gereken hususlar. (http://tap.org.tr/tasinabilir_pillerin_kullaniminda_dikkat_edilmesi_gereken_hususlar-185.html)

[7] N. Karasar, Bilimsel araştırma yöntemleri. (22th ed). Ankara: Nobel, 2011.

# 26

# Mental models which influence the attitudes of science students towards recycling

**Gonca Harman, Zeynep Aksan, Dilek Çelikler**

Department of Elementary Science Education, Faculty of Education, Ondokuz Mayıs University, Samsun, Turkey

**Email address:**

gonca.harman@omu.edu.tr (G. Harman), zeynep.axan@gmail.com (Z. Aksan), dilekc@omu.edu.tr (D. Çelikler)

**Abstract:** The aim of this study was to evaluate the mental models which influence the attitudes towards recycling of students attending the Department of Science Teaching. 31 first-year university students who are enrolling in the department of science education participated in the study. Based on the study results, it was determined that only a limited number of mental models influenced the students' attitudes towards recycling. It was also determined that the students' mental models focused on 8 categories, which were the "benefits of recycling," the "sorting of wastes at their source," the "promotion of recycling," the "use of different recycling containers for different types of wastes," "recyclable wastes," "recycling facilities," the "recycling process" and the "logo of recycling".

**Keywords:** Recycling, Mental Models, Drawing, Science Students

## 1. Introduction

The products we use in our daily lives for a variety of purposes are generally marketed and sold inside packages composed of paper, cardboard, glass, metal, plastic and composites. For this reason, the use of daily products is leading to a constant increase in the amount of waste generated. However, since materials used in package production are generally reusable, it is possible to convert package-related wastes once again into a usable form through collection, sorting and recycling [1].

Recycling is the process by which reusable waste materials such as glass, metals, paper, cardboard and composites are converted into raw materials or byproducts by undergoing various physical, chemical and biological treatments and processes, thus allowing them to be reused in production processes [2; 3]. Although they are extensively used and consumed, materials such as metals, glass, plastics and paper are among the easiest materials to recycle [4]. For instance, glass can be recycled many times, over and over again. As result, glass can be used repeatedly in the production of glass-based products. Similarly, new metal products can be obtained through the recycling of waste metals. Newspapers, packaging paper and cardboards can be recycled to produce second quality paper and packaging materials. Through recycling, objects and materials such as plastic bags, plastic

pipes and multiple purpose plastics can be recycled to produce plastic buckets, bowls, blinds, bottles, watertight objects/materials, automobile parts, construction materials, fibers, textile products composed of synthetic fibers, toys and office materials [5].

The recycling of waste materials is not only important in terms of contributing to the environment, public health and the economy; but it also represents an effective type of environmental protection activity for preventing pollution and the depletion of natural resources [6]. Recycling has a crucial role in in the protection of the environment and natural resources; in reducing the demand for raw materials; in preventing unnecessary energy consumption; and in reducing the amount of waste produced [7; 8; 4]. Recycling also represents a resource protection mechanism that is very beneficial for both the economy and the environment [9], as well as a prerequisite for a sustainable future. Recycling entails important tasks and responsibilities at the level of individuals. It is essential for individuals, and especially for the teachers who will raise these individuals, to embrace the principles relating to the reuse and recycling of waste materials; to be knowledgeable about these principles; and to be able to convey these principles to other individuals. The reduction of the generation of waste, the sorting of wastes at their source, and the systematic recycling of wastes can only be achieved through the efforts of environmentally conscious, sensitive and aware individuals.

Schools play an important role in instilling behavior patterns that focus on the protection of the environment, such as the consistent use of recycled products. The importance of recycling, as well as environmental awareness, should be taught and conveyed in schools through classes on the environment [2]. In addition to this, it is both necessary and important to identify the mental models which shape individuals' behavior regarding recycling, and which effectively illustrate the internal and cognitive representations in the minds of individuals on this subject [10; 11]. As individuals are not able to comprehend and grasp nature directly, they tend to form mental models through internal representations [12]. Individuals expressed their mental models through speech and writing [13]. Using their preliminary knowledge and the scientific information they acquired during their education [14], individuals structure their mental models based on their own expressions and behaviors [15]. What is interesting about mental models is their ability to influence the activities of individuals [16]. Thus, the mental models students have regarding the concept of recycling will certainly influence their behaviors towards recycling. In addition to this influence, it was also noted that there are no studies in both the national and international literature which attempted to identify the mental models of science students regarding recycling. These considerations were instrumental in determining the subject of the current study.

In this study, the mental models of the students were illustrated with the aid of drawings. The drawing method is based on an open-ended questioning approach, which, when used in combination with other methods or processes, is effective in ensuring that the thoughts, learning and the level of understanding of individuals which might otherwise remain concealed can be revealed and identified without the limitation of words. Thus, this method which can be applied to any age group is useful in allowing students to reflect their learning and thoughts [17].

The aim of this study was to evaluate the mental models which influence the attitudes towards recycling of students attending the Department of Science Teaching in Turkey.

## 2. Methods

The study was conducted using the general screening model. The general screening model is a screening approach conducted on populations consisting of a large number of individuals in order to reach a general conclusion regarding the population [18]. It is performed by screening the population as a whole, or a certain group or sample within the population. A sample selection was performed based on the suitability sample, which is defined as the group of individuals who could be reached/contacted for the study [19]. Students' mental models about recycling were identified with drawing in the study.

The study participants included 31 first-year university students who are enrolling in the Faculty of Education,

Department of Science Teaching of a public university in Turkey.

The assessment tool used for identifying the mental models of the students regarding the concept of recycling consisted of a single open-ended question. The students were asked to answer the question "What does the word 'recycling' bring to your mind?" in both drawing and writing.

Science students' answers to the questions were analyzed by using the qualitative content analysis method. By doing content analysis, similar data collected by each researcher was gathered under the same categories, and then after combining these categories they were finalized. For the better understanding of the resulting data, the data was organized and interpreted. Content analysis of the data were achieved by following the steps: (1) data coding, (2) the creation of categories, (3) the regulation of the codes and categories, (4) the identification and interpretation of the findings [20]. The drawing-related and writing-related data obtained during the study were evaluated and analyzed separately as two different categories. The response categories and the frequencies placed in the categories were calculated. In addition, in order to identify common categories of science students' answers, they were constantly compared with each other. The reply codes belonging to the science students, common categories and frequencies of these categories have been created.

## 3. Results

The study data were structured and shown with tables based on 8 different categories, which were the "benefits of recycling," the "sorting of wastes at their source," the "promotion of recycling," the "use of different recycling containers for different types of wastes," "recyclable wastes," "recycling facilities," and the "recycling process."

*Table 1. Students' mental models about recycling: First category: the benefits of recycling*

| Student's answers | Explain f | Drawing f |
|---|---|---|
| to preserve trees | 11 | 10 |
| to use of recycled products | 10 | 1 |
| to prevent of environmental pollution | 8 | 1 |
| to leave a clean environment for future generations. | 7 | 2 |
| to reduce damage that was given nature | 4 | - |
| to preserve natural resources | 2 | - |
| to contribute to the economy | 2 | - |
| to provide the recycling of wastes | 2 | - |
| to prevent damage to living | 2 | - |
| to preserve the balance of nature | 1 | - |
| to prevent disposal of waste | 1 | - |
| to prevent air pollution | 1 | 1 |
| to reduce waste generation | 1 | - |
| to create new products with less cost | 1 | - |
| to live healthy | 1 | - |

Table 1 shows the mental models of the science students regarding the benefits of recycling category. Regarding the benefits of recycling; the students provided drawings and written descriptions which mainly focused on the preservation of trees, the use of recycled products, the prevention of environmental pollution, and the necessity to leave a clean environment for future generations.

*Table 2. Students' mental models about recycling: Second category: the sorting of wastes at their source*

| Student's answers | | Explain | Drawing |
|---|---|---|---|
| | Types of waste | f | f |
| Separate recycle | Paper | 2 | 1 |
| | Plastic | 2 | 1 |
| | Glass | 2 | 1 |
| | Battery | 1 | - |
| Same recycle | Paper, plastic, glass | - | 2 |

Table 2 shows the mental models of the science students regarding the sorting of wastes at their source category. Concerning this category, it was observed that some of the students expressed the view that each type of waste should be disposed in separate recycling containers, while other students expressed the view that a common recycling container should be used for all wastes. In their drawings and written descriptions, the students mentioned wastes such as batteries, paper, plastic and glass, as well as other types of wastes.

*Table 3. Students' mental models about recycling: Third category: the promotion of recycling*

| Student's Answers | Explain | Drawing |
|---|---|---|
| | f | f |
| to ensure that individuals fulfil their recycling-related responsibilities | 1 | - |
| to abide to the applicable rules imposed by the state | 1 | - |
| to ensure that individuals become knowledgeable | 1 | 1 |
| to use public announcements on TV in order to promote recycling | - | 1 |

Table 3 shows the mental models of the science students regarding the promotion of recycling category. In their drawings and written descriptions, the students emphasized that in order to promote recycling, it is necessary to ensure that individuals fulfil their recycling-related responsibilities and abide to the applicable rules imposed by the state, and also to ensure that individuals become knowledgeable about recycling. However, one of the students emphasized in his drawings the strength of the media, and described the necessity of using public announcements on TV in order to promote recycling.

*Table 4: Students' mental models about recycling: Fourth category: the "use of different recycling containers for different types of wastes*

| Student's Answers | Drawing |
|---|---|
| | f |
| Paper | 8 |
| Plastic | 6 |
| Glass | 4 |
| General | 3 |
| Battery | 2 |
| Metal | 1 |

Table 4 shows the mental models of the science students

regarding the use of different recycling containers for different types of wastes category. Some of the students provided drawings which made generalizations regarding the types of wastes to be disposed in recycling containers. In other words, these students overlooked the different types of wastes that exist and which need to be disposed differently. Such an approach was, of course, not compatible with the requirements of recycling processes, since different types of package wastes require different recycling processes. In addition, two of the students erroneously described in their drawings that batteries can be thrown in recycling containers, whereas batteries are, in fact, disposed in special battery disposal containers. On the other hand, four of the students correctly described in their drawings that batteries should be thrown in battery disposal containers.

Table 5 shows the mental models the science students illustrated regarding the recyclable wastes category through their drawings and written descriptions. Both the drawings and descriptions of the students indicated that they mainly focused on paper, cardboard and plastic wastes. In addition, one of the students erroneously described flue gas as a recyclable waste, while flue gases are in fact pollutants for which recycling is not possible.

*Table 5. Students' mental models about recycling: Fifth category: the recyclable wastes*

| Student's Answers | Explain | Drawing |
|---|---|---|
| | f | f |
| Paper-cardboard | 3 | 12 |
| Plastic | 4 | 9 |
| Glass | 2 | 4 |
| Battery | 2 | 9 |
| Oil | 1 | 4 |
| Phone battery | 1 | 2 |
| Plastic bag | - | 2 |
| Flue gases are in fact pollutants | - | 1 |

*Table 6. Students' mental models about recycling: Sixth category: the recycling facilities*

| Student's Answers | Drawing |
|---|---|
| | f |
| Paper | 3 |
| Glass | 1 |
| Plastic | 1 |
| Battery | 1 |
| General | 1 |

Table 6 shows the mental models that the science students illustrated through drawings regarding the recycling facilities category. Some of the student made drawings showing specific recycling facilities for each type of waste, while one of the students erroneously made a drawing showing that all types of wastes can be processed by a single type of facility.

*Table 7. Students' mental models about recycling: Seventh category: the recycling process*

| Student's Answers | Explain f | Drawing f |
|---|---|---|
| Paper | 1 | 4 |
| Plastic | - | 1 |
| Oil | - | 1 |

Table 7 shows the mental models of the science students regarding the recycling process category. The drawings and written descriptions of the students regarding the recycling processes of paper, plastic and oil wastes indicated that their mental models about the types of recyclable wastes were fairly limited.

*Table 8. Students' mental models about recycling: Eighth category: the logo of recycling*

| Student's Answers | Drawing f |
|---|---|
| logo of recycling | 10 |

It was noted that 10 of the students used the möbius strip, the logo of recycling, in their drawings.

# 4. Conclusions and Recommendations

Based on the study results, it was determined that the mental models which influenced the attitudes towards recycling of students receiving education at the department of science teaching were structured according to 8 categories. Evaluation of the drawings and written descriptions provided by the students showed that these categories were the "benefits of recycling," the "sorting of wastes at their source," the "promotion of recycling," the "use of different recycling containers for different types of wastes," "recyclable wastes," "recycling facilities," the "recycling process" and the "logo of recycling".

With regards to the benefits of recycling category, most students mentioned the preservation of trees in both their drawings and written descriptions. This indicated that the students placed importance on the recycling paper and cardboard. The recycling of paper and cardboard is a very effective approach for preserving trees. The students also expressed in their descriptions a willingness/tendency to use recycled products, which indicated that they paid attention to the type of package used in the products they buy. This was an important finding which reflected that the students were conscious consumers. Another noteworthy point which the students expressed in their descriptions was the perceived necessity to prevent environmental pollution and to leave a clean future for future generations. This was significant in that indicated the students had an environmental model in their minds based on sustainability. As such, recycling was considered essential for a sustainable future. In the literature, students tend to rather describe the "formation of new products" and the "utilization of waste materials" as the main advantages of recycling. In parallel to the results of this study, one of the noteworthy and important results observed in the studies from the literature is the finding that students generally have a good grasp of the benefits of recycled products to the environment [2].

Recycling can be performed by sorting of wastes centrally or at their source. With regards to the sorting of wastes

category, the students provided drawings and written descriptions which especially placed emphasis on the sorting of wastes at their source. It was observed that some of the students expressed the view that each type of waste should be disposed in their separate and specific recycling containers, while other students expressed the view that wastes should be disposed in a common recycling container. It was noteworthy that the drawings and written descriptions provided by the students regarding the sorting of wastes at their sources especially made mention of battery, plastic and glass wastes, along with other types of wastes. However, recyclable wastes are not limited to paper, plastic and glass packages; in addition to these, it is also possible to recycle packages composed of cardboard, composite and metal.

All individuals have a centrally important role and responsibility in ensuring the proper recycling of wastes. Fulfilling this role and responsibility requires individuals who are environmentally conscious, sensitive and aware. While instilling such consciousness, sensitivity and awareness to individuals, it also important and necessary to encourage them to act in an environmentally responsible manner. In their drawings and written descriptions regarding the promotion of recycling category, the students described the need to implement measures that would encourage individuals to fulfill their responsibilities, and to abide to the existing rules concerning recycling. In addition, as the media is nowadays an effective tool for rapidly and easily reaching the masses, organizing public announcement on TV regarding recycling (as suggested by one of the students in his drawings) will be a useful approach for encouraging individuals to assume their share of responsibilities. In addition to these, it was noteworthy that only a limited of number of students had mental models about encouraging individuals to perform recycling.

It is possible to recycle glass, paper, cardboard, plastic, composite and metal packages. In their drawings showing different types of recycling containers depending on the types of wastes, most students did not include composite package materials, indicating that they lacked any mental models regarding this type of package. Although used batteries are normally collected in battery disposal containers, two of the students erroneously showed in their drawings that used batteries can be thrown in recycling containers. Four of the students, on the other hand, made drawings emphasizing the necessity to dispose batteries in battery disposal containers. Some of the students provided drawings which made generalizations regarding the types of wastes to be disposed in recycling containers. In other words, these students overlooked the different types of wastes that exist and which need to be disposed differently. Such an approach was, of course, not compatible with the requirements of recycling processes, since different types of package wastes require different recycling processes.

With regards to the recyclable wastes category, both the drawings and the written descriptions of the students mainly illustrated paper, cardboard and plastic wastes. This finding probably stemmed from the fact that, during their daily lives, the students frequently encountered paper and cardboard collectors in their urban environment and/or the media, as well as various campaigns for the collection plastic bottles and caps. On the other hand, one of the students erroneously described flue gas as a recyclable waste, while flue gases are in fact pollutants for which recycling is not possible.

With regards to the recycling facilities category, some of the student made drawings showing specific recycling facilities for each type of waste, while one of the students erroneously made a drawing showing that all types of wastes can be processed by a single type of facility – an inaccurate generalization that runs contrary to the actual requirements of recycling processes. This indicated that the student in question lacked a mental model which considered the fact that each type of waste undergoes different processing, and that different recycling facilities consequently require different machinery and equipment. For this reason, in order to remedy this lack of knowledge, it would probably be beneficial to organize tours at recycling facilities and to have the students informed by specialists.

With regards to the recycling process category; it was observed that many of the students' drawings and written descriptions focused on the recycling of waste paper, plastic and oils, and that the knowledge of most students regarding recyclable wastes was limited to these materials. This observation indicated that these students' mental models on this subject were fairly limited in scope.

With regards to the recycling logo category, it was observed that 10 of the students drew the möbius strip in their drawings. This symbol is commercially used to indicate that a package is produced from recyclable or reusable materials. The students most likely encountered this symbol through the products they normally purchase. The fact that the students were knowledgeable of this symbol was significant, in that it indicated the students' knowledge and awareness of the recyclability of package wastes. Students who pay attention to this symbol are more likely contribute to recycling, by disposing packages with the logo into recycling containers specific for each type of waste.

In a previous study conducted by [21], the level of awareness regarding solid wastes and recycling was identified as being very low among teacher candidates. In the current study, which was performed using similar methods to Karatekin's study, the students' mental models that influence their behaviors towards recycling were also fairly limited. Furthermore, the literature describes that the lack of knowledge regarding recycling processes, the benefits of recycling, and the types of products which can recycled is one of the most important obstacles for adequate recycling practices [22]. However, despite their relatively limited scope, it was nevertheless important that the students had developed various different mental models regarding the benefits of recycling, the sorting of wastes at their source, the promotion of recycling, the use of different recycling containers for different types of wastes, recyclable wastes, recycling facilities, the recycling process, and the recycling logo. These mental models are likely to significantly influence the students' behaviors regarding recycling.

# Appendix

Examples from Students' Drawings

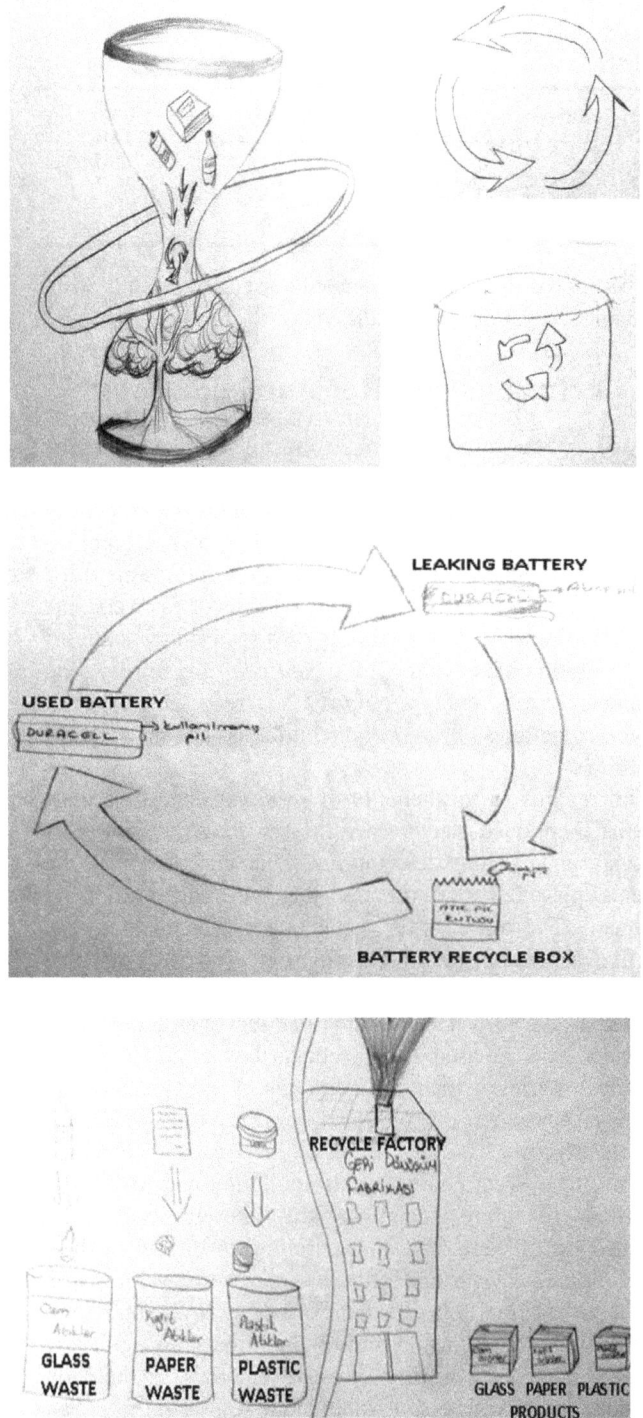

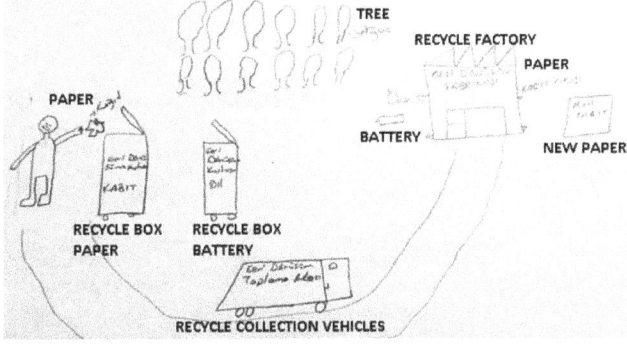

# References

[1]  M. Mert, "Lise öğrencilerinin çevre eğitimi ve katı atıklar konusundaki bilinç düzeylerinin saptanması", Yüksek Lisans Tezi, Hacettepe Üniversitesi, Ankara, 2006.

[2]  O. Çimen, and M. Yılmaz, M, İlköğretim öğrencilerinin geri dönüşümle ilgili bilgileri ve geri dönüşüm davranışları. Uludağ Üniversitesi Eğitim Fakültesi Dergisi, 2012, 25(1), 63-74.

[3]  K. Yücel, "Türkiye'de katı atık yönetimi ve geri kazanım", Bilim Uzmanlığı Tezi, Yıldız Teknik Üniversitesi, Fen Bilimleri Enstitüsü, İstanbul, 1997.

[4]  Ş. Özbay, "Fen ve teknoloji programı içinde kompost hakkında verilen etkinliklerin öğrencilerin akademik başarılarına ve çevre tutumlarına etkisi", Çanakkale Onsekiz Mart Üniversitesi Fen Bilimleri Enstitüsü, Çanakkale, 2010.

[5]  N. Ejder, Atık peyzajının ev konumundaki görünümleri ve yeniden düzenlenmesi. Verimlilik Dergisi, 1995, 149-156.

[6]  H. Spiegelman, and B. Sheehan, The future of waste. BioCycle, 2004, 45(1), 59.

[7]  J. R. Hopper, and J. M. Nielsen, Recycling as altruistic behavior: Normative and behavioral strategies to expand participation in a community recycling program. Environment and Behavior, 1991, 23,195-220.

[8]  S. Oskamp, Resource conservation and recycling: Behavior and policy. Journal of Social Issues, 1995, 51, 157–177.

[9]  P. Valle, E. Reis, J. Menezes, and E. Rebelo, Behavioral determinants of household recycling participation. Environment and Behavior, 2004, 36(4), 505-540.

[10]  A. G. Harrison, and D. F. Treagust, Secondary students' mental models of atoms and molecules: Implications for teaching chemistry. Science Education, 1996, 80(5), 509-534.

[11]  S. Ritchie, K. Tobin, and K. S. Hook, Teaching referents and the warrants used to test the viability of students' mental models: Is there a link?. Journal of Research in Science Teaching, 1997, 34(3), 223–238.

[12]  I. M. Greca, and M. A. Moreira, Mental models, conceptual models and modeling. International Journal of Science Education, 2000, 22(1), 1-11.

[13]  J. K. Gilbert, C. Boulter, and M. Rutherford, Models in explanations, Part 1: Horses for courses?. International Journal of Science Education, 1998, 20(1), 83-97.

[14]  A. G. Harrison, and D. F. Treagust, A typology of school science models. International Journal of Science Education, 2000, 22(9), 1011- 1026.

[15]  M. A. Kurnaz, and A. Değermenci, 7. sınıf öğrencilerinin Güneş, Dünya ve Ay ile ilgili zihinsel modelleri. Elementary Education Online, 2012, 11(1), 137-150.

[16]  S. Vosniadou, Capturing and modelling the process of conceptual change. Learning and Instruction, 1994, 4, 45–69.

[17]  M. Aydoğdu, and T. Kesercioğlu, İlköğretimde Fen ve Teknoloji Öğretimi. Ankara: Anı Yayıncılık, 2005.

[18]  N. Karasar, Bilimsel Araştırma Yöntemleri, (22. baskı.). Ankara: Nobel, 2011.

[19]  J. R. Fraenkel, and N. E. Wallen, How to Design and Evaluate Reseach in Education, (5th ed.). New York: McGraw Hill, 2003.

[20]  A. Yıldırım, and H. Şimşek, Sosyal Bilimlerde Nitel Araştırma Yöntemleri. Ankara: Seçkin Yayıncılık, 2008.

[21]  K. Karatekin, Öğretmen adayları için katı atık ve geri dönüşüme yönelik tutum ölçeğinin geliştirilmesi: Geçerlik ve güvenirlik çalışması. Uluslararası Avrasya Sosyal Bilimler Dergisi, 2013, 4(10), 71-90.

[22]  R. J. Gamba, and S. Oskamp, Factors influencing community residents` participation in commingled curbside recycling programs. Environment and Behavior, 1994, 26, 587-612.

# 27

# Biodiesel from green alga *Scenedesmus acuminatus*

**Yuwalee Unpaprom[1], Sawitree Tipnee[1], Ramaraj Rameshprabu[2], ***

[1]Program in Biotechnology, Faculty of Science, Maejo University, Sansai, Chiang Mai-50290, Thailand
[2]School of Renewable Energy, Maejo University, Sansai, Chiang Mai-50290, Thailand

**Email address:**

rrameshprabu@gmail.com (Ramaraj R.), rameshprabu@mju.ac.th (Ramaraj R.), yuwalee@mju.ac.th (Unpaprom Y.)

**Abstract:** Renewable fuels for alternative energy sources have been paid a great attention in recent years. Biodiesel has been gaining worldwide popularity as an alternative energy source. The production of biofuels from microalgae, especially biodiesel, has gained huge popularity in the recent years, and it is assumed that, due to its eco-friendly and renewable nature, it can replace the need of fossil fuels. *Scenedesmus* genus was discussed by phycologists as promising microalgae for biofuel production based on its biomass and fatty acid productivity. In the present study, *S. acuminatus* was cultivated in piggery wastewater effluent to couple waste treatment with biodiesel production. The batch feeding operation by replacing 10% of algae culture with piggery wastewater effluent every day could provide a stable net biomass productivity of $3.24 \ g \ L^{-1} \ day^{-1}$. The effect of acid hydrolysis of lipids from *S. acuminatus* on FAME (fatty acid methyl esters) production was investigated. Direct transesterification (a one-stage process) of the as harvested *S. acuminatus* biomass resulted in a higher bio-diesel yield content than that in a two-stage process. This study results revealed that it is feasible to produce biodiesel from wet microalgae biomass directly without the steps of drying and lipid extraction.

**Keywords:** Biodiesel, Fresh Water, *Scenedesmus acuminatus*, Piggery Wastewater Effluent

## 1. Introduction

Biofuel is a renewable energy, which may be instead of the fossil fuel resources in the future with decreasing of the fossil fuel on a daily basis. Biodiesel is a renewable fuel alternative for diesel engines [1]. It can be produced in any climate. Biodiesel is biodegradable, nontoxic and a low emission profiles, environmentally friendly biofuel, also contributes no net carbon dioxide or sulfur to the atmosphere and emits less gaseous pollutants than conventional diesel fuel [2–4]. Due to these merits, biodiesel fuel has received considerable awareness in recent years. Biodiesel is a mix of monoalkyl esters of long-chain fatty acids, obtained by chemical reaction (transesterification), coming from renewable feedstock such as vegetable oil or animal fats, and alcohol with a catalyst [5]. It is called the biodiesel fuel, which consists of the simple alkyl esters of fatty acids, is presently making the transition from a research topic and demonstration fuel to a marketed commodity. Various biomasses can be used to produce biodiesel. Traditional feedstock of biodiesel contains plant oils and animal fats [6,7]. Nevertheless, such raw materials may compete with food supply, increase the utilization of limited farmland, and require long time to harvest which is hard to satisfy the large and long-term global energy demand; the most commonly used are rapeseed, canola, corn, soybean, oil palm, coconut and soybean, but also other crops such as mustard, hemp and waste vegetable oil animal fats [8]. Even algae are promising source of biodiesel in nature, primarily to highlight the order-of-magnitude differences present in the oil yields from algae when compared with other oilseeds. For example, the Tallow tree could yield significantly higher quantities of oil than current crops, and microalgae offer the potential for triglyceride production rates some 200 times higher than terrestrial biomass [9]. Hence, microalgae are considered a promising alternative and a renewable feedstock source for biofuels.

In addition, microalgae are believed to be excellent candidates for fuel production because of their high photosynthetic efficiency, high growth rate, and high area-specific yield. Moreover, microalgae can be cultivated in saline/brackish water and on non-arable land; therefore, this precludes competition for the conventional crop land [10]. Consequently, microalgae have received more attention in the recent decades.

Biodiesel, which is produced from biomass by transesterification of triacylglycerols, is one of the most prominent renewable energy sources [5]. Microalgae are emerging as one of the most promising resources of biodiesel, with a projected yield of 58,700 to 136,900 liter ha$^{-1}$ year$^{-1}$ [9]. For microalgae cultivation, the huge consumption of water resources and inorganic nutrients is costly. Addition of organic carbon, though found highly stimulatory for microalgal growth, increases the feedstock cost. Thus, an economically acceptable and environmentally sustainable carbon source for alga-based biodiesel is currently needed. The use of micro-algae for biodiesel in itself is particularly attractive because it is up to 30 times [10], more efficient in producing oil for biodiesel compared to conventional methods, and micro-algae can be cultured in poor quality salty water or nutrient loaded water. Micro-algae are much more efficient converters of solar energy than any known terrestrial plant, because they grow in suspension where they have unlimited access to water and more efficient access to $CO_2$ and dissolved nutrients [11–14].

Continued only increasing use of petroleum will intensify local air pollution and magnify the global warming problems caused by $CO_2$ [14]. One of the most serious environmental problems today is that of global warming, caused primarily by the heavy use of fossil fuels. Photosynthetic microalgae are potential candidates for utilizing excessive amounts of $CO_2$ [13], since when cultivated these organisms are capable of fixing $CO_2$ to produce energy and chemical compounds upon exposure to sunlight [12]. The derivation of energy from algal biomass is an attractive concept in that unlike fossil fuels, algal biomass is rather uniformly distributed over much of the earth's surface [5]. In a particular case, such as the emission of pollutants in the closed environments of underground mines, biodiesel fuel has the potential to reduce the level of pollutants and the level of potential or probable carcinogens [4].

Biodiesel is a renewable fuel that is produced by chemically reacting algal oil with an alcohol such as methanol. The reaction requires a catalyst, usually a strong base, such as sodium or potassium hydroxide, and produces new chemical compounds called methyl esters. It is these esters that have come to be known as biodiesel. There are four primary ways to make biodiesel, direct use and blending, microemulsions, thermal cracking (pyrolysis) and transesterification. The most common way is transesterification as the biodiesel from transesterification can be used directly or as blends with diesel fuel in diesel engine [8–10].

One promising approach is to couple biodiesel production with wastewater treatment, as algae can be successfully cultivated in wastewaters. The algal cultivation process was focused on in this study for reducing the cost. Microalgae can be cultivated in wastewaters because they can utilize the nutrients contained in most wastewaters. By using wastewater as a nutrient source, the cost of nutrients and water in biodiesel production can be reduced, and also the quality of wastewater discarded after treatment can be improved simultaneously. Cultivation of microalgae in swine wastes, dairy manure, and other animal residues has been reported by several literatures

[7,15,16]; by using *Scenedesmus* Sp., which is part of our ecosystem and is very accessible, a more environmental friendly and renewal fuel can be produced. In terms of oil production, of the published algal species, members of the *Scenedesmus* genus have been identified as potential oil-producing species, with both rapid growth, as well as relatively high lipid content [17, 18].

The present article reports simultaneous biodiesel production and waste recycling with the green microalga *S. acuminatus* isolated from the fresh water fish pond. During the study the alga was subjected and compared with different methods of transesterification for maximum FAME yield.

## 2. Materials and Methods

### 2.1. Microalgal Isolation, Purification and Identification

Algae samples were collected by plankton net (20-μm pore size) from freshwater fish pond (18° 55′4.2″N; 99° 0′41.1″E) at a location near Maejo University, Sansai, Thailand. The collected samples were samples of about 5 ml were inoculated into 5-ml autoclaved Bold Basal Medium (BBM) [19] in 20-ml test tubes and cultured at room temperature (25±1°C) under 67.50 ± 2 μmol$^{-1}$m$^2$ sec$^{-1}$ intensity with 16:8 h photoperiod for 10 days. After incubation, individual colonies were picked and transferred to the same media for purification in 250 mL conical flask. The culture broth was shaken manually for five to six times a day. The pre-cultured samples were streaked on BBM medium-enriched agar plates and cultured for another 10 days with cool white fluorescent light using the same light intensity.

The single colonies on agar were picked up and cultured in liquid BBM medium, and the streaking and inoculation procedure was repeated until pure cultures were obtained. The purity of the culture was monitored by regular observation under microscope. The isolated microalgae were identified microscopically using light microscope with standard manual for algae [20, 21]. Green microalga *S. acuminatus* structure shown in figure 1.

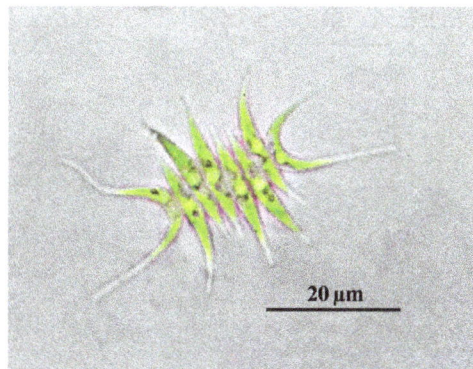

**Figure 1.** *Scenedesmus acuminatus.*

### 2.2. Inoculums Preparation

Isolated and purified microalgae were inoculated in 250-ml Erlenmeyer flasks containing 125 ml culture medium (BBM).

Flasks were placed on a reciprocating shaker at 120 rpm for 7 d at room temperature of 25±1 °C. Light was provided by cool white fluorescent lamps at an intensity of 37.5 μmol$^{-1}$m$^2$ sec$^{-1}$. The inoculums were then transferred to 1000-ml Erlenmeyer flasks (photo-bioreactor). All experiments were carried out in triplicate.

### 2.3. Medium Preparation

The raw piggery wastewater effluent was collected from the Faculty of Animal Science and Technology (18°54'55.36"N; 99° 1'6.61"E), Maejo university near the laboratory was used as a substrate to cultivate *S. acuminatus*. Pre-treatment was carried out by sedimentation and filtration with a filter cloth to remove large, non-soluble particulate solids. After filtration the substrate was autoclaved for 20 min at 121°C, after which the liquid was stored at 4 °C for 2 days for settling any visible particulate solids and the supernatant was used for microalgae growth studies. The characteristics and features of the autoclaved wastewater are summarized in Table 1.

*Table 1. Characteristics of autoclaved piggery wastewater effluent (means ± SD).*

| Parameter | Autoclaved |
|---|---|
| pH | 7 ± 0.0 |
| COD (mg L$^{-1}$) | 3200± 63 |
| TN (mg L$^{-1}$) | 748.0 ± 3.0 |
| TP (mg L$^{-1}$) | 128 ± 43 |
| Suspended solid (mg L$^{-1}$) | 288 ± 43 |

### 2.4. Photo-Bioreactor Set Up

The standard reactor of continuous stirred tank reactor (CSTR) was used. The batch-fed algal cultures were grown in photo-bioreactor (CSTR), cultured at room temperature (25±1°C) under 67.50 ± 2 μmol$^{-1}$m$^2$ sec$^{-1}$ intensity with 16:8 h photoperiod. The triplicate reactors were operated at 10 days detention time and other operational factors were list in Table 2 and methods were presented in Table 3.

*Table 2. Operational parameters*

| Operational parameter | Photo-bioreactor |
|---|---|
| Scale | Lab |
| Detention time | 10 days |
| Reactor design | 1L up-flow flask |
| Water volume | 1L |
| Feeding | Batch feed daily |
| Filter size | 0.45 μm |
| Mixing speed | Magnetic mixer |
| Light intensity | 67.05 ± 2 μmol s$^{-1}$ m$^{-2}$ |
| Operation period | 30 days |

### 2.5. Analytical Methods

All the indices including pH, chemical oxygen demand (COD), total nitrogen (TN), total phosphorous (TP) and algal biomass of total suspended solids were continuously monitored throughout the study, following the standard protocols of APHA [20].

*Table 3. Environmental and algal biomass measurements*

| Parameter | Equipment or method |
|---|---|
| Light intensity | LI-COR light meter (LI-250) |
| Water temperature | Thermometer |
| Settleable solid | Imhoff cone |
| Species | Microscope |
| Operation period | 30 days |
| pH | Method 423 (Standard Methods) |
| COD | Method 508B (Standard Methods) |
| NH$_4^+$-N | Method 417D (Standard Methods) |
| TKN | Method 420A (Standard Methods) |
| NO$_2^-$-N | Method 419 (Standard Methods) |
| NO$_3^-$-N | Method 418A (Standard Methods) |
| TN | = Total of N species |
| TP | Method 424D (Standard Methods) |
| TSS | Method 209 (Standard Methods) |

### 2.6. Nutrient Removal

Every day 50 mL microalgae culture was decanted from the reactors then by centrifugation at 4000 rpm, 10°C for 10 minutes and filtration by 0.45 μm membranes. After these preprocessing, the supernatant was used to monitor the concentration of NO$_3$-N, NH$_4$-N, TN and TP following the standard testing methods [20]. The removal efficiency can be calculated following formula:

$$\text{Removal efficiency} = (C_i - C_0)/C_0 * 100\%$$

where $C_o$ and $C_i$ are defined as the mean values of nutrient concentration at initial time $t_0$ and time $t_i$, respectively..

### 2.6. Lipid Extraction

Total lipids from 100 mg microalgae were extracted using 2mL chloroform/methanol (v/v: 2/1) [21], ultrasonic treatment for 10 min and centrifugation at 4000 rpm for 5 min. The supernatants were then collected into pre-weighted centrifuge tubes. This process was repeated three times. The collected supernatants were dried under nitrogen flow and then at 60°C until the weight of samples remained constant.

### 2.7. Fatty acid Methyl Ester (FAME) Content Analysis and Transesterification

Biodiesel samples were analyzed quantitatively and qualitatively to determine the biodiesel yield and FAME composition. The samples were weighed and moved to 10-mL flasks. Then, 5 mL H$_2$SO$_4$–methanol (v/v H$_2$SO$_4$/methanol) was added, and the flask was stirred at a specific temperature for a specific amount of time with refluxing. After the specific time period, the flask was cooled to room temperature. Next, 2 mL of hexane and 0.75 mL of distilled water were added to the flask and mixed for 30 s on a vortex mixer. The mixture formed two phases, and the upper hexane layer contained the fatty acid methyl esters (FAMEs). The hexane layer was transferred to a new vial and mixed with the internal standard C17-ME for analysis by gas chromatography (GC). The fatty acid methyl esters (FAME) were then extracted with hexane and analyzed by GC-MS as described by Thomæus et al. [22]. FAME analysis was performed using GC-MS (Agilent 6890 -HP5973 model, Australia). All the measurements of the values used in the tables and figures represent the average ± SD of four individual replicates during the whole experiment.

# 3. Results and Discussion

## 3.1. Microalgal Growth in Piggery Wastewater Effluent Medium

Algal growth is directly affected by the availability of nutrients and light, the pH stability, temperature, and the initial inoculum density [11,13]. Microalgae can grow an abundantly under suitable conditions and with sufficient nutrients. They often double their biomass within 3.5 h or, at the longest, 24 h during their exponential growth phase. The green microalga, *S. acuminatus* was chosen for the study because, it was found to be dominant among other algal species in their natural environment. Instead of huge differences in the climatic conditions of the places where collection has been done and the place where all the experimental work was carried out, the microalga showed luxurious growth, which reveals its flexible nature to adapt the wide range of the environmental condition. There were several indexes for algal biomass measurement and roughly we could classify into two groups, (1) direct index such as dry weight and (2) indirect index such as chlorophyll, so-called proxy index [11–14]. According to Ramaraj et al. [12], TSS was applied to be an index of algal biomass measured in this study. The average biomass was 3.24 and ranged 3.07–3.42 g/L (Table 4).

Culturing of microalgae in wastewater also substantially reduces the need of chemical fertilizers and their related burden on life cycle. Through the utilization of wastewater, the zero-waste concept is further implemented, and thus stimulates a more sustainable practice for the microalgae

biofuel industry. It has even been proposed that integrated phyco-remediation and biofuel technology appears to be the only source of sustainable production of biofuels [15, 16, 23].

*Table 4.* Algal biomass and lipid measurements

| Parameter | Average | Minimum | Maximum |
|---|---|---|---|
| Total biomass (g/L) | 3.24 | 3.07 | 3.42 |
| Lipid production (mg/L) | 710.11 | 543.5 | 844 |
| Productivity of lipid (mg/L/d) | 71.01 | 54.35 | 84.4 |

## 3.2. Characterization of Piggery Wastewater Effluent and Nutrient Removal Efficiency of S. scuminatus

Growing algae depends on the availability of principal nutrients like nitrogen, phosphorus, carbon, sulphur and micronutrients including silica, calcium, magnesium, potassium, iron, manganese, sulphur, zinc, copper, and cobalt. Algal cells have the capability to uptake nitrogen and phosphorus from water. Nitrogen and phosphorus are the two important nutrient compounds to analyze a water source for potential algae growth [23, 24]. Many treatments of piggery wastewater by microalgae have been investigated, as a means of providing environmental protection from and recovery of nutrients. Neglecting the cost, processing of piggery wastewater by microalgae with the simultaneous production of oil would seem to be a good choice. The removal of nitrogen, phosphorus, calcium and inorganic carbon from piggery wastewater by microalgae cultivation as a function of incubation time has been studied [23, 24].

The algal species such as *C. mexicana*, *M. reisseri*, *C. vulgaris*, *N. pusilla*, *S. Obliquus*, and *O. multisporus* shown the maximum nitrogen, phosphorus and inorganic removal (62%, 28%, and 29%) were obtained with C. mexicana, respectively, while the maximum calcium removal (71%) was obtained with *C. vulgaris*. The lowest nitrogen, phosphorus and inorganic carbon removal (8%, 3%, and 1.3%) were obtained with *M. reisseri* after 20 days of cultivation [23].

*Table 5.* Nutrient removal Piggery wastewater and effluent

| Microalgal species | Wastewater type | Nitrogen | Phosphorus | Reference |
|---|---|---|---|---|
| *C. Mexicana* | Piggery wastewater | 62% | 28% | 23 |
| *S. obliquus* | Piggery effluent | 41% | 59% | 24 |
| *S. acuminatus* | Piggery effluent | 75% | 88% | This study |

The nutrient removal efficiency of *S. acuminatus* was analyzed in this study and results shown in Table 5. Microalgae can be efficiently used to remove significant amount of nutrients because they need high amounts of nitrogen and phosphorus for protein (45–60% of microalgae dry weight), nucleic acid, and phospholipid synthesis [16]. The nitrogen in sewage effluent arises primarily from metabolic interconversions of extra derived compounds, whereas 50% or more of phosphorus arises from synthetic detergents [23]. Consequently, this study results demonstrated that *S. acuminatus* was highly utilized the macronutrients

from the piggery in wastewater effluent medium.

## 3.3. Lipid Yield and Fatty acid Methyl Ester of S. Acuminatus

Algae grown on wastewater media are a potential source of low-cost lipids for production of liquid biofuels. This study investigated lipid productivity and nutrient removal by green algae grown during treatment of wastewaters effluent without supplemented $CO_2$. The lipid productivity and lipid content were presented in Table 4. The major fatty acid composition of each isolate was determined using GC analysis. Table 6 shows the fatty acids (FA) profile of *S. abundans* grown under large scale cultivation using indigenously made photobioreactor. The FA profile of alga was determined by the quantification of FAME content which reveals the abundance of FA with carbon chain length of C16 and C18. Oleate (C18:1), palmitate (C16:0), linolenate (C18:3), linoleate (C18:2), palmitoleate (C16:1) and stearate (C18:0) were contributing over 90% of the total FAME content. The properties of biodiesel are highly influenced by the FA profile of the algae.

**Table 6.** *Fatty acid methyl ester profile of S. acuminatus*

| Fatty acids | Contents (% of total fatty acids) |
|---|---|
| Capric acid (C10:0) | 0.13 |
| Myristic acid (C14:0) | 1.87 |
| Pentadecylic acid (C15:0) | 0.53 |
| Palmitic acid (C16:0) | 19.34 |
| Palmitoleic acid (C16:1) | 11.73 |
| Hexadecadienoic acid (C16:2) | 3.81 |
| Strearic acid (C18:0) | 19.55 |
| Oleic acid (C18:1) | 23.2 |
| Linoleic acid (C18:2) | 9.75 |
| Linolenic acid (C18:3) | 4.95 |
| Others | 5.1 |
| Total saturated fatty acids | 34.45 |
| Total unsaturated fatty acids | 54.65 |
| Unsaturated/saturated fatty acid ratio | 1.55 |

### 3.4. Algal Biodiesel Production

Algal biodiesel production consists, primarily of five steps. They are: (a) algae production; (b) algae harvesting; (c) oil extraction; (d) transesterification or chemical treatment; and (e) separation and purification [25]. The oil extraction step includes cell disruption by mechanical, chemical, or biological methods and oil collection by solvent. Major bottlenecks of oil extraction are that the extraction of internal oils is energetically demanding because the cell walls of some species of microalgae are strong and thick and that the oil extraction yield is negatively affected in case of a wet biomass [26]. Extracted microalgal oils are typically converted to biodiesel by transesterification using alcohols and catalysts.

Recently, the combination of oil extraction and biodiesel conversion, called direct (in-situ) transesterification has been studied. Direct transesterification refers to the conversion of algal oils present in biomass to biodiesel. Here, direct transesterification includes both the esterification of free fatty acid and the transesterification of triglyceride from microalgae. This process simplifies the production process and improves the biodiesel yield compared with conventional extraction because of the elimination of an oil extraction step that incurs oil loss. The reactions are simple and comprise the addition of alcohols, catalysts, and biomass and sometimes co-solvents [27]. So far, direct transesterification was carried out with chemical catalysts; lesser reaction time and high

yields are the advantages of the chemical method of direct transesterification, though, high energy requirements, difficulties in the recovery of the catalysts and glycerol, and pollution to the environment by these catalysts, are the major disadvantages of this process. Accordingly, the direct transesterification of the oleaginous biomass resulted in a higher biodiesel yield and fatty acid methyl ester (FAME) content than the extraction–transesterification method.

### 3.5. Analysis of S. acuminatus Biodiesel

Biodiesel consists largely of fatty acid methyl esters, which are produced by the transesterification of biologically-derived lipids [31], and the quality of biodiesel is considerably affected by the composition of the fatty acids in the biodiesel. In Table demonstrated that the palmitic, stearic, oleic, linoleic and linolenic acid were recognized as the most common fatty acids in biodiesel. In addition, the fatty acids profiles of the isolates indicated the presences of lauric (C12:0), myristic (C14:0), palmitic (C16:0), heptadecanoic (C17:0), stearic (C18:0), palmitoleic (C16:1), oleic (C18:1n9c), α-linolenic (C18:3n3), and γ-linolenic acid (C18:3n6). For the green alga *Scenedesmus*, the fatty acid compositions of 14:0, 16:0, 16:1, 16:2, 16:3, 18:0, 18:1, 18:2, α-18-3 have been confirmed under many conditions including photoautotrophic and heterotrophic cultivation, nitrogen starvation, and outdoors in a photobioreactor. Biodiesel fuels enriched in methyl oleate are desirable, relatively small percentages of saturated fatty esters can wreck the cold flow properties of biodiesel. Our finding shows that most of these strains contain 34% of saturated fatty acids (C16 and C18). Among the tested microalgal species, *S. acuminatus* showed the highest oleic acid content, making it the most suitable isolate for the production of good quality biodiesel. Moreover, *S. acuminatus* grown in piggery wastewater effluent showed a rise in palmitic acid content, which is desirable for good-quality biodiesel.

## 4. Conclusions

Microalgae, *Scenedesmus acuminatus* was feed batch -cultured in a photo-bioreactor to facilitate better culture control and higher biomass productivity. A one-step direct transesterification of microalgal cells was successfully performed which has great significance for fatty acid composition analysis of micro-scale samples in applications such as strain screening. Moreover, this method can be used for direct transesterification of microalgal cells without dehydration beyond centrifugation. Direct transesterification of microalga paste greatly simplifies the process of fatty acid analysis while completely eliminating the drying and oil extraction steps, and thus, the developed method has great potential for a variety of applications. The results of this study indicate that the naturally isolated microalgal strain *S. acuminatus* may be a valuable candidate for biodiesel production.

# References

[1] B. R. Moser, "Proposed technological improvements to ensure biodiesel's continued survival as a significant alternative to diesel fuel", Biofuels, 2014, 5: 5–8.

[2] L. C. Meher, D. Vidya Sagar, S. N. Naik, "Technical aspects of biodiesel production by transesterification – a review", Renewable and Sustainable Energy Reviews, 2006, 10: 248–268.

[3] K. Bozbas, "Biodiesel as an alternative motor fuel: production and policies in the European Union", Renewable and Sustainable Energy Reviews, 2008, 12: 542–552.

[4] E. M. Shahid, Y. Jamal, "Production of biodiesel: a technical review", Renewable and Sustainable Energy Reviews, 2011, 15: 4732–4745.

[5] Á. Sánchez, R. Maceiras, Á. Cancela, A Pérez, "Culture aspects of Isochrysis galbana for biodiesel production", Applied Energy, 2013, 101: 192–197.

[6] X. Yu, P. Zhao, C. He, J. Li, X. Tang, J. Zhou, Z. Huang, "Isolation of a novel strain of Monoraphidium sp. and characterization of its potential application as biodiesel feedstock", Bioresource Technology, 2012, 121: 256–262.

[7] G. Zaimes, V. Khanna, "Microalgal biomass production pathways: evaluation of life cycle environmental impacts", Biotechnol Biofuels, 2013, 6:88.

[8] H.-Y. Ren, B.-F. Liu, C. Ma, L. Zhao, N.-Q. Ren, "A new lipid-rich microalga Scenedesmus sp. Strain R-16 isolated using Nile red staining: effects of carbon and nitrogen sources and initial pH on the biomass and lipid production", Biotechnology for Biofuels, 2013, 6:143.

[9] A. Demirbas, M. F. Demirbas, "Importance of algae oil as a source of biodiesel", Energy Conversion and Management, 2011, 52: 163–70.

[10] X. Zhang, J. Rong, H. Chen, C. He. Q. Wang, "Current status and outlook in the application of microalgae in biodiesel production and environmental protection", Frontiers in Energy Research, 2014, 2: 32.

[11] R. Ramaraj, D. D-W. Tsai, P. H. Chen, "Algae Growth in Natural Water Resources", Journal of Soil and Water Conservation, 2010, 42: 439–450.

[12] R. Ramaraj, D. D-W. Tsai, P. H. Chen, "Chlorophyll is not accurate measurement for algal biomass", Chiang Mai Journal of Science, 2013, 40: 547–555.

[13] R. Ramaraj, D. D-W. Tsai, P. H. Chen, "An exploration of the relationships between microalgae biomass growth and related environmental variables", Journal of Photochemistry and Photobiology B: Biology, 2014, 135: 44–47.

[14] R. Ramaraj, D. D-W. Tsai, P. H. Chen, "Freshwater microalgae niche of air carbon dioxide mitigation", Ecological Engineering, 2014; 68: 47–52.

[15] M. K. Kim, J. W. Park, C. S. Park, S. J. Kim, K. H. Jeune, M. U. Chang, J. Acreman, "Enhanced production of Scenedesmus spp. (green microalgae) using a new medium containing fermented swine wastewater", Bioresource Technology, 2007, 98: 2220–2228.

[16] J. K. Pittman, A. P. Dean, O. Osundeko, "The potential of sustainable algal biofuel production using wastewater resources", Bioresource Technology, 2011, 102: 17–25.

[17] L. Xin, H. Hong-Ying, Y. Jia, "Lipid accumulation and nutrient removal properties of a newly isolated freshwater microalga, Scenedesmus sp. LX1, growing in secondary effluent", New Biotechnology, 2009, 27(1): 59–63.

[18] L. Rodolfi, G. C. Zittelli, N. Bassi, G. Padovani, N. Biondi, G. Bonini, M. R. Tredici, "Microalgae for oil: strain selection, induction of lipid synthesis and outdoor mass cultivation in a low-cost photobioreactor", Biotechnology and Bioengineering, 2009, 102: 100–112.

[19] S. Kant, P. Gupta, "Algal Flora of Ladakh". Scientific Publishers, Jodhpur, India, 1998, p.341.

[20] APHA, AWWA, WPCF, "Standards Methods for the Examination of Water and Wastewater, 21$^{st}$ ed. APHA-AWWA-WPCF, Washington, DC, 2005.

[21] E. G. Bligh, W. J. Dyer, "A rapid method of total lipid extraction and purification", Canadian Journal of Biochemistry and Physiology, 1959, 37: 911–917.

[22] S. Thomæus, A. S. Carlsson, S. Stymne, "Distribution of fatty acids in polar and neutral lipids during seed development in Arabidopsis thaliana genetically engineered to produce acetylenic, epoxy and hydroxyl fatty acids", Plant Science, 2001, 161: 997–1003.

[23] R. A. Abou-Shanab, M. K. Ji, H. C. Kim, K. J. Paeng, and B. H. Jeon, "Microalgal species growing on piggery wastewater as a valuable candidate for nutrient removal and biodiesel production", Journal of Environmental Management, 2013, 115: 257–264.

[24] L. Zhu, Z. Wang, Q. Shu, J. Takala, E. Hiltunen, P. Feng, "Nutrient removal and biodiesel production by integration of freshwater algae cultivation with piggery wastewater treatment", Water Research, 2013, 47: 4294–4302.

[25] L. Lardon, A. Helias, B. Sialve, J. Steyer, O. Bernard "Life-cycle assessment of biodiesel production from microalgae", Environmental Science and Technology, 2009, 43: 6475–6481.

[26] P. Hidalgo, C. Toro, G. Ciudad, R. Navia, "Advances in direct transesterification of microalgal biomass for biodiesel production", Reviews in Environmental Science and Biotechnology, 2013, 12: 179–199.

[27] T. R. S. Baumgartner, J. A. M. Burak, D. Baumgartner, G. M. Zanin, P. A. Arroyo, "Biomass production and ester synthesis by in situ transesterification/esterification using the microalga Spirulina platensis", International Journal of Chemical Engineering, 2013, 425604: 1–7.

# 28

# Nickel – the ultimate substitute of Coal, Oil and Uranium

**U. V. S. Seshavatharam[1], S. Lakshminarayana[2]**

[1]Honorary faculty, I-SERVE, Alakapuri, Hyderabad-35, AP, India
[2]Dept. of Nuclear Physics, Andhra University, Visakhapatnam-03, AP, India

**Email address:**

seshavatharam.uvs@gmail.com (U. V. S. Seshavatharam), lnsrirama@gmail.com (S. Lakshminarayana)

**Abstract:** During E-CAT test run some hidden and unknown energy is being coming out in the form of heat energy in large quantity. Based on the principle of conservation of energy and from the well known nuclear fusion and fission reactions it is possible to guess that, the E-CAT hidden energy may be in the form of binding of protons and neutrons of the Nickel and Lithium atomic nuclei. By considering the nuclear binding energies of $^{58}_{28}Ni$, $^{62}_{28}Ni$ and $^{7}_{3}Li$ an attempt is made to understand the energy liberation mechanism in E-CAT. With reference to the net energy production of (5825 ± 10%) Mega Joules liberated from one gram Ni of the E-CAT's 32 days third party test run, it can be suggested that, for every transformation of $^{58}_{28}Ni$ to $^{62}_{28}Ni$ via $^{7}_{3}Li$, liberated heat energy is 3.64 MeV and for one gram of $^{58}_{28}Ni$ liberated energy is 5984 Mega Joules. For each transformation of $^{58}_{28}Ni$ to $^{62}_{28}Ni$ via $^{7}_{3}Li$, 3 hydrogen atoms can be expected to be emitted. Note that, energy liberated for one gram of $^{58}_{28}Ni$ in cold fusion is 1.66 MWh and energy liberated for one gram of $^{235}_{92}U$ in nuclear fission is 22.6 MWh. Clearly speaking, energy released in Nickel based E-CAT is just 13.6 times less than the energy released in Uranium fission.

**Keywords:** Cold Fusion, Low Energy Nuclear Reactions, E-CAT ( Energy Catalyzer)

## 1. Introduction

With reference to the current [1-6] and old [7] review reports, one can understand the current 'golden status' and old 'Pathetic status' of Cold fusion or Low energy nuclear interactions (LENR). Since 1989 scientists proposed several interesting proposals for understanding the observed excess heat generation with various experimental setups [8-18]. Many researchers and scientists around the world have reported successful experiments at a number of international conferences, and selected articles are collected in an on-line data base. The results, however, have not been taken seriously by main stream science, even after full support by two Nobel laureates, Julian Schwinger and Brian Josephson. Julian Schwinger (1918-1994), a Nobel prize winner in Physics, 1965, who also worked with Oppenheimer, was a strong advocate of cold fusion [19]. Brian Josephson, a Nobel prize winner in Physics, 1973, is a discoverer of the Josephon effect in the field of superconductivity. He is presently a strong supporter of cold fusion [20].

These new interactions are exclusively based on 'excess heat generation' and are absolutely free from the currently believed alpha, beta and gamma radiations. Experts believe that, LENR could use 1% of the Nickel mined to produce current world energy at a price four times cheaper than coal. Joseph Zawodny, a senior research scientist with NASA 's Langley Research Center says: "It has the demonstrated ability to produce excess amounts of energy , cleanly , without hazardous ionizing radiation, without producing nasty waste". It is not a surprise to say that, very soon LENR will dominate all the current leading research areas of physics in the near future.

Andrea Rossi says: "one gram of Nickel is equivalent to 5,00,000 liters of oil". As there is no emission of the alpha, beta and gamma radiation, as there is no emission of other gasses like $SO_2$, $CO_2$ etc, as Nickel abundance is high in the earth core and as the technical risk of power generation is less, E-CAT seems to be the best and ultimate green power generator in coming future. Very interesting information is that Mr and Mrs Bill Gates are planning to fund Italy's ENEA for LENR/Cold fusion technology. In this paper authors made an attempt to understand the working mechanism of heat generation in the E-CAT.

## 2. History of the E-CAT

Kanarev [21] and Japanese researcher Dr. T. Mizuno [22] provided measurable proof of fusion and fission products. In Italy, the cold fusion research pioneered by Francesco Piantelli in 1989 has been extended and supported by the local inter-university centers in Bologna (Focardi, Campari) and Sienna (Piantelli, Gabbani, Montalbano, Veronesi). A detailed report about this research was published by the Italian National Agency for New Technology, Energy and Environment in 2008. Piantelli filed two patents WO9520816 (1997) and WO2010058288 (2010), describing different methods, and published an article ITSI920002 about cold fusion of nickel with deuterium or hydrogen. Recently, interest in cold fusion as an alternative to nuclear energy was raised by the successful demonstration of the Rossi cold fusion device called E-cat. E-CAT (Energy Catalyzer) seems to be the most promising apparatus in this regard [1-6]. It is invented and being developed by Andrea Rossi. The Focardi-Rossi method of nuclear reaction Ni + H -> Cu is based on the preliminary research of Focardi and colleagues. Sergio Focardi is an emeritus professor at the University of Bologna while Andrea Rossi is a skilled researcher and inventor. After years of successful collaboration, they gave on January 14, 2011, the first public demonstration of a nickel-hydrogen fusion reactor, called E-cat, capable of producing more than 10 kilowatts of heat power, while only consuming a fraction of that. In 2008 Rossi filed International patent application WO 2009/125444 A1 entitled Method and Apparatus for Carrying out Nickel and Hydrogen Exothermal Reaction. Ignoring the skepticism in the main stream science, Rossi proceeded further with the development and manufacturing of his E-cat generator. Public demonstrations of the E-cat reactor with some invited experts were made on January 14, March 29, April 19 and 28, September 7, October 6 and October 28, 2011. During the larger public demonstration on October 28, 2011, Rossi invited a few dozen people, including a group of engineers from an unnamed potential US customer, as well as a handful of journalists. According to Rossi, each module received an initial energy input of 400 watts and produced a self-sustaining, continuous output of about 10 kilowatts per hour for the next few hours.

## 3. E-CAT – Recent Test Run Observations and Understanding its Hidden Mechanism

In a third party inspection, the E-CAT subject to testing was powered by 360 W for a total of 96 hours, and produced in all 2034 W thermal [3]. In this context experts say:
1. Something "REAL" is happening and we are certainly dealing with a new source of energy.
2. There are efforts ongoing to explore the validity of the theories and the weak interaction theories suggest what the physics might be.

Proceeding further, authors of this paper request the readers to go through the detailed and interesting report of 50 pages carefully prepared by the third party, titled "Observation of abundant heat production from a reactor device and of isotopic changes in the fuel" [2]. In the report one can see the experimental setup, photos of the reactor, experimental data, observations, comments & opinions of the authors along with clear cut information on the non-emission of generally believed nuclear radiation. The utmost important point to be noted is that, as per the report, "The isotope composition in Lithium and Nickel was found to agree with the natural composition before the run, while after the run it was found to have changed substantially. Nuclear reactions are therefore indicated to be present in the run process, which however is hard to reconcile with the fact that no radioactivity was detected outside the reactor during the run".

After successfully conducting the 32 day test run with E-CAT, at Switzerland [3], authors concluded in the following way.

1. "A 32-day test was performed on a reactor termed E-Cat, capable of producing heat by exploiting an unknown reaction primed by heating and some electro-magnetic stimulation. In the past years, the same collaboration has performed similar measurements on reactors operating in like manner, but differing both in shape and construction materials from the one studied here. Those tests have indicated an anomalous production of heat, which prompted us to attempt a new, longer test. The purpose of this longer measurement was to verify whether the production of heat is reproducible in a new improved test set-up, and can go on for a significant amount of time. In order to assure that the reactor would operate for a prolonged length of time, we chose to supply power to the E-Cat in such a way as to keep it working in a stable and controlled manner. For this reason, the performances obtained do not reflect the maximum potential of the reactor, which was not an object of study here".

2. "In summary, the performance of the E-Cat reactor is remarkable. We have a device giving heat energy compatible with nuclear transformations, but it operates at low energy and gives neither nuclear radioactive waste nor emits radiation. From basic general knowledge in nuclear physics this should not be possible. Nevertheless we have to relate to the fact that the experimental results from our test show heat production beyond chemical burning, and that the E-Cat fuel undergoes nuclear transformations. It is certainly most unsatisfying that these results so far have no convincing theoretical explanation, but the experimental results cannot be dismissed or ignored just because of lack of theoretical understanding. Moreover, the E-Cat results are too conspicuous not to be followed up in detail. In addition, if proven sustainable in further tests the E-Cat invention has a large potential to become an important energy source. Further investigations are required to guide the

interpretational work, and one needs in particular as a first step detailed knowledge of all parameters affecting the E-Cat operation. Our work will continue in that direction".

From current known physics point of view it is quite shocking, quite bitter and demands the need of review and revision of our known physical laws and concepts. Based on the principle of conservation of energy it is clear that, during LENR and Cold fusion some hidden and unknown energy is being coming out in the form of heat energy. From the well known nuclear fusion and fission reactions it is possible to guess that, the hidden energy may be in the form of binding of protons and neutrons of the Nickel and Lithium atomic nuclei.

## 4. Estimating the Possible Liberated Heat Energy in E-CAT

Authors of the recent E-CAT test run say [2] :

1. "The fuel generating the excessive heat was analyzed with several methods before and after the experimental run. It was found that the Lithium and Nickel content in the fuel had the natural isotopic composition before the run, but after the 32 days run the isotopic composition has changed dramatically both for Lithium and Nickel. Such a change can only take place via nuclear reactions. It is thus clear that nuclear reactions have taken place in the burning process. This is also what can be suspected from the excessive heat being generated in the process".

2. "The unused fuel shows the natural isotope composition from both SIMS and ICP-MS, i.e. 58Ni (68.1%), 60Ni (26.2%), 61Ni (1.1%), 62Ni (3.6%), and 64Ni (0.9%), whereas the ash composition from SIMS is: 58Ni (0.8.%), 60Ni (0.5%), 61Ni (0%), 62Ni (98.7%), 64Ni (0%), and from ICP-MS: 58Ni (0.8%), 60Ni (0.3%), 61Ni (0%), 62Ni (99.3%), 64Ni (0%).We note that the SIMS and ICP-MS give the same values within the estimated 3% error in the given percentages. Evidently, there is also an isotope shift in Nickel. There is a depletion of the 58Ni and 60Ni isotopes and a buildup of the 62Ni isotopes in the burning process. We note that 62Ni is the nucleus with the largest binding energy per nucleon. The origin of this shift cannot be understood from single nuclear reactions involving protons".

3. "The Lithium content in the fuel is found to have the natural composition, i.e. 6Li 7 % and 7Li 93 %. However at the end of the run a depletion of 7Li in the ash was revealed by both the SIMS and the ICP-MS methods. In the SIMS analysis the 7Li content was only 7.9% and in the ICP-MS analysis it was 42.5 %. This result is remarkable since it shows that the burning process in E-Cat indeed changes the fuel at the nuclear level, i.e. nuclear reactions have taken place. It is notable, but maybe only a coincidence, that also in Astrophysics a 7Li depletion is observed".

4. "Our measurement, based on calculating the power emitted by the reactor through radiation and convection, gave the following results: the net production of the reactor after 32 days' operation was (5825 ± 10%) [MJ], the density of thermal energy (if referred to an internal charge weighing 1 g) was (5.8 · $10^6$ ± 10%) [MJ/kg], while the density of power was equal to (2.1 · $10^6$ ± 10%) [W/kg]. These values place the E-Cat beyond any other known conventional source of energy. Even if one conservatively repeats the same calculations with reference to the weight of the whole reactor rather than that of its internal charge, one gets results confirming the non-conventional nature of the form of energy generated by the E-Cat, namely (1.3 · $10^4$ ± 10%) [MJ/kg] for thermal energy density, and (4.7 · $10^3$ ± 10%) [W/kg] for power density".

From above points and with reference to the net energy production (5825 ± 10%) MJ (of the reactor for 32 days run with one gram of Ni) - quantitatively it can understood in the following way. Binding energy of $^{58}_{28}Ni$ is 506.6 MeV and binding energy of $^{62}_{28}Ni$ is 544.41 MeV. Similarly binding energy of $^{7}_{3}Li$ is 41.45 MeV. For a moment guess that, $^{7}_{3}Li$ joins with $^{58}_{28}Ni$ forming $^{62}_{28}Ni$ and emits 3 hydrogen atoms. Clearly speaking, $^{7}_{3}Li$ transforms to 4 neutrons and 3 hydrogen atoms. 4 neutrons joins with $^{58}_{28}Ni$ forming $^{62}_{28}Ni$. To have stability $^{62}_{28}Ni$, must gain an effective binding energy (544.41-506.6)=37.81 MeV [23,24]. It can be gained from the binding energy 41.45 MeV of $^{7}_{3}Li$. If so the remaining binding energy of $^{7}_{3}Li$ is [41.45-(544.41-506.6)]=3.64 MeV and it may be liberated out in the form of heat energy. For every transformation of $^{58}_{28}Ni$ to $^{62}_{28}Ni$ via $^{7}_{3}Li$, liberated heat energy is 3.64 MeV and for one gram of $^{58}_{28}Ni$ liberated energy is 5984 Mega Joules. This can be compared with the observed E-CAT's 32 day output energy with one gram Ni fuel. If it is possible to design the E-cat to have mole transformations of $^{58}_{28}Ni$ to $^{62}_{28}Ni$ via $^{7}_{3}Li$ for second, then for every second, 0.3512 Tera Joules of energy can be liberated.

In general, the number of successful transformations $^{58}_{28}Ni$ to $^{62}_{28}Ni$ will depend on the quantity of Ni powder, fineness of the nickel powder, working temperature, working pressure and volume of the E-CAT, kinetic energy of Nickel and Hydrogen, melting points of nickel and lithium, unknown catalyst, efficiency of the E-CAT etc. Keeping all these parameters, instead of mass of Nickel, it is possible to consider number of $^{58}_{28}Ni$ to $^{62}_{28}Ni$ transformations per second. Design capacity of E-CAT can be fixed in this way for different large scale, medium scale and small scale applications. So, number of transformations, $n \cong f \cdot N_A$ where $f \approx 10^{-6}$ to 1 and can be called as the working factor. Now in a simplified view, proposed mechanism can be expressed in the following way.

If chosen time unit is One second,

$$\text{Energy liberated/sec} \cong f \times N_A \times 3.64 \text{ MeV/sec}$$

$$\cong f \times 0.3512 \text{ Tera J/sec}$$

where $f \approx 10^{-6}$ to $1$.

If chosen time unit is One hour,

$$\text{Energy liberated/hour} \cong f \times N_A \times 3.64 \text{ MeV/hour}$$

$$\cong f \times 0.3512 \text{ Tera J/hour}$$

where $f \approx 10^{-6}$ to $1$. E-CAT reactor speed can be defined as the number of transformations of $^{58}_{28}Ni$ to $^{62}_{28}Ni$ per second. Note that for one gram of Ni fuel, number of transformations can be understood as follows.

$$n \cong \frac{1 \times 10^{-3} \text{ (kg)}}{58.69334 \times 1.66053892 \times 10^{-27} \text{(kg)}} \cong 1.02603 \times 10^{22}$$

$$\Rightarrow f \cong \frac{n}{N_A} \cong \frac{1}{58.69334} \cong 0.01703768$$

Note that, energy liberated for one gram of $^{58}_{28}Ni$ in cold fusion is 1.66 MWh and energy liberated for one gram of $^{235}_{92}U$ in nuclear fission is 22.6 MWh. Clearly speaking, energy released in E-CAT is just 13.6 times less than the energy released in $^{235}_{92}U$ fission.

## 5. Understanding Nickel–Hydrogen fusion in E-Cat

Andrea Rossi's stimulating experiments and inventions [5,6] on cold fusion suggest and confirm that under pressure and temperature, Hydrogen atom's proton joins with 62Nickel nucleus and electron joins with 62Nickel electronic orbits forming stable 63Cu atom and liberates heat energy of the order of MeV. It can be understood as follows.

1) Binding energy of 63Cu is 549.96 MeV and binding energy of 62Ni is 544.41 MeV. To have stability, 63Cu requires 549.96-544.41=5.55 MeV.
2) With reference to 'average binding energy per nucleon' and unified atomic mass unit, (within the nucleus) effective mass of each nucleon can be assumed to be 931.5-0.511 = 931.0 MeV. When proton joins with 62Ni nucleus, 938.3-930.0=7.3 MeV seems to be excess. 63Cu takes 5.55 MeV from 7.3 MeV and releases 1.75 MeV in the form of heat.
3) For 63Cu, average biding energy per nucleon is 8.73 MeV and for 62Ni, average biding energy per nucleon is 8.78 MeV. When suddenly 62Ni transforms to 63Cu, 8.78-8.73=0.05 MeV per nucleon will be liberated. Thus for 63 nucleons, 63*0.05 =3.15 MeV will also be released.
4) Thus a total of 1.75+3.15=4.9 MeV may be coming out in each Ni-Hi fusion of E-Cat. If so for one gram of 62Ni, liberated energy may be close to $7.625.10^9$ J.

5) Based on the above point-2, fusion of four protons (into a stable 4He) may liberate of 4*7.3=29.2 MeV and this can be compared with currently believed PP chain reactions in astrophysics.

## 6. Characteristic Applications of E-Cat

1) Mini power plants for 5 to 10 Villages or two Towns or one City
2) Medium scale industrial power generation
3) Power generation for Busses, Lorries, Trucks, Cars and Bikes
4) Power generation for Trains, Ships, Submarines and Aero planes (if possible)
5) Power generation for medium and big residential and commercial apartments
6) Cold room heating, hot water generation and direct food cooking with hot water.
7) Farm field mini alternators for 2 to 3 successful crops per year with high yield.
8) Hose hold and special purpose laboratory mini alternators
9) De-centralized, Uninterrupted, Pollution free and Risk free power supply
10) Slow and gradual stopping of coal and oil usage for power generation and minimizing their transportation charges.
11) Based on the Environmental safety, Economical conditions and Safe& ease operating techniques, in future,
   A) Closed and intermittent running biomass based power plants,
   B) Coal based thermal power plants and
   C) Uranium based nuclear power plants can be converted to Ni based E-CAT power plants.
12) Travelling charges, transportation charges and power consumption charges can be reduced.

## 7. Discussion

From the 32 day experimental run of E-CAT, (measured) liberated energy for one gram of Nickel-58 is 5825 MJ with ±10% error. With the proposed method, (estimated) liberated energy for one gram of Nickel-58 is 5984 MJ. This is an excellent fit. If one is willing to consider the proposed methodology, E-CAT working mechanism and isotopic change mechanism both can be understood. Not only that, by considering the number of transformations of $^{58}_{28}Ni$ to $^{62}_{28}Ni$ per hour it is possible to decide the design capacity of E-CAT. Thus it may be helpful in designing the future E-CAT with all possible controls like E-CAT reactor volume, E-CAT working temperature, E-CAT working temperature, frequency of addition quantity of Nickel, frequency of addition quantity of Lithium etc.

On the E-CAT's recent third party test run, one can see various positive and appreciating comments in Andrea Rossi's blog, http://www.journal-of-nuclear-physics.com/.

Andrea Rossi personally appreciated authors current approach on E-CAT mechanism and published the basic idea of this paper as a comment in the blog. No doubt, E-CAT can be considered as the most promising equipment for future power generation with plenty of available nickel, low working temperatures, no nuclear radiations, no pollution and no risk. Currently Andrea Rossi is seriously concentrated on developing E-CAT in all respects. Mean while Indian government, scientists, professors, industrialists and engineers may focus their attention on E-CAT design for fulfilling the infinite demand of electric power generation in India.

## 8. Conclusions

So far no model is successful in understanding and estimating the energy liberated in Cold fusion phenomenon. Considering the proposed concepts, with ease and efficiency, it is possible to fit, estimate and design a cold fusion based high power density apparatus like E-CAT for different energy level applications. With further research and analysis basics of 'cold fusion' can be established. In view of the recently developed compact 1MW E-CAT heat power plant designed by the Leonardo corporation, Nickel can certainly be considered as the ultimate substitute of Coal, Oil and Uranium in near future.

## Acknowledgements

Authors are very much thankful to Dr. Andrea Rossi for his valuable comments, suggestions, guidance and encouragement in preparing this paper. Authors humbly request the Noble committee to kindly honor Dr. Andrea Rossi and Dr.Sergio Focardi for developing the E-CAT. Authors would like to thank Dr. Stoyan Sarg Sargoytchev for his excellent review on the history of Cold fusion.

The first author is indebted to professor K. V. Krishna Murthy, Chairman, Institute of Scientific Research on Vedas (I-SERVE), Hyderabad, India and Shri K. V. R. S. Murthy, former scientist IICT (CSIR) Govt. of India, Director, Research and Development, I-SERVE, for their valuable guidance and great support in developing this subject. Both the authors are very much thankful to the anonymous referees for their valuable comments and kind suggestions in improving and bringing this subject into current main stream physics research.

## References

[1] Bo Hoistad, Lars tegner, Roland Petterson, Hanno Essen, Giuseppe Levi, Evelyn Foschi: " Independent Third Party Report" on http://www.elforsk.se/LENR-matrapport-publicerad

[2] Giuseppe Levi et al. Indication of anomalous heat energy production in a reactor device. arXiv:1305.3913

[3] Giuseppe Levi et al. Observation of abundant heat production from a reactor device and of isotopic changes in the fuel. http://www.sifferkoll.se/sifferkoll/wpcontent/uploads/2014/10/LuganoReportSubmit.pdf (Oct 2014) http://www.e-catworld.com/2014/10/08/e-cat-report-released/ 8th October 2014

[4] http://animpossibleinvention.com/2014/10/08/new-scientific-report-on-the-e-cat-shows-excess-heat-and-nuclear-process/#comments. 8th October 2014

[5] Andrea Rossi. Method and Apparatus for carrying out nickel and hydrogen exothermal reactions. Patent application. WO2009125444 (A1) and US 201110005506 Al.

[6] Andrea Rosi, Hydrogen Nickel LENR (Low Energy Nuclear Reaction) Andrea Rossi Cold Fusion - The E-Cat Energy Revolution. http://www.xvi-ncbc.com/news/hydrogen-nickel-lenr-low-energy-nuclear-reaction-andrea- rossi-cold-fusion-the-e-cat-energy-revolution.

[7] Bruce V. Lewenstein. Cornell Cold Fusion Archive, collection no. 4451. Division of Rare and Manuscript Collections, Cornell University Library. 5th edition 31 August 1994.

[8] Hugh G. Flyn, Method of generating energy by acoustically induced cavitation fusion and reactor therefore, US Patent 4,333,796 (filed 1978, issued 1982)

[9] M. Fleischmann and S. J. Pons, Electroanal. Chem, 261, 301, (1989)

[10] Marwan, J. et al. A new look at low-energy nuclear reaction (LENR) research: a response to Shanahan. Journal of Environmental Monitoring 12.9 (2010): 1765.

[11] Yeong E. Kim et al. Optical theorem formulation of low-energy nuclear reactions. PHYSICAL REVIEW C VOLUME 55, NUMBER 2 FEBRUARY 1997

[12] Y.N. Srivastava et al. A primer for electroweak induced low-energy nuclear reactions. Pramana. Vol. 75, No. 4 pp.617-637 (2010)

[13] V.I. Dubinko. Low energy nuclear reactions driven by discrete breathers. To be published in Journal of Condensed nuclear science.

[14] Storms, E.K., The science of low energy nuclear reaction (World Scientific, Singapore, 2007).

[15] Z. Sun and D. Tomanek, Cold Fusion: How Close Can Deuterium Atoms Come inside Palladium?, Phys. Rev. Letters 63 (1989) 59-61.

[16] E. N. Tsyganov, Cold nuclear fusion, Physics of Atomic Nuclei, 75 (2012) 153–159

[17] Cold Fusion. The history of research in Italy. Italian National Agency for New Technologies, Energy and Environment.

[18] Stoyan Sarg Sargoytchev. Theoretical Feasibility of Cold Fusion According to the BSM - Supergravitation Unified Theory. Journal of nuclear physics. (Dec. 2013) http://www.journal-of-nuclear-physics.com/

[19] J. Schwinger, A progress Report: Energy Transfer in Cold Fusion and Sonoluminecsence, Infinite Energy, Issue 24, p. 19,(1999)

[20] http://www.tcm.phy.cam.ac.uk/~bdj10/ (B. Josephson web page), http://www.youtube.com/watch?v=EDv6phew-ck

[21] Ph. M. Kanarev, and Tadahiko Mizuno, COLD FUSION BY PLASMA ELECTROLYSIS OF WATER. New Energy Technologies, Issue #1 January - February 2003 http://www.free-energy-info.tuks.nl/Issue10.pdf

[22] Tadahiko Mizuno, Experimental Confirmation of the Nuclear Reaction at Low Energy Caused by Electrolysis in the Electrolyte, Proceedings for the Symposium on Advanced Research in Energy Technology 2000, Hokkaido University, March 15, 16 and 17, 2000, pp. 95-106

[23] W.D. Myers and W.J. Swiatecki. Table of Nuclear Masses according to the 1994 Thomas-Fermi Model. LBL-36803.1994.

[24] W.D. Myers & W.J. Swiatecki, Nuclear Properties According to the Thomas-Fermi Model, LBL-36557.1994.

# 29

# Notes on the Boussinesq integrable hierarchy

**O. Dafounansou[1], D. C. Mbah[2], A. Boulahoual[3], M. B. Sedra[3, 4]**

[1]Department of Physics, Faculty of Science, Douala University, Douala, Cameroun
[2]CEPAMOQ, Douala University, Douala, Cameroun
[3]LHESIR, Faculty of Science of Kenitra, Ibn Toufail University, Kenitra, Morocco
[4]ENSAH, Mohammed First University, Al Hoceima, Morocco

**Email address:**

daf.osman@gmail.com (O. Dafounansou), mbahdavidc@yahoo.fr (D. C. Mbah), boulahoual@yahoo.com (A. Boulahoual),
sedramyb@gmail.com (M. B. Sedra)

**Abstract:** This work is dedicated to some notes on the Moyal momentum algebras applied to the sl₃ Boussinesq integrable hierarchy. Starting from a brief review of the Moyal momentum algebra structures, we establish in detail the Non-commutative Boussinesq hierarchy by using the Lax pair Generating Technique. Then we shows that these equations can be obtained as 3-reduction of Non-commutative KP hierarchy in a similarly form via some conformal realizations.

**Keywords:** Moyal Momentum Algebra, Moyal KP Hierarchy, Non-Commutative Boussinesq Hierarchy

## 1. Introduction

The origin of integrable system dates back to the 19$^{\text{th}}$ century with the KdV equation, which describe the long solitary wave in the shallow water [1]. Since the study of integrability of nonlinear system, has taken more consideration [2]. For such systems integrability means the existence of an infinite number of conserved quantities in involutions. A definition given by Ward is that such system, more precisely few of them, can be derived from the anti-self-dual Yang Mills equations by reduction with gauge groups [3, 4].

These studies yield exacts solutions in many problem in theoretical high energy physics and mathematics. It appears that the geometry of integrable system is crucial for understanding many aspects of field theories [5]. E. Witten has conjectured that the energy of 2-dimensional gravity coincide with the Tau-function of KdV hierarchy [6].In addition the integrable systems can be linked to the infinite-dimensional conformal algebra and its extensions. From their Poisson bracket structure it turns out that Boussinesq and KP hierarchy are respectively isomorphic to $W_3$ and $W_{1+\infty}$ algebra. In current days, there are deep interest in the non-commutative aspect of different soliton equations [7, 8, 9], with successful applications to string theories [10]... It appears that the Moyal momentum algebra $\widehat{\Sigma}_m^{(r,s)}$ via its $sl_{n-1}$ − momentum Lax operators provides an interesting

tools for the study of $\theta$ −deformed integrable systems.

We will study integrable systems of $(1 + 1)$ and $(2 + 1)$ dimensional evolution equation namely the Boussinesq and KP equation respectively. We starts with some basic properties of the Moyal$\star$ Product, introducing the Moyal Momentum Algebra $\widehat{\Sigma}_m^{(r,s)}$. Then we adopt the Lax Pair Generating Technique to study the evolution equations of Non-commutative Boussinesq hierarchy. By the way we establish the Non-commutative KP hierarchy before discussing the 3-reduction of NC KP hierarchy and the link with the previous Boussinesq hierarchy.

## 2. Moyal Product $\star$ and Operators Algebra $\widehat{\Sigma}_m^{(r,s)}$

Our formulation will be based on star product $\star$ called the Moyal product. Given a smooth manifold $M$ with $x = (x_1, x_2, ..., x_n)$ coordinates system. This manifold will be endowed with the skew-symmetric bilinear bracket defined on $C^\infty(M)$ by [11, 12]:

$$\{f, g\} = \omega^{ij} \frac{\partial f}{\partial x^i} \frac{\partial g}{\partial x^j} \quad f, g \in C^\infty(M) \qquad (1)$$

$\{,\}$ verifies the Jacobi identity, if $\omega^{ij}$ is a non-degenerate skew-symmetric matrix, hence $M$ is symplectic manifold with and even dimension. We consider extend tensorial manifold $F = M \otimes T$ with $t = (t1, t2, \ldots, tn)$ denoting the extra coordinates system of $T$. The Moyal product will not affect $t$ and it is given by [2, 13]:

$$f(x,t) \star g(x,t) = \exp\left(\theta \, \omega^{ij} \frac{\partial}{\partial x^i} \frac{\partial}{\partial x^j}\right) f(x,t) g(\tilde{x},t)\Big|_{x=\tilde{x}} \quad (2)$$

Expanding this equation we find:

$$f(x,t) \star g(x,t) = \sum_{s=0}^{\infty} \frac{\theta^s}{s!} \omega^{i_1 j_1} \ldots \omega^{i_s j_s} \frac{\partial^s f(x,t)}{\partial x^{i_1}\ldots\partial x^{i_s}} \frac{\partial^s g(x,t)}{\partial x^{j_1}\ldots\partial x^{j_s}} \quad (3)$$

The Moyal bracket is defined as follow:

$$f(x,p,t) \star g(x,p,t) =$$

$$\sum_{s=0}^{\infty} \sum_{i=0}^{s} \frac{\theta^s}{s!} (-1)^i \, C_s^i \left(\partial_x^i \partial_p^{s-i} f(x,p,t)\right)\left(\partial_x^{s-i} \partial_p^i g(x,p,t)\right) \quad (5)$$

and the Moyal bracket :

$$\{f(x,t), g(x,t)\} = \frac{f \star g - g \star f}{2\theta}$$

$$\{f(x,t), g(x,t)\} = \sum_{s=0}^{\infty} \frac{\theta^{2s}}{(2s+1)!} \sum_{i=0}^{2s+1} (-1)^i \binom{2s+1}{i} \times$$

$$\left(\partial_x^{2s+1-i} \partial_p^i f(x,t)\right)\left(\partial_x^i \partial_p^{2s-i} g(x,t)\right)$$

$$f \text{ and } g \in C^{\infty}(F) = C^{\infty}(M \otimes T) \quad (6)$$

The point of introducing the above properties is to define the Moyal momentum algebra. The Moyal momentum was introduced first by authors [14], and systematically studied later with some applications to conformal field theory and $\theta$-deformed integrable models [15]. This algebra is a pseudo momentum operators algebra denoted by $\hat{\Sigma}(\theta)$.

$\hat{\Sigma}(\theta)$ consist of the object of the form $u(x,\tau) \star f(p)$ where $f(p)$ is polynomial in momentum $p, \tau = (t_1, t_2, \ldots, t_n)$. The Moyal momentum algebra is isomorphic to the ordinary pseudo differential operator $L_m = \sum_{i \in Z} u_{m-i} \star \partial^i$. The construction of $\hat{\Sigma}(\theta)$ consist of replacing the ordinary pseudo differential lax operators by the Lax momentum operators:

$$L_m = \sum_{j \in Z} u_{m-j}(x,\tau) \star p^j \quad (7)$$

$L_m$ is a $C^{\infty}(F)$ function of ordinary spin $m$ living in a non-commutative space parameterized by $\theta$. The conformal dimensions are given as follow:

$$[u_i] = i, [\theta] = 0, [p] = [\partial_x] = -[x] = 1, [\partial_{t_k}] = -[t_k] = k. \quad (8)$$

$\hat{\Sigma}(\theta)$ can be decomposed as:

$$\hat{\Sigma}(\theta) = \oplus_{r \leq s} \oplus_{m \in Z} \hat{\Sigma}_m^{(r,s)}(\theta) \quad (9)$$

where $\hat{\Sigma}_m^{(r,s)}$ denotes the space of momentum lax operators of conformal spin $m$ and degrees start from $r$ to $s$:

$$L_m^{(r,s)} = \sum_{j=r}^{s} u_{m-j} \star p^j \quad (10)$$

$L_m^{(r,s)}$ involving zero value $u_{m-k}$ ($r \leq k \leq s$) term belong

$$\{f,g\} = \frac{f \star g - g \star f}{2\theta} \quad (4)$$

With $\lim_{\theta \to 0} f \star g = fg$.

If we consider the 2d-phase space $M$, with $x(x_1 = x, x_2 = p)$ coordinate, the matrix $\omega^{ij}$ becomes:

$$\omega^{ij} = \begin{pmatrix} 0 & 1 \\ -1 & 0 \end{pmatrix}$$

hence expression (3) can be written as [13]:

to the space $\Sigma_m^{(r,s)} \big/ \Sigma_m^{(k,k)}$.

$\Sigma_m^{(0,0)}$ is the space of operators of degree 0 denoting function coefficient of conformal spin $m$:

$$u_l \star u_m = u_l u_m.$$

The Moyal bracket of two operators $\in \Sigma_m^{(r,s)}$ gives rise to an operator $\in \Sigma_m^{(r,2s-1)}$. To perform all the forthcoming calculations, the formulae (3) will be use its more simplified way. This has be done in several papers [8, 15].
We have

$$p^n \star f(x,p) = \sum_{s=0}^{n} \theta^s \, C_n^s \, f^{(s)}(x,p) \, p^{n-s}. \quad (11)$$

$$p^{-n} \star f(x,p) = \sum_{s=0}^{\infty} (-1)^s \, \theta^s C_{n+s-1}^s \, f^{(s)}(x,p) \, p^{-n-s}. \quad (12)$$

$$\{p^n, f\}_\theta = \sum_{k=0}^{\infty} \theta^{2k} \, C_n^{2k+1} \, f^{(2k+1)} \, p^{n-2k-1}. \quad (13)$$

$$\{p^{-n}, f\}_\theta = -\sum_{k=0}^{\infty} \theta^{2k} \, C_{2k+n}^{2k+1} \, f^{(2k+1)} \, p^{-n-2k-1}. \quad (14)$$

## 3. Moyal Boussinesq Hierarchy

The $sl_n$ − moyal hierarchy is defined by the lax equation [16]

$$\frac{\partial L}{\partial t_k} = \left\{\left(L^{1/n}\right)_+^k, L\right\}_\theta \quad (15)$$

Where

$$\left(L^{1/n}\right)_+^k = \underbrace{\left(L^{\frac{1}{n}} \star L^{\frac{1}{n}} \star \ldots \star L^{1/n}\right)_+}_{k}$$

It follows that the coefficient in order $n - 1$ vanishes, we have the special form of $L \in \Sigma_n^{(0,n)}\Big/ \Sigma_n^{(n-1,n-1)}$ called $L -$ hierarchy.

$$L = L_n = p^n + \sum_{i=0}^{n-2} u_{n-i} \star p^i \qquad (16)$$

and

$$L^{1/n} = p + \sum_{i=1} \omega_{i+1} \star p^{-i} \qquad (17)$$

is the $n^{th}$ root of $L$. Thus the $sl_3$-Boussinesq moyal momentum Lax operator we will deal with is :

$$L_3 = p^3 + u_2 \star p + u_3 \qquad (18)$$

The explicit expression of $L^{1/3}$ and the straightforward calculations gives the Boussinesq hierrarchy. This has been done by many authors [8, 15].

Instead of the above approach in this section, we will adopt the Lax pair generating technique to determine the Non-commutative Boussinesq hierarchy [4]. Briery, the Lax pair generating technique consist of finding for a given $L_m$, the operator $T$ such that:

$$\{L_m, T + \partial_t\}_\theta = 0 \qquad (19)$$

The equation (19) is equivalent to (15) where $T$ is the analogue of $\left(L^{k/m}\right)_+$ [4, 8]. This technique is based on the following ansatz:

$$T = p^\alpha \star L^\beta + \tilde{T} \qquad (20)$$

with $\alpha$ and $\beta \in Z$.
where $p^\alpha$ is a monome of momentum operator. Actually the clue of the problem is to determine the expression of the operator $\tilde{T}$ ; keeping in mind that $T$ and $\tilde{T}$ have the same degree.

- The $t_1$ flow $\partial_{t_1} u = \dot{u}$ :

$$L = p^3 + u_2 \star p + u_3 = p^3 + u_2 p + u_3 - \theta u_2'$$

$$T = p^{-2} \star L + \tilde{T} \qquad (21)$$

The equation (19) leads to trivial equations with $\tilde{T} = A \in C$:

$$\frac{\dot{u}_2}{2\theta} = -u_2' , \qquad (22)$$

$$\frac{\dot{u}_3}{2\theta} = -u_3' , \qquad (23)$$

where we denote by $\frac{\partial u}{\partial x} = u'$ and $\frac{\partial u}{\partial t} = \dot{u}$ We can obtain the ordinary form of the Boussinesq hierarchy via the correspondence $\frac{1}{2\theta}\frac{\partial}{\partial t} \longrightarrow -\frac{\partial}{\partial t}$.

- The $t_2$ flow $\partial_{t_2} u = \dot{u}$ :
We consider the ansatz :

$$T = p^{-1} \star L + \tilde{T}, \qquad (24)$$

where $\tilde{T} = A \in \hat{\Sigma}_2^{(0,0)}$.
Considering the differential part of $p^{-1} \star L$, we get:

$$\{L, (p^{-1} \star L)_+ + \partial_t\}$$
$$= (3u_2' - 2u_2')p^2 - \left(2(u_3' - \theta u_2'') + \frac{\dot{u}_2}{2\theta}\right)p$$
$$+ u_2 u_2' - \frac{\dot{u}_3}{2\theta} + \frac{\dot{u}_2'}{2} + \theta^2 u_2'''$$

By identifying with:

$$-\{L, \tilde{T}\} = -3 A'p^2 - u_2 A' - \theta A''' ,$$

one gets :

$$A = -\frac{1}{3}u_2 + a,$$

taking $a = 0$, then:

$$A = -\frac{1}{3}u_2. \qquad (25)$$

Therefore:

$$T = p^2 + \frac{2}{3}u_2 \qquad (26)$$

$$\frac{\dot{u}_2}{2\theta} = -2u_3 + 2\theta u_2''. \qquad (27)$$

Substitute $A$ in:

$$u_2 u_2' - \frac{\dot{u}_3}{2\theta} + \frac{\dot{u}_2'}{2} + \theta^2 u_2''' = -u_2 A' - \theta A''',$$

we get:

$$\frac{\dot{u}_3}{2\theta} = \frac{2}{3}u_2 u_2' + \frac{8}{3}\theta^2 u_2''' - 2\theta u_3''. \qquad (28)$$

We recognize the pair of equations (27) and (28) is nothing but the non-commutative Boussinesq equation.

- The $t_4$ ow $\partial_{t_4} u = \dot{u}$
Here we consider the following ansatz:

$$T = p \star L + \tilde{T} \qquad (29)$$

with:

$$\tilde{T} = a \star p^2 + b \star p + c$$

$$\tilde{T} = Ap^2 + Bp + C \qquad (30)$$

where coefficients of polynome in $p$ belong to $\hat{\Sigma}^{(0,0)}$. To find the Lax pair of equation

$$\{L, T + \partial_t\}_\theta = 0,$$

we start by calculating the following terms:$\{L, p \star L\}_\theta$ and $\{L, \tilde{T}\}_\theta$ :

$$\{L, p \star L\}_\theta = -u_2' p^4 + (-u_3' + 4\theta u_2'')p^3$$
$$+ (3\theta u_3'' - 6\theta^2 u_2''' - u_2 u_2')p^2$$
$$+ \left[-3\theta^2 u_3''' - (u_2 u_3)' + 2\theta u_2 u_2''\right.$$
$$\left. + 4\theta^3 u_2^{(4)}\right]p + \theta^3\left(u_3^{(4)} - \theta u_2^{(5)}\right)$$
$$+ \theta^2 u_2' u_2'' + \theta u_2(u_3'' - \theta u_2''')$$
$$- u_3(u_3' - \theta u_2'').$$

$$-\{L, \tilde{T}\}_\theta = -3A' p^4 - 3B' p^3$$
$$- (\theta^2 A''' + 3C' + u_2 A' - 2u_2' A)p^2$$
$$- [\theta^2 B''' + u_2 B' - u_2' B$$
$$- 2A(u_3' - \theta u_2'')]p - \theta^2 C''' - u_2 C'$$
$$- \theta^2 u_2'' A' + B(u_3' - \theta u_2'').$$

Then by identifying the order 4, 3, 2 in $p$, we obtain:

$$A = \frac{1}{3} u_2, \tag{31}$$

$$B = \frac{1}{3}(u_3 - 4\theta u_2'), \tag{32}$$

$$C = \frac{17}{9}\theta^2 u_2'' - \theta u_3' + \frac{2}{9}(u_2)^2, \tag{33}$$

With these values, the identification in order 1 leads to:

$$\frac{\dot{u}_2}{2\theta} = \frac{4}{3}\left[\theta(u_2 u_2'' + (u_2')^2) - (u_2 u_3)' - 2\theta^2 u_3''' + 2\theta^3 u_2^{(4)}\right] \tag{34}$$

Finally, the term of the order $0 \in \hat{\Sigma}^{(0,0)}$ fields:

$$\frac{\dot{u}_3 - \theta \dot{u}_2'}{2\theta} = \frac{4}{3}\left[-u_3 u_3' + \frac{1}{3}u_2^2 u_2' + \frac{2}{3}\theta^4 u_2^{(5)} + \theta\ (u_3 u_2'' + u_2' u_3') + \right.$$
$$\left. \theta^2\ u2' u2'' + u2\ u2'''\right]. \tag{35}$$

Hence:

$$T = p \star L + \tilde{T},$$

$$T = p^4 + \frac{4}{3} u_2 p^2 + \frac{4}{3}(u_3 - \theta u_2')p + \frac{8}{9}\theta^2 u_2'' + \frac{2}{9}(u_2)^2. \tag{36}$$

Equations (34) and (35) correspond to the $t_4$ evolution equations of the non-commutative Boussinesq hierarchy.

# 4. Moyal KP Hierarchy

In this section, we drop the $-\frac{1}{2\theta}\frac{\partial}{\partial t_k}$. time derivation we start with a more familiar notations similar to Lax representation for a hierarchy in Sato's framework. We consider the KP Lax operator:

$$L_{KP} = L = p + \sum_{i=1}^{\infty} u_{i+1} \star p^{-i} \in \left.\hat{\Sigma}_1^{(-\infty,1)}\right/_{\hat{\Sigma}_1^{(0,0)}}$$

$$L = p + \sum_{i=1}^{\infty} v_{i+1} p^{-i}. \tag{37}$$

Then the non-commutative KP evolution equations take the lax form:

$$\frac{\partial L}{\partial t_k} = \{B_k, L\}_\theta = \{(L^k)_+, L\}_\theta \tag{38}$$

We use the later to determine KP hierarchy in a simpler way by using the Moyal $\star$ product and recover the hierarchy similar to the one found by using the supershmidt-Manin $\star$ product, just by a conformal realization of fields $v_i$. It turns out that the KP hierarchy consists of an infinite set of differential equation for each time $t_k$ [13, 17].

** The $t_1$ flow $\partial_{t_1} u = \dot{u}$ :

$$\dot{v}_{i+1} = v_{i+1}', \tag{39}$$

** The $t_2$ flow $\partial_{t_2} u = \dot{u}$ :

$$\{B^2, L\}_\theta = \{p^2 + 2v_2, p + \sum_{i=1}^{\infty} v_{i+1} p^{-i}\}_\theta$$

$$= \sum_{i=1}^{\infty} 2v_{i+1}' p^{-i+1} - 2v_2' - 2\sum_{i=1}^{\infty} v_{i+1}\{p^{-i}, v_2\},$$

we keep the terms in $p^{-1}, p^{-2}$ and $p^{-3}$ then we get the following equations:

$$\dot{v}_2 = v_3'$$
$$\dot{v}_3 = 2v_4' + 2v_2 v_2'$$
$$\dot{v}_4 = 2v_5' + 4v_2' v_3$$
$$\vdots$$

$$\tag{40}$$

by a conformal realization the field $v_i$ is expressed in term of $u_i$:

$$v_2 = u_2, v_3 = u_3 + \theta\ u_2'$$
$$v_4 = u_4 + 2\theta\ u_3' + \theta^2 u_2'',$$
$$v_5 = u_5 + 3\theta\ u_4' + 3\theta^2 u_3'' + \theta^3 u_2''', \tag{41}$$

we find the previous hierarchy in the following form:

$$\dot{u}_2 = 2(u_3' + \theta\ u_2'),$$
$$\dot{u}_3 = 2u_4' + 2\theta\ u_3'' + 2u_2 u_2'$$
$$\dot{u}_4 = 2u_5' + 2\theta\ u_4'' + 4u_3 u_2' - 4\theta u_2 u_2'', \tag{42}$$
$$\vdots$$

** The $t_3$ flow $\partial_{t_3} u = \dot{u}$:

$$\{B^3, L\}_\theta = \{p^3 + 3v_2 p + 3v_3, p + \sum_{i=1}^{\infty} v_{i+1} p^{-i}\}_\theta$$

keeping the term up to $p^{-2}$, we find:

$$\dot{v}_2 = 6v_2 v_2' + 3v_4' + \theta^2 v_2'''$$
$$\dot{v}_3 = 6(v_2 v_3)' + 3v_5' + \theta^2 v_3''' \tag{43}$$

using the conformal realization (41), we get:

$$\dot{u}_2 = 6u_2 u_2' + 4\theta^2 u_2''' + 3u_4' + 6\theta\ u_3'',$$
$$\dot{u}_3 = 6(u_2 u_3)' + 4\theta^2 u_3''' + 3u_5' + 6\theta\ u_4'' \tag{44}$$
$$\vdots$$

It appears that if one takes the first two equations of (42) and eliminating $u_3$ and $u_4$ in the fisrt equation of (44) we get the non-commutative KP equation where $t_2 \equiv y$ and $t_3 \equiv t$.

## 5. Boussinesq Hierarchy as 3-Reduction of Moyal KP Hierarchy

This approach pictures the link between the KP Lax operator and others integrables models. Let's rewrite the KP Lax operator $L = p + \sum_{i=1}^{\infty} u_{i+1} \star p^{-i}$ or in the form $L = p + \sum_{i=1}^{\infty} v_{i+1} p^{-i}$. For Boussinesq equation, we denote the Lax operators by $\mathcal{L} = p^3 + u_2 \star p + u_3$ or $\mathcal{L} = p^3 + v_2 p + v_3$ .Then the Boussinesq hierarchy obtained by 3-reduction is given by the following Lax equation.

$$\frac{\partial \mathcal{L}}{\partial t_k} = [B_k, L^3] \tag{45}$$

Where $B_k = (L^k)_+$ with the contrain $L^3 = B_3$. The $t_1$ flows are trivial.

For The $t_2$ flows we have:

$$B_2 = (L^2)_+ = p^2 + 2v_2 \tag{46}$$

$$\frac{\partial \mathcal{L}}{\partial t_2} = [B_2, L^3]_\theta. \tag{47}$$

With:

$$L^3 = p^3 + 3v_2 p + 3v_3 + (3v_2^2 + 3v_4 + \theta^2 v_2'')p^{-1} + \cdots \tag{48}$$

and the contrain $L^3 = B_3 \Rightarrow Res(L^3) = 0$, we find:
The term $[B_2, L^3]_\theta$ yields:

$$v_4 = -v_2^2 - \frac{\theta^2}{3} v_2'' \tag{49}$$

$$[B_2, L^3]_\theta = 6v_3' p - 2(\theta^2 v_2''' + 3v_2 v_2').$$

Finally with the lax equation (47) we obtain:

$$\dot{v}_2 = 6v_3' \tag{50}$$

$$\dot{v}_3 = -2 (\theta^2 v_2''' + 3 v_2 v_2') \tag{51}$$

it turns out that the $\partial_{t_2}$ time derivation of equation (50) yields:

$$\ddot{v}_2 = 6\dot{v}_3' = 6\big(-2 (\theta^2 v_2''' + 3 v_2 v_2')\big)',$$

Therfore we get the Non-commutative Boussinesq equation:

$$\ddot{v}_2 = -12(3 v_2 v_2' + \theta^2 v_2''')'. \tag{52}$$

Taking classical limit $\theta = \frac{1}{2}$ we obtain the Boussinesq equation in the ordinary form [18]. Notice that the map (41) doesn't change equation (52).

The $t_4$ flows are given as follow: we start by calculating:

$$L^4 = B_4 = p^4 + 4v_2 p^2 + 4v_3 p + 4\theta^2 v_2'' + 6v_2^2 + 4v_4.$$

The condition $Res(L^3) = 0$ yields :

$$L^4 = B_4 = p^4 + 4v_2 p^2 + 4v_3 p + 2v_2^2 + \frac{8}{3}\theta^2 v_2''. \tag{53}$$

then the equation :

$$\frac{\partial \mathcal{L}}{\partial t_4} = [B_4, L^3]_\theta \tag{54}$$

gives rise to:

$$\dot{v}_2 = 4( 3(v_2 v_3)' + 2\theta^2 v_3'''), \tag{55}$$

$$\dot{v}_3 = 4 \left( 3v_3 v_3' - 3v_2^2 v_2' - 6\theta^2 v_2' v_2'' - 3\theta^2 v_2 v_2''' - \frac{2}{3}\theta^4 v_2^{(5)} \right). \tag{56}$$

Notice that the presence of the term $-6\theta^2 v_2' v_2''$ in the last equation doesn't matter, since when applying the conformal map (41) to the terms $v_i$ that coming from the KP Lax operator $L$ we recover the following equations.

$$\ddot{u}_2 = 4 \left( 3(u_2 u_3)' + 3\theta\big(u_2 u_2'' + u_2'^2\big) + 2\theta^2 \left(u_3''' + \theta u_2^{(4)}\right)\right), \tag{57}$$

$$\dot{u}_3 - \theta \dot{u}_2' = 12 \left( u_3 u_3' - u_2^2 u_2' + \theta(u_3 u_2'' + u_2' u_3') - \theta^2(u_2' u_2'' + u2u2''' - 29\theta4u25. \tag{58}$$

## 6. Conclusions

We have presented by two different methods how to obtain the deformed Boussinesq hierarchy. Of course there are several versions of theory and each has its advantages and flaws. In this work, the results found in the first method show the consistency of Lax pair generating technique. Where by rescaling time derivation we recover the ordinary form of Boussinesq hierarchy. We also got a look to the KP hierarchy which has been simplified by using a conformal realization that shows the equivalence between the Moyal $\star$ product and the Kupershmidt-Manin $\star$ product. We have also shown that the Boussinesq hierarchy obtained by the 3-reduction of KP hierarchy using the same conformal map gives rise to equations similar to that obtained by Lax Pair Generating Technique. We hope our discussion will make the Moyal momentum be more accessible in the study of some integrable models.

## References

[1]    A.B. Zamolodchikov, Integrable field theory from conformal field theory, Proceedings of the Taniguchi Symposium, Kyoto, (1988); Int. J. Mod. Phys. A3 (1988) 743;

[2]    A. Das and Z. Popowicz, Phys. Lett. A272 (2000) 65. [3] Szablikowski B.M. and Blaszak M., Meromorphic Lax representations of (1+1)-dimensional multi-Hamiltonian dispersionless systems, J. Math. Phys. 47 paper 092701 (2006);

[3]    M. Hamanaka and K. Toda, Phys. Lett. A 316 (2003) 77;

[4]    J. Madore An Introduction to Non-commutative Geometry and its Physical Applications Second Edition LMS 257 (1999);

[5]    Kontsevich M., Intersection theory of the moduli space of curves and the matrix Airy function, Comm. Math. Phys. 147 1-23 (1992);

[6]    M. T. Grisaru, L. Mazzanti, S. Penati, L. Tamassia, JHEP 0404:057, 2004;

[7] M.B. Sedra, Moyal non-commutative integrability and the BurgersKdV mapping, Nuclear Physics B 740 [PM] (2006) 243270;

[8] A. F. Dimakis and F. Muller-Hoissen, Rep. Math. Phys. 46 (2000) 203; Non-Commutative Kortewegde-Vries equation, hep-th 0007074;

[9] A. Connes, Non-commutative geometry, Academic Press (1994);

[10] B. A. Kupershmidt, Phys. Lett. A 102, 213 (1984);

[11] M. H. Tlili AFST 6e srie, Tome 9, No 3 (2000), P. 551-564;

[12] Strachan, I.A.B., The Moyal bracket and the dispersionless limit of the KP hierarchy, J. Phys. A. 20 (1995) 1967-1975;

[13] A. Das and Z. Popowicz, J. Phys. A, Math.Gen. 34, 6105 (2001) and [hep- th/0104191]; B. A. Kupershmidt, Lett. Math. Phys. 20, 19 (1990);

[14] A. Boulahoual and M. B. Sedra, hep-th/0208200, Chin. J. Phys 43, 408 (2005); A. Das and Z. Popowicz, Properties of Moyal-Lax Representation Phys. Lett. B 510 (2001) 264270 ; O. Dafounansou, A. El Boukili and M. B. Sedra,Some Aspects of Moyal Deformed Integrable Systems Chin. J. Phys 44, 274 (2006);

[15] O. Babelon, D. Bernard, M. Talon, Introduction to Classical Integrable System Cambridge University Press (2003) and references therein;

[16] A. F. Dimakis and F. Muller-Hoissen, J. Phys. A: Math. Theor. 40 (2007) 7573 - 7596; O. Lechtenfeld and A. D. Popov, Non-commutative Multi- solitons in (2+1)dimensions, JHEP 0111(2001)040;

[17] Dai Zheng-De, Jiang Mu-Rung, Dai Qing-Yun, Li Shao-Lin; Chin.Phys.Lett. Vol.23, No 5 (2006)1065.

# PERMISSIONS

# LIST OF CONTRIBUTORS

**Salih Mohammed Salih, Osama Ibrahim Abd and Kaleid Waleed Abid**
Renewable Energy Research Center, University of Anbar, Ramadi, Iraq.

**Nagaraj Banapurmath, T. Narasimhalu and Rohan Kittur**
Department of Mechanical Engineering, B. V. B. College of Engineering and Technology, Hubli-580031, India.

**Anand Hunshyal**
Department of Civil Engineering, B. V. B. College of Engineering and Technology, Hubli-580031, India.

**Radhakrishnan Sankaran**
Nehru College of Engineering and Research Centre Pampady, Trissure (Dist), Kerala, India.

**Mohammad Hussain Rabinal**
Department of Physics, Karnatak University, Dharwad, Karnataka, India.

**Narasimhan Ayachit**
Department of Physics, Rani Channamma University, Belgaum, Karnataka, India.

**Amit Shor, Md. Nazmus Sakib Khan and GM. Ahteshamul Haque**
Dept. of Mechanical Engineering, Rajshahi University of Engineering & Technology (RUET), Rajshahi, Bangladesh.

**Keith Openshaw**
Retired forest/energy economist. Formerly with the International Resource Group (IRG ENGILITY) of Washington D.C..

**Natthawut Dutsadee, Nigran Homdoung and Rameshprabu Ramaraj**
School of Renewable Energy, Maejo University, Sansai, Chiang Mai-50290, Thailand.

**Khamatanh Santisouk and Shangphuerk Inthavideth**
Department of Mechanical Engineering, Faculty of Engineering, National University of Laos, Vientiane, Laos.

**Rameshprabu Ramaraj and Natthawud Dussadee**
School of Renewable Energy, Maejo University, Sansai, Chiang Mai-50290, Thailand.

**O. Christopoulou, M. Fountoukidou, St. Sakellariou, St. Tampekis, F. Samara A. Sfoungaris and I. Sfoungaris**
University of Thessaly, School of Engineering, Department of Planning and Regional Development, Thessaly, Greece.

**A. Stergiadou**
Aristotelian University of Thessaloniki, School of Agriculture, Forestry and Natural Environment, Department of Forestry and Natural Environment, Thessaloniki, Greece.

**G. Tsantopoulos and K. Soutsas**
Democritus University of Thrace, Department of Forestry Management of the Environment and Natural Resources, Komotini, Greece.

**Stergiadou Anastasia**
Aristotle University of Thessaloniki, Faculty of Forestry and Natural Environment, Institute of Forest Engineering and Topography, Thessaloniki, Greece.

**Kolkos Georgios**
Forester-Enviromentalist AUTH, Ariadnis, Thessaloniki, Greece.

**Nattaporn Chaiyat and Natthawud Dussadee**
School of Renewable Energy, Maejo University, Chiang Mai, Thailand.

**Wirawut Temprasit, Alounxay Pasithi, Suthida Wanno, Supannee Suwanpakdee,**
Sudaporn Tongsiiri and Niwooti Whangchai

Faculty of Fisheries Technology and Aquatic Resources, Maejo University, Sansai, Chiang Mai 50290, Thailand.

**Natthawud Dussadee**
School of Renewable Energy, Maejo University, Sansai, Chiang Mai 50290, Thailand.

**Purnima Dashora, Chetna Ameta, Rakshit Ameta and Suresh C. Ameta**
Department of Chemistry, PAHER University, Udaipur, (Raj.) India.

**Hashim A. Hussain**
Electromechanical. Eng. Dept, University of Technology, Baghdad, Iraq.

**Qusay Jawad**
Elect. Eng. Dept, University of Technology, Baghdad, Iraq.

**Khalid F. Sultan**
Electromechanical. Eng. Dept, University of Technology, Baghdad, Iraq.

**Yousif I. Al-Mashhadany**
Electrical Engineering Dept., College of Engineering, University of Al-Anbar, Al-Anbar, Iraq.

**Hussain A. Attia**
Electronics and Communications Eng. Dept., American University of Ras Al Khaimah, Dubai, UAE.

**Yousif Ismail Al Mashhadany**
Electrical Engineering Department, Engineering College, University of Anbar, Anbar, Iraq.

**Jasim Abdulateef**
Mechanical Engineering Department, Diyala University, Diyala, Iraq.

**Jonathan Richard Raush and Terrence Lynn Chambers**
Department of Mechanical Engineering, University of Louisiana at Lafayette, Lafayette, U. S. A..

**Md. Niaz Murshed Chowdhury and Sumaiya Saleh**
Research Assistant, Department of Economics, South Dakota State University, Brookings, USA.

**Samim Uddin**
Department of Economics, University of Chittagong, Chittagong, Bangladesh.

**Ahmed Fakhri**
School of Electronic and Electrical Engineering, University of Leeds, Leeds, UK.

**Waleed Al-sallami**
School of Mechanical Engineering, University of Leeds, Leeds, UK.

**Nawar H. Imran**
School of Chemical and Process Engineering, University of Leeds, Leeds, UK.

**Ramaraj Rameshprabu**
School of Renewable Energy, Maejo University, Sansai, Chiang Mai-50290, Thailand.

**Rungthip Kawaree and Yuwalee Unpaprom**
Program in Biotechnology, Faculty of Science, Maejo University, Sansai, Chiang Mai-50290, Thailand.

**A. G. Stergiadou**
Institute of Forest Engineering and Topography, Department of Agriculture, Forestry and Natural Environment, Aristotle University of Thessaloniki, Thessaloniki, Greece.

**V. Drosos**
Department of Forestry and Management of Natural Recourses, Democritus University of Thrace, Orestiada, Greece.

**A. K. Douka**
Law Faculty, Aristotle University of Thessaloniki, Thessaloniki, Greece.

**Dilek Çelikler and Zeynep Aksan**
Department of Elementary Science Education, Faculty of Education, Ondokuz Mayıs University, Samsun, Turkey.

**Zeynep Aksan, Gonca Harman and Dilek Çelikler**
Department of Elementary Science Education, Faculty of Education, Ondokuz Mayıs University, Samsun, Turkey.

**Rameshprabu Ramaraj, Natthawud Dussadee**
School of Renewable Energy, Maejo University, Sansai, Chiang Mai-50290, Thailand.

**Rameshprabu Ramaraj and Natthawud Dussadee**
School of Renewable Energy, Maejo University, Sansai, Chiang Mai-50290, Thailand.

**Niwooti Whangchai**
Faculty of Fisheries Technology and Aquatic Resources, Maejo University, Sansai, Chiang Mai 50290, Thailand.

**Yuwalee Unpaprom**
Program in Biotechnology, Faculty of Science, Maejo University, Sansai, Chiang Mai-50290, Thailand.

**Dilek Çelikler and Filiz Kara**
Department of Elementary Science Education, Faculty of Education, Ondokuz Mayıs University, Samsun, Turkey.

**Gonca Harman, Zeynep Aksan and Dilek Çelikler**
Department of Elementary Science Education, Faculty of Education, Ondokuz Mayıs University, Samsun, Turkey.

**Yuwalee Unpaprom and Sawitree Tipnee**
Program in Biotechnology, Faculty of Science, Maejo University, Sansai, Chiang Mai-50290, Thailand.

**Ramaraj Rameshprabu**
School of Renewable Energy, Maejo University, Sansai, Chiang Mai-50290, Thailand.

**U. V. S. Seshavatharam**
Honorary faculty, I-SERVE, Alakapuri, Hyderabad-35, AP, India.

**S. Lakshminarayana**
Dept. of Nuclear Physics, Andhra University, Visakhapatnam-03, AP, India.

**O. Dafounansou**
Department of Physics, Faculty of Science, Douala University, Douala, Cameroun.

**D. C. Mbah**
CEPAMOQ, Douala University, Douala, Cameroun.

**A. Boulahoual**
LHESIR, Faculty of Science of Kenitra, Ibn Toufail University, Kenitra, Morocco.

**M. B. Sedra**
LHESIR, Faculty of Science of Kenitra, Ibn Toufail University, Kenitra, Morocco

ENSAH, Mohammed First University, Al Hoceima, Morocco..

# Index